Handbook of Business Analytics

Editors:
Ravindra Sharma, PhD
School of Management Studies
Swami Rama Himalayan University
Dehradun, India

Geeta Rana, PhD
School of Management Studies
Swami Rama Himalayan University
Dehradun, India

CRC Press
Taylor & Francis Group
Boca Raton London New York

CRC Press is an imprint of the
Taylor & Francis Group, an **informa** business
A SCIENCE PUBLISHERS BOOK

First edition published 2026
by CRC Press
2385 NW Executive Center Drive, Suite 320, Boca Raton FL 33431

and by CRC Press
4 Park Square, Milton Park, Abingdon, Oxon, OX14 4RN

Library of Congress Cataloging-in-Publication Data (applied for)

ISBN: 978-1-032-82272-3 (hbk)
ISBN: 978-1-032-84644-6 (pbk)
ISBN: 978-1-003-51428-2 (ebk)

DOI: 10.1201/9781003514282

Typeset in Times New Roman
by Prime Publishing Services

Preface

Business analytics is a rapidly evolving field that involves the use of data analysis techniques and statistical methods to help businesses make better decisions.

This book offers insights into business analytics, a crucial analytical method used by several organizations to assess risk, forecast future market trends, and identify when upgrades are required. The book will help analysts and researchers to find trends and patterns in the data, which uses past information as a source, and a variety of analytical algorithms for forecasting future events with a high level of accuracy, is the fundamental aim of business analytics. The book includes real-world examples and case studies that demonstrate how analytics techniques can be applied to solve business problems. This could help readers to understand the practical application of analytics in different industries and contexts. Further, this book highlights the importance of ethical and responsible use of data. Topics such as data privacy, security, and bias, will help readers to understand the ethical implications of data-driven decision-making.

Happy Reading!

Editors
Dr. Ravindra Sharma
Dr. Geeta Rana

Acknowledgements

Business analytics are widely used in several sectors, including healthcare, manufacturing, and any other company that monitors or carefully analyzes its users or clients. Many high-end business analytics software solutions and platforms have been developed to ingest and process large data sets. One of the highly appreciated examples of Business Analytics is working with available data to figure out and assess how and why the tastes and preferences change of customers who visit a particular restaurant regularly.

The edited book "Handbook of Business Analytics" provides an insight into business analytics as a key analytical technique that can be utilized by several companies to evaluate risk, project future commercial trends, and determine when upgradation is needed. This book will help analysts and researchers to discover trends and patterns in the data, which uses past information as the source, and a variety of analytical algorithms for forecasting future events with a high level of accuracy is the fundamental aim of business analytics.

This book is a perfect example of the unwavering commitment and perseverance that made the endeavor possible. Maa Daat Kali and Lord Hanuman ji bestowed us with incessant wisdom and courage during the journey of this book.

The book's content matter reflects the curiosity and attention of academicians and students towards it. We express our heart-filled gratitude towards the author's contribution in bringing valuable and appreciating subject matter to rejoice and use.

A large number of academic and industry friends and colleagues' support made this book successful. The multi-facet support includes review and re-review of chapters, suggestions for revisions, modifications, and critical comments to make the chapters more thorough and meaningful. We acknowledge their tremendous support and are thankful for their valuable comments.

We thankfully acknowledge all the support, inspiration, and motivation we received from our faculty colleagues. We would remain indebted to their outstanding intellectual efforts. Our gratitude is due to Dr. Vijay Dhasmana the President of Swami Rama Himalayan University, Jolly Grant, Dehradun, India; Dr. Rajendra Dobhal, the Vice Chancellor of Swami Rama Himalayan University, Dr. Vijendra Chauhan, Director General Academic Swami Rama Himalayan University, and Dr. Bindu Dey, Director Research, Swami Rama Himalayan University, for their faith, trust, and responsibility empowered us and extended full support and being a constant source of inspiration and pillars of strength throughout the pursuit.

We express our thanks to our publisher CRC (Taylor & Francis Publication, USA), and the entire editorial team who lent their wonderful support throughout and ensured the timely processing of the manuscript and bringing out the book.

Finally, it would not be justified if we did not acknowledge the continuous and unstained support from our families. Their belief in us facilitated accomplishing our task and bringing forth this compiled project in book form for scholars, readers, and content enthusiasts.

Editors

Dr. Ravindra Sharma

Dr. Geeta Rana

Contents

Preface *iii*
Acknowledgements *v*
List of Contributors *ix*

1. Foundation Concepts for Business Analytics 1
2. Unraveling the Dimensions of HR Analytics and Sustainable HR Analytics: Paving to a New Paradigm 34
3. Data Science for FinTech Innovations: Trends, Tools, and Applications 58
4. Marketing Analytics to Identify Success and Impact 81
5. Harnessing Data for Sustainability: Applications and Challenges in Environmental Analytics 99
6. Application of Artificial Intelligence-based Technologies for Evaluating a Healthcare Provider's Maturity in the Healthcare Worker Onboarding Process 119
7. Utilizing Analytic Hierarchy Process (AHP) to Prioritize Factors Impacting Financial Inclusion for Economic Advancement 134
8. Digital Transformation in Family-Owned Enterprises: Challenges and Opportunities for Business Analytics Adoption 151
9. Role of Business Intelligence and Big Data in Higher Education 165
10. Can FoMO Enhance Continuous Intention to use ChatGPT?: Mediation of Innovative Capability of using Data Analytics 180

11. Smart Tourism Supply Chains: Integrating AI, Advanced Analytics, and Robotics for Enhanced Customer Experiences 196

12. Emerging Trends and the Future of Business Analytics 219

13. Analysis and Forecasting Consumer Price Index of Indian Economy via Predictive Modeling Algorithms 230

14. The Meme Revolution: Integrating Cultural Phenomenon and Analytics in Branding Strategy 244

15. Ethical and Legal Challenges in Business Analytics and Digital Technologies 269

Index 283

Editors Biography 285

List of Contributors

Arpana Kumari
Symbiosis Centre for Management Studies Noida, (Symbiosis International University), India.

Anuja Shukla
Swiss School of Business Management, Switzerland.

Arun Bhatia
Central University of Himachal Pradesh, India.

Ashima Agrawal
Jaipuria Institute of Management, Jaipur, India.

Amrita Chaurasia
School of Commerce, finance and accountancy, Christ University, India.

Aditi Sharma
HPKV Business School, School of Commerce and Management Studies, Central University of Himachal Pradesh.

Ashutosh Kumar Choudhary
DASH & HSST, Swami Rama Himalayan University, Dehradun, India.

Bhakti Parashar
VIT-Business School, VIT-Bhopal University, India.

Deepak Halan
Jaipuria Institute of Management, Noida, India.

Devvandya Mishra
Jaipuria Institute of Management, Jaipur, India.

Deepak Srivastava
Swami Rama Himalayan University (SRHU) India.

Geeta Rana
School of Management Studies, Swami Rama Himalayan University, Dehradun, India.

H.H.D.P.J. Opatha
Faculty of Management Studies and Commerce, University of Sri Jayewardenepura, Srilanka.

Imali Fernando
University of Sri Jayawardenapaura, Srilanka.

Kalai Arasi Mehavarnam
PSG Institute of Management, Coimbatore, India.

Lokesh Vijayvargy
Jaipuria Institute of Management Jaipur, India.

M. Manoprabha
Department of Statistics and Applied Mathematics, Central University of Tamilnadu, India.

A.Mansurali
Department of Management, Central University of Tamil Nadu. India

Mridul
Amity Centre for Behavioural Sciences (ACBS), Amity University, Noida, India.

N.W.K.D.K. Dayarathna,
University of Sri Jayewardenepura, Sri Lanka.

Naga Ravi Shankar Mulagaleti
Swiss School of Business and Management Geneva.

Nidhi Srivastava
GL Bajaj Institute of Management Research, Greater Noida, India.

Nidhi S. Natrajan
Symbiosis Centre for Management Studies, NOIDA
Constituent of Symbiosis International (Deemed University), Pune, India

Park Thaichon
University of Southern Queensland, Australia

Poornima Jirli
Swiss School of Business and Management Geneva.

Ravindra Sharma
School of Management Studies, Swami Rama Himalayan University, Dehradun, India.

Rutuparan Anil Pandey
Uniting NSW (Equifax), North Sydney, New South Wales, Australia.

Ramzan Sama
Jaipuria Institute of Management, Jaipur, India.

Sanjay Sharma
Shri Guru Ram Rai University (SGRRU), Dehradun, India.

Shuchita Singh
GL Bajaj Institute of Management Research, Greater Noida, India.

Shivani Agarwal
Galgotias University, Greater Noida, India.

Vipul Jain
Shri Guru Ram Rai University (SGRRU), Dehradun, India.

V. Harish
PSG Institute of Management, Coimbatore, India.

V. Srividya
PSG Institute of Management, Coimbatore. India.

1 Foundation Concepts for Business Analytics

V. Harish[1*], Ravindra Sharma[2], Geeta Rana[3] and Bhakti Parashar[4]

"Foundation of Business Analytics" is an attempt to provide a detailed review and consolidation of the various core concepts and significant progress in the emerging domain of Business Analytics. The chapter attempts to provide the reader with an overview of the term "Business Analytics" and its association with the term "Data Analytics". The chapter then highlights the significant milestones that have critically evolved the dynamic field, tracing its evolution from its early stages to its present status as a vital and indispensable tool for an efficient decision-making process that is being backed by data across organizations. It sets the stage for a comprehensive topic exploration by outlining the chapter's scope and objectives. The next section details the transformative impact of technological advancements on Business Analytics and discusses the importance of data collection and pre-processing techniques to maintain data quality and integrity. The next section details different forms of Business Analytics, including descriptive, diagnostic, predictive, and prescriptive analytics, where each category is thoroughly explained. The chapter concludes by emphasizing the diverse applications of Business Analytics across various industries, underlining its importance as a valuable asset in driving data-driven decisions and optimizing business performance. In summary, "Foundation of Business Analytics" is an insightful book chapter that provides a detailed analysis of the core principles, techniques, and applications of Business Analytics and it attempts to equip readers with a comprehensive understanding of this dynamic field, enabling them to leverage data effectively and make informed decisions to succeed in today's data-driven business landscape.

1. Introduction

Business Analytics (BA) uses statistical techniques with a massive volume of data to solve day-to-day business problems (Shah and Murthi, 2021) and in simple terms refers to the gathering, analysing, and understanding of data to provide valuable and different perspectives that can be utilized to make improved and

[1] 0000-0002-3913-9419
0000-0001-6432-2680
0000-0002-3733-5176
0000-0002-2561-117X
* Corresponding author: harish@psgim.ac.in

DOI: 10.1201/9781003514282-1

faster decisions (Wixom et al., 2013, Sarker, 2021). One of the major applications of BA is that it empowers managers to make timelier and well-informed choices by analysing data, classifying trends, and looking at various patterns of data that might not otherwise be missed (Mikalef et al., 2020). The ability of BA to assist organizations in identifying and mitigating risks is another essential aspect of the discipline where firms can look for possible risks and plan for mitigation strategies by analyzing data effectively (Liu et al., 2020).

Businesses can adopt BA to improve operational efficiency by analyzing data on key performance indicators, and this allows them to identify areas for improvement and implement necessary changes to enhance performance (Bag et al., 2020, Sharma et al., 2019). As a result, companies can reduce costs, boost productivity, and improve customer satisfaction. BA encompasses a wide range of services, which can be primarily divided into four categories: descriptive, diagnostic, predictive, and prescriptive. Each of these categories relies on various tools and technologies and among these are statistical software applications, data mining tools, and visualization tools, with the choice of tools depending on the specific needs of the organization and the completeness of the data being analyzed (Jaggia et al., 2020).

2. Review of Literature

For businesses to succeed, gaining access to the right data at the right time has become a crucial factor, particularly in the current digital age, where competition is fierce (Winter et al., 2012). This requirement is catered to by the technique of business analytics. The evolution of the term ‘Business Analytics’ initially began with the introduction of decision support systems (Sprague, 1980). Following that, concepts like Machine Learning (Samuel, 1959), DBMS (Codd, 1970), Data Science (Cleveland, 2001), and Big Data (Mayer-Schnberger, 2013) came into the picture. Though it is claimed that Business Analytics gained popularity in 2012 (Chen et al., 2012), it is claimed that Hans Peter Luhn coined the term in the 1950s (Hassan, 2019).

Researchers have different perceptions of business analytics and have coined different definitions. Business analytics is a tool that helps to study the past and make informed decisions in the future (Power, 2018). Business analytics is also seen as the use of models, analysis, and management to make data-driven decisions (Davenport, 2006; Harish et al. 2023). Business analytics is also seen as a tool to extract value out of data (Pape, 2016). Authors also define Business Analytics as the usage of analytical resources to gain insights and make informed decisions (Seddon et al., 2016).

Firms of recent times require constant growth and innovation to thrive (Jantunen et al., 2018) and business analytics provides a platform to constantly innovate (Yoshikuni et al., 2023). Innovation further leads to competitive advantage (Li et al., 2023). The amount of advantage that a business can gain from the usage of business analytics depends on many factors like the accuracy of data, target selection methods, top management adaptability, etc (Davenport et al., 2009). Business analytics also finds application in the field of the supply chain, leading to optimized supply chain performance, ultimately leading to increased profits (Kakhki and Sajadi, 2024; Rana et al., 2024). Business analytics also finds application in the field of marketing, allowing companies to gain more information on customers and hence, take a targeted marketing approach (Agag et al., 2023). Authors also put forward that building an informed decision-making culture in an organization would bring about long-term benefits and this can be achieved with the help of the adoption of business analytics (Szukits and Móricz, 2023). Business analysis also plays a part in the financial decision-making process of a company as it helps make informed investment decisions and other important financial decisions (Charles et al., 2023), such as risk analysis (Zhang et al., 2022). Business Analytics also helps in HR processes like staff selection, hiring, etc (Di Lauro et al., 2024). Business Analytics is also useful for firms to make strategic decisions that make a huge impact on the dynamics of the company (Pröllochs and Feuerriegel, 2018). On the whole, Business analytics plays a crucial role in various segments of a company like finance, supply chain, HR, etc, and hence, is capable of making a huge impact on any business.

Business analytics also finds application in a variety of industrial sectors. The healthcare sector is one of the most beneficial sectors of business analytics as it helps in identifying and controlling medical variations (Stadler et al., 2016). The retail industry also utilises business analytics in making decisions like inventory management decisions, product bundling, deciding store layout, etc (Kaur and Kang, 2016). Business analytics also finds applications in niche sectors like sports, in applications like team analysis, player analysis, ticket pricing, etc (Troilo et al., 2015).

Authors have identified three pillars that lay the foundation for the technological development of business analytics, namely, data visualisation, statistical modelling, and machine learning (Raghupathi and Raghupathi, 2021). Visualisation is the process of presenting the data gathered in a visually appealing way, in the form of graphs, tables, etc (Raghupathi and Raghupathi, 2018) and there are numerous business analytics tools available for this purpose (Dill et al., 2012). The statistical model involves the creation of a mathematical model of the data under study and analyzing it using statistical techniques (Ohri, 2012). Machine learning is an advanced form of statistical modelling where imperfect data (missing values, null values, etc) can be handled without any hurdles (Barga et al., 2015).

There are five major types of business analytics, as identified by researchers. These are descriptive analytics, real-time analytics, predictive analytics, prescriptive analytics, and diagnostic analytics (Wolniak & Grebski, 2023). Descriptive analytics provides the current state of the data (W.R.V. Raghupathi, 2013). Real-time analytics helps process data and gain insights at the moment it arrives (W. Chen et al., 2023). Diagnostic analytics helps identify the reason for the data patterns or trends identified (Ridzuan and Zainon, 2022). Predictive analytics moves a step further from a diagnostic model by making predictions of future data (Broby, 2022). Prescriptive analytics explains the next steps that have to be taken and as the name suggests, prescribes future actions (Lepenioti et al., 2019).

Researchers have identified a four-stage technique that is used to solve business analytics problems. These stages include concept design, proposal, implementation, and presentation and evaluation (W.R.V. Raghupathi, 2013). Researchers also suggest the use of different techniques and tools like traditional and modern visualisation tools, analysis techniques, statistical techniques, Optimization, and Heuristic algorithms, etc. at different stages of the project depending on the need (Liu et al., 2023).

Researchers have also identified a few challenges with business analytics, which include data challenges such as data ownership and data privacy (Schwartz, 2011). Other challenges include top management challenges and strategic challenges (W. Raghupathi and Raghupathi, 2014). Hence, a whole lot of scope lies in solving these problems and there lies a research gap in these areas where more research can be done to bring in a fitting solution.

2.1 Objectives

1. To comprehend the evolution of "Business Analytics".
2. To examine the key technological advancements in "Business Analytics".
3. To understand the types of "Business Analytics".
4. To examine the techniques used in "Business Analytics".
5. To understand the application and future directions of "Business Analytics".

3. Methodology

Social science research has predominantly been classified into two competing paradigms namely qualitative and quantitative research generally qualitative research is adopted to recognize the perceptions and perspectives of the respondents (Beaudry and Miller, 2016), and hence qualitative research was adopted due to its nature of interpretive nature which assists in understanding the field of BA. PRISMA framework (Fig. 1) was adopted a detailed study on the research topic. Prevalent databases, namely "Web of science" and "Scopus", for literature were identified from 2017 to 2023. The subjects identified for shortlisting the articles were "Business management and accounting. Social sciences, Decision Science, Economics, econometrics and Finance and multi-disciplinary" with keywords of Business AND Analytics. As a result, in July 2023, 12,513 articles were identified and duplicates were removed and filtered adopting the "inclusion and exclusion criteria", as an outcome of the same, 184 relevant articles were recognized and a final list of 76 titles was considered for the study.

It can be inferred from the above-given Fig. 2, that while early documents appeared on the topic as early as 1987, there has been an exponential growth of articles since 2011 indicating the growing prominence of the subject.

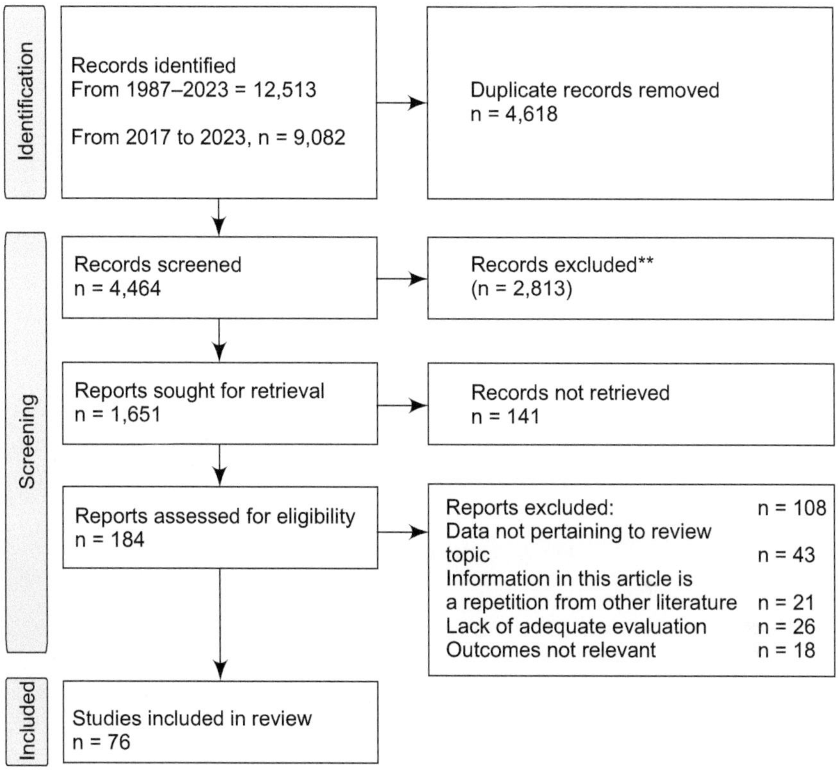

Fig. 1 PRISMA framework.
Sources: Author's own

It can be observed from Fig. 3 that a majority of research on the research topic was done in the United States, followed by the U.K., Australia, China, Canada, Germany, France, and India.

A keyword analysis on the mentioned topic area and the identified articles was done as shown in Fig. 4, indicated that the top keywords used were "Business Analytics, Big data, Decision Making, Advanced Analytics, Business Intelligence and Data Analytics".

4. Analysis

4.1 Business Analytics and Data Analytics

Data Analytics is a field that centres on the investigation of data, with a primary emphasis on deriving meaningful insights from unprocessed data, and it encompasses the course of gathering, refining, and converting the vast amount of data to conduct diverse analyzes to uncover patterns, trends, correlations,

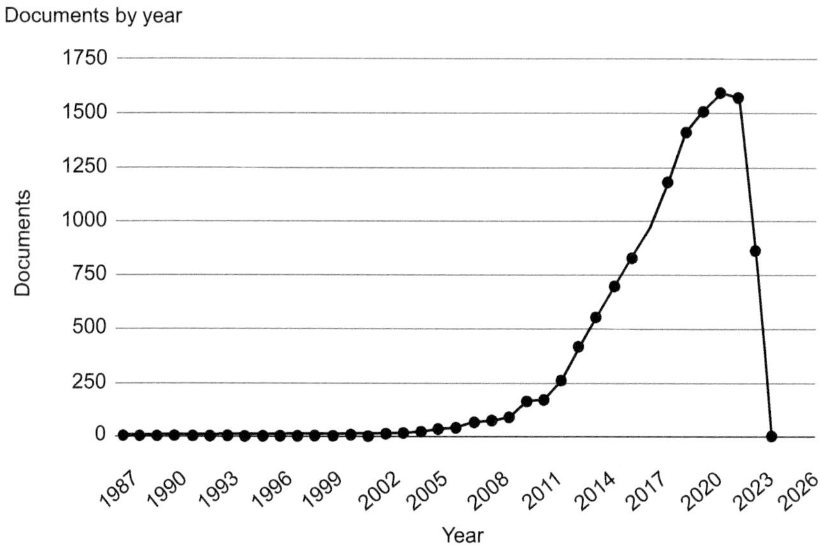

Fig. 2 List of documents by year.
Sources: Authors's own

and anomalies. Data Analytics is essential for making informed decisions based on data, gaining insights into customer behaviour, streamlining processes, and pinpointing areas that require enhancement. It primarily focuses on historical data, allowing businesses to develop a more comprehensive comprehension of previous events and patterns through descriptive analytics and commonly utilises various statistical methods, data mining, and data visualisation techniques to effectively explore and interpret data.

However, BA extends beyond the realm of DA, as it is influenced by the specific requirements of the business and the key focus of BA is to offer valuable understandings based on which can be used for strategic planning, resource allocation, risk management, and overall business optimisation.

DA is typically carried out by data analysts, data scientists, and statisticians who possess the necessary expertise to handle and analyze data proficiently, and they collaborate closely with subject matter experts to analyze the findings and deliver customised insights that cater to specific business requirements. On the other hand, BA serves a broader range of individuals, such as business executives, managers, and stakeholders, and the findings derived from BA are effectively communicated to guide business strategies and operations in a concise and actionable manner.

The primary goal of DA is to discover significant informative trends and patterns in the data, leading to a deeper comprehension of previous events and enhancing current processes to increase operational efficiency, whereas Business analytics focuses on enhancing business performance and results through facilitating strategic, well-informed, and proactive decision-making. It assists organisations in

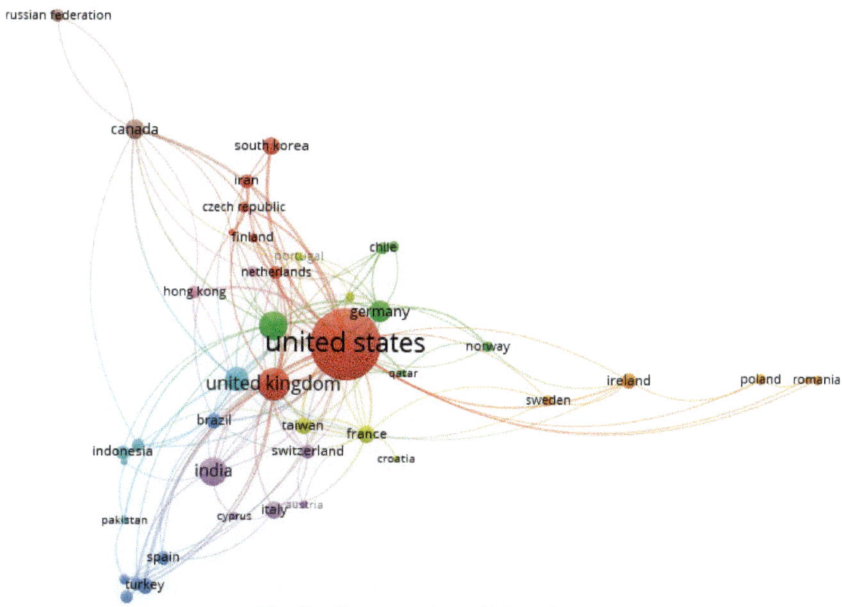

Fig. 3 Country-wise collaboration.
Sources: Authors's own

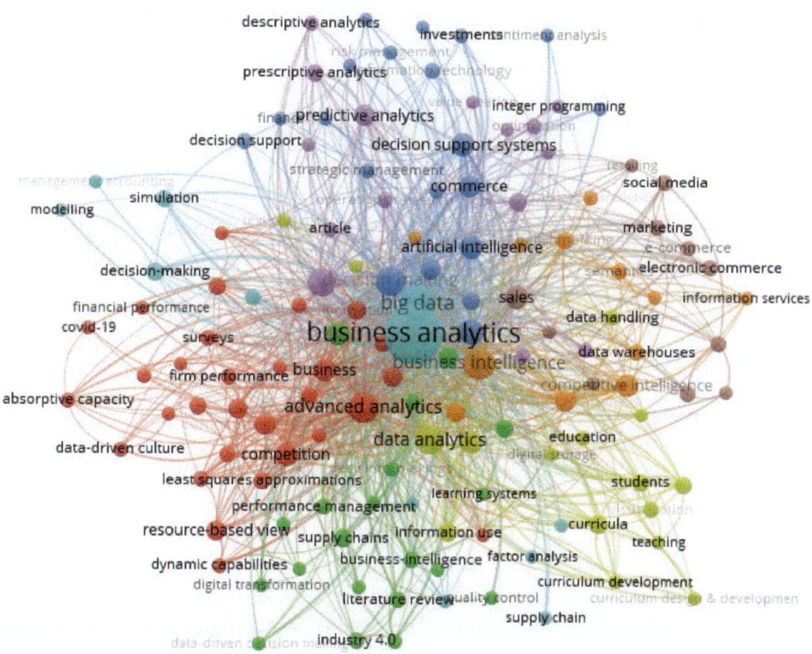

VOSviewer

Fig. 4 Keyword analysis of the articles identified.
Source: Author's own

attaining higher revenue, lower expenses and provides a competitive edge for the firm by utilising insights derived from data analysis.

DA is the process of reading out information from the data that is available in different formats, including formal data from companies' databases and informal data from sources like webpages, the internet, social media sites, and weblogs and it can concentrate on a particular dataset or merge various datasets to obtain a holistic understanding of the information.

In summary, DA and BA are two distinct fields that work together to support an organization's data-driven decision-making process while data Analytics serves as the basis, offering valuable insights and comprehension of past events and patterns, BA leverages these insights, utilising sophisticated techniques and models to forecast future outcomes, recommend actions, and strategically propel the business forward.

4.2 Evolution of Business Analytics

The evolution of Business analytics has evolved rapidly over the past few decades and it can be divided into three distinct periods: the early years, the emergence of business intelligence, and the current era where each of these eras has contributed to the field's evolution and had a substantial influence on the mode in which firms work and make decisions.

Early Years (1950s–1970s)

The origins of Business analytics can be looked back to the early 50s, when the first business computers were developed and implemented, and it was during this period, that businesses began to gather and investigate sales, customer, and financial data from assorted sources (Klimberg, 2022). One of the most prominent tools was using basic statistics for inference and statistical analysis was one of the earliest forms of BA, which involved using statistical methods to analyze data and classify patterns and inclinations and this was typically accomplished using costly mainframe computers and difficult-to-use specialized software which as augmented by the development of more sophisticated computer systems and software in the 1970s, allowing them to collect and analyze greater quantities of data (Davenport, 2018).

Emergence of Business Intelligence (1980s–1990s)

In the 1980s and 1990s, business intelligence (BI) emerged, involving the analyzing the vast amount of data to gain valuable information to improve business performance, and specific BI tools were created to assist businesses in obtaining an insightful understanding of their operations and making better-informed decisions (Chen and Lin, 2021). During this phase of development, businesses began to develop more sophisticated data warehouses and data consolidation points, enabling them to acquire and store vast quantities of data, and the info thus obtained is then used to generate business performance reports and dashboards.

The emergence of online analytical processing (OLAP), enabled businesses to analyze data from multiple perspectives and dimensions (Patel and Sharma, 2020), was one of the most significant developments of this era as the OLAP tools were interactive and user-friendly which made it simpler for business users to make decisions grounded on DA.

Rise of Business Analytics (2000s–Present)

The turn of the century witnessed the development of BA to drive business performance and strategy, whereas BA is just an advanced form of BI that employs more complex analytical techniques, such as data mining to obtain valuable information (Božič and Dimovski, 2019).

The emergence of big data, which denotes the large volumes of data captured by enterprises and other organizations, was one of the defining features of this era (Zaki, 2019) as big data has created both challenges and opportunities for businesses, as collecting, storing, and analyzing large volumes of data requires new technologies and techniques. One of the most important technologies was the advent of cloud computing, leading to simplifying and reducing the cost of storage and analyzing data by enterprises (Li and Zhang, 2021). Additionally, cloud-based analytics tools and services have made it simpler for businesses to access advanced analytics capabilities without investing in costly hardware and software.

In the last few years, the importance of specific technologies such as artificial intelligence (AI) in BA has grown. AI and Machine learning (ML) are utilized to develop predictive models and recognize patterns in data, used to drive business performance and strategy (Rathore, 2023 A).

4.2.1 Key Technological Advancements in Business Analytics

Since its inception in the early years of computation, Business analytics has made great strides where technological advancements have played a substantial role in determining the field and enabling businesses to collect, process, and investigate immense quantities of data to make informed decisions. In this section, we will examine some of the most significant technological developments that have revolutionized BA.

- **Big Data (BD):** The proliferation of BD is a noteworthy technological advancement in BA where in today's data-driven society, businesses produce enormous quantities of data daily. This information includes customer data, financial data, supply chain information, and social media data and the process of consolidation of data, storage, and analysis capability has enabled businesses to acquire a sense of their operations and make better and faster choices (Sheng et al., 2021) and to handle this enormous amount of data, firms have adopted various tools such as Hadoop and Spark, which can process massive amounts of data and enable real-time analysis (Rodrigues et al., 2021).
- **Predictive Analytics:** By utilizing algorithms and machine learning techniques, business analytics can help analyze data to forecast future events

(Ngiam and Khor, 2019), and this capability is crucial for companies to identify trends, seize opportunities, and mitigate risks. For instance, IBM's Watson Analytics uses natural language processing to make complex data analysis accessible to users without technical expertise, making it a widely used tool (Schuff et al., 2018). Additionally, Microsoft Azure ML serves as a predictive analytics platform that allows businesses to develop and implement predictive models on a large scale (Barga et al., 2015).

- **Artificial Intelligence and Machine Learning:** AI and ML are two closely connected technological advancements that are reshaping BA (Das et al., 2015) where both of these technologies employ sophisticated algorithms to analyze massive quantities of data that would be difficult or impossible for humans to discover (Dixit, 2022) and AI-powered data visualization tools make it easier for firms to extract valuable information from complex data and communicate them in a more accessible manner.
- **Data Visualization:** Data Visualization is another important technological advancement in business analytics where visualization tools enable the transformation of unprocessed data into interactive, graphical representations, allowing users to gain insights swiftly and easily (Li and Li, 2020). Popular among businesses, data visualization tools such as Tableau and Qlikview empower firms to generate dashboards and reports that simplify the identification of patterns and trends (Sharma, 2020).

Technological advancements have been instrumental in shaping the field of BA and as businesses generate more data and face new challenges, they will continue to rely on these technologies to obtain a more in-depth understanding of their operations, enhance their decision-making, and stimulate growth.

5. Data in Business Analytics

Data is the new oil in today's digital era, and businesses are increasingly leveraging it to obtain a competitive advantage, and with the emergence of big data, BA has become an integral component of decision-making across industries. The effectiveness of BA depends heavily on the caliber of the data used for analysis (Paradza and Daramola, 2021). In this section, the various data-related aspects of BA, such as data types, data sources, data accumulation and storage, and data pre-processing will be discussed.

- **Types of data:** "Structured and unstructured data" are the two broad classifications applicable to BA data where structured data is information that is prepared in a predetermined arrangement and is readily searchable and analyzed (Ajah and Nweke, 2019) and data stored in relational databases, spreadsheets, and transactional data are instances of structured data and unstructured data, on the other hand, is data that is not organized in a predetermined format and is difficult to search or analyze (Batko and Ślęzak, 2022) and cases of unstructured data comprise of social media data, emails, videos, pictures, and auditory files and are more tough to analyze than structured data (Khan et al., 2022) but it can also provide essential customer behavior and sentiment insights. Data that lies somewhere between structured and unstructured data is referred to as "semi-structured data" and it is somewhat structured, but not in a predetermined format. XML files and JSON data are some examples (Berndt, 2020).

- **Data Sources:** Data sources refer to the numerous locations from which data is gathered for analysis and can be obtained from both internal and external sources for BA (Ferraris et al., 2019).

Internal data sources include transactional data, customer data, employee data, and financial data generated within the organization. Data generated outside the organization, such as market data, social media data, government data, and third-party data, are examples of external data sources and the selection of data sources is heavily influenced by business requirements and the type of analysis being performed (Burgess et al., 2019), for instance, if a company wishes to undertake sentiment analysis on customer feedback, social media data would be a valuable source of information.

- **Data Collection and Storage:** Data gathering indicates a gathering of information for analysis from a variety of sources and the procedure for collecting data varies depending on the nature of data and its sources whereas, in contrast, unstructured data may necessitate manual data entry or data mining, and once collected, data must be retained in an appropriate data storage system (Wang et al., 2020). In general, there are two categories of data storage systems: "On-premises and cloud-based". "On-premises data storage" denotes data storage systems located on the premises of an organization. Relational databases, data repositories, and file servers are examples of these systems. "On-premises data storage" gives organizations full control over their data, but it can be expensive to implement and maintain for small and medium firms while, "cloud-based data storage" refers to data storage systems that are hosted on cloud servers (Mohammad and Pradhan, 2021) and some of the "cloud-based data storage" systems are Amazon S3, Google Cloud Storage, and Microsoft Azure and "Cloud-based data storage" is ideally more economical and offers scalability and adaptability (Alam, 2022).
- **Data Pre-processing:** The process of cleansing, transforming, and preparing data for analysis is termed data pre-processing and is a key stage in the data analysis, as it ensures the accuracy and dependability of the results (Bekar et al., 2020) by eradicating inconsistencies, errors, and absent values that could skew an analysis constitutes data cleansing (Mirnaghi and Haghighat, 2020).

5.1 Types of Business Analytics

BA as a subject has different kinds and this section examines the four primary groups of BA: "descriptive, diagnostic, predictive, and prescriptive" (El Morr et al., 2019).

- **Descriptive Analytics:** The fundamental type of data analysis, it entails summarizing and aggregating past data to obtain insight into historical performance, the primary objective of descriptive analytics is to answer the query of the past (Greasley, 2019) and this type of analytics is used to track key performance indicators (KPIs) such as revenue, profit, and consumer retention and in general, firms can adopt descriptive analytics to comprehend their historical performance and identify trends, patterns, and outliers in their data.
- **Diagnostic Analytics:** It attempts to explain why something occurred and this is adopted to identify a problem's root cause or to explain a particular trend or pattern (Muneeswaran et al., 2021) and diagnostic analytics can be used to investigate issues related to customer behavior, product quality, supply chain problems, and any other area where understanding the root causes of a specific issue is necessary.
- **Predictive Analytics:** It is a form that predicts upcoming events using algorithms and ML techniques, and it aims to provide an answer to the query of forthcoming events (Menezes et al., 2019), such as forecasting sales, detecting customer churn, and predicting equipment malfunction, and much more.

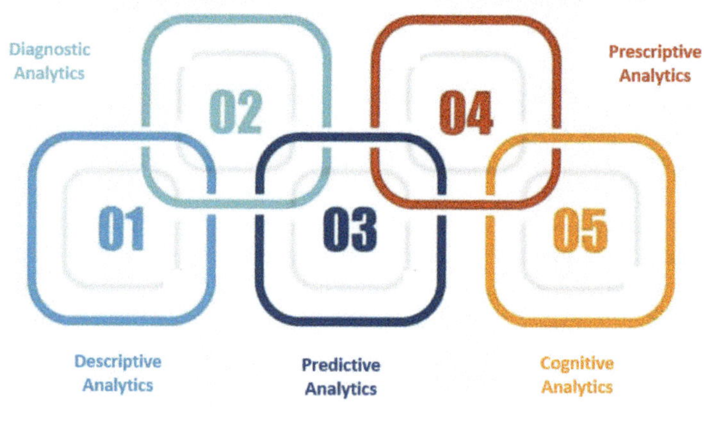

Fig. 5 Types of Business Analytics.
Source: Author's own

- **Prescriptive Analytics:** It employs algorithms and ML techniques to recommend specific actions to improve business outcomes, and fundamentally it focuses on what should be done to attain a specific objective (Frazzetto et al., 2019) such as enhancing pricing recommendations, improving supply chain efficiency, and increasing consumer engagement.
- **Cognitive Analytics:** It is a subset of BA that analyzes data using AI and ML algorithms, and this type of analytics concentrates on comprehending the thought processes and decision-making abilities of humans and then recreating them in machines (Gudivada et al., 2016). Cognitive Analytics can assist businesses in better comprehending consumer behavior, enhancing

decision-making, and automating mundane tasks, and this form of analytics is frequently employed in industries such as healthcare, finance, and retail, where decision-making is crucial and data volumes are large. Cognitive analytics is an emerging discipline that is anticipated to continue expanding and transforming over the next few years.

Table 1 Comparison of the types of Business Analytics.

Points of Comparison	*Descriptive Analytics*	*Diagnostic Analytics*	*Predictive Analytics*	*Prescriptive Analytics*	*Cognitive Analytics*
Focus	Historical data	Identifying root cause	Future predictions	Recommending solutions	Understanding cognition
Purpose	Describe what happened	Explain why it happened	Predict what will happen	Recommend what to do	Insight generation
Data Type	Past data	Past data	Past data and current	Past data and present	Past and present
Techniques	Aggregation	Drill down	Regression, machine learning	Optimization, simulation	ML, Natural Language Processing, Neural Networks
Example	Sales reports	Customer churn analysis	Demand forecasting	Supply chain optimization	Analyzing customer sentiment to improve product design
Key Questions	What happened?	Why did it happen?	What will happen?	What should we do about it?	What insights can we gain
Timeframe	Past	Past	Future	Future	Real-time analysis

Table 2 5 W 2 H for types of analytics.

5W	*Descriptive Analytics*	*Diagnostic Analytics*	*Predictive Analytics*	*Prescriptive Analytics*	*Cognitive Analytics*
Why?	To summarize data	To find root causes	To predict future outcomes	To prescribe actions based on insights	To provide insights using ML
What?	Historical data analysis	Analysis of performance	Analysis of trends and patterns	Optimization and decision-making	Advanced data analysis
Who?	Business Analysts	Data Analysts	Data Scientists	Business Analysts, Data Analysts, Data Scientists	Data Scientists, ML Experts

Contd.

Table 2 *Contd.*

Where?	On-premise, Cloud-based	On-premise, Cloud-based	On-premise, Cloud-based	On-premise, Cloud-based	Cloud-based
When?	Historical data analysis	Post-event analysis	Real-time analysis	Real-time analysis	Real-time analysis
How?	Visualization, Reporting	Statistical Analysis	Data Mining, ML	Optimization Models	ML
How Much?	Cost-effective	Cost-effective	More expensive than descriptive/ diagnostic analytics	More expensive than predictive analytics	The most expensive type of analytics

These are some of the most significant distinctions between the four types of analytics, emphasizing their focus, objective, data type, techniques, examples, key questions, and timeframe. DA focuses on past data and provides an overview of past events, whereas diagnostic analytics identifies the underlying cause of a specific outcome.

6. Advanced Analytics Techniques

With the advent of big data and more sophisticated computational skills, businesses now have access to more sophisticated analytical tools and techniques, and implementing advanced analytics techniques requires more sophisticated tools and software due to their complexity (Bănărescu, 2015) and these techniques can aid organizations in gaining a deeper understanding of their operations, predicting future trends, and making more informed decisions and in this section, we will examine the most prevalent advanced analytics techniques employed by businesses today.

- **Cluster analysis:** It is a statistical technique for categorizing objects according to their similarities and the focus is to group comparable objects and separate dissimilar objects (Blashfield and Aldenderfer, 1978) and this method is frequently employed to segment consumers based on their behaviour or to group products according to their attributes by utilizing cluster analysis, businesses can gain insight into consumer preferences and behaviour and develop targeted marketing strategies.
- **Data Mining:** Data mining is the discovery of patterns, trends, and relationships within large data sets and this method is frequently employed to uncover insights that are not readily apparent through simple data analysis (Hasan et al., 2020) and can be used to anticipate consumer behavior, detect fraud, and optimize supply chain operations (Rathore, 2023 B).
- **Machine learning:** It is a form of AI that enables computers to learn without being programmed (Janiesch et al., 2021) to predict future outcomes based on

historical data and can be used to predict customer attrition, identify fraudulent transactions, and optimize pricing strategies (AL-Najjar et al. 2022).

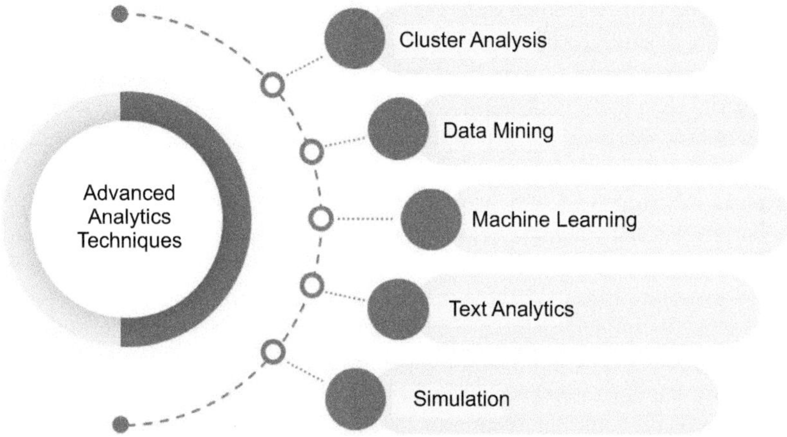

Fig. 6 Advanced analytics techniques

- **Text Analytics:** It examines unstructured text data to discover patterns and insights (Marzouk and Enaba, 2019) and this method is frequently employed to analyze customer feedback, social media posts, and other unstructured data types and text analytics can be employed to determine consumer sentiment, comprehend brand perception, and track public opinion (Kumar et al., 2021).
- **Simulation:** Simulation is a method involving the creation of a virtual model of a real-world environment to analyze its behaviour under varying conditions (Banks, 1999) and can be utilized to augment the supply chain, foretell equipment failure, and pinpoint process improvement opportunities (El-Haik and Al-Aomar, 2006) by simulating various scenarios, businesses can make more informed decisions.

In conclusion, sophisticated analytics techniques can provide businesses with a greater comprehension of their operations and enable them to make more knowledgeable decisions, such as cluster analysis can assist in grouping customers or products based on their similarities, and data mining can be used to unearth hidden insights, ML can be used to predict, text analytics to analyze unstructured data, and simulation can be used to test various scenarios.

7. Tools and Technologies for Business Analytics

The following are some of the most significant tools used in BA:

- **Statistical Tools:** Analyzing data and deriving insights from it requires the use of statistical tools where these instruments can assist in recognizing relationships and associations within the various variables, some examples of statistical tools are SPSS, SAS, R, and Python, where these tools are capable of data cleansing, hypothesis testing, regression analysis, and clustering, among other duties (Mattos et al., 2021). SPSS is a popularly utilized statistical tool in the field of BA which offers an intuitive interface for data analysis and is capable of handling both small and large datasets, SAS is a prevalent statistical tool in the business world, particularly in the healthcare, finance, and retail sectors (Xia and Gong, 2014), R and Python are open-source programming languages that are gaining popularity for data analysis due to their adaptability and potency.

- **Data Mining and Machine Learning Tools:** The relationships in the variables can be identified along with the future outcomes, and some tools are IBM SPSS Modeller, RapidMiner, and KNIME. While IBM SPSS Modeller is a great tool for establishing predictive models, RapidMiner also specializes in creating models using a friendly interface. KNIME, on the other hand, is an open-source application capable of working with large datasets and flexible enough to perform different works like text mining and image analysis (Kadaru and UmaMaheswararao 2017).
- **Big Data Tools:** These are the ones that have been used in handling data sets beyond what traditional tools can handle (Hariri et al., 2019) such as Hadoop, Spark, and Hive, which are applied for different purposes, including storage, processing, and analysis of data. Hadoop is an open-source framework that has been primarily designed for storing and analyzing huge data sets while spark is a tool of big data, which is a general use tool, that has great capability to manage large databases in high volume sets, including different functions such as data streaming and machine learning operations and on the other end, Hive acts as a data warehouse application that allows users to query the vast datasets of Hadoop (Sharma et al., 2021; Eckroth, 2018).

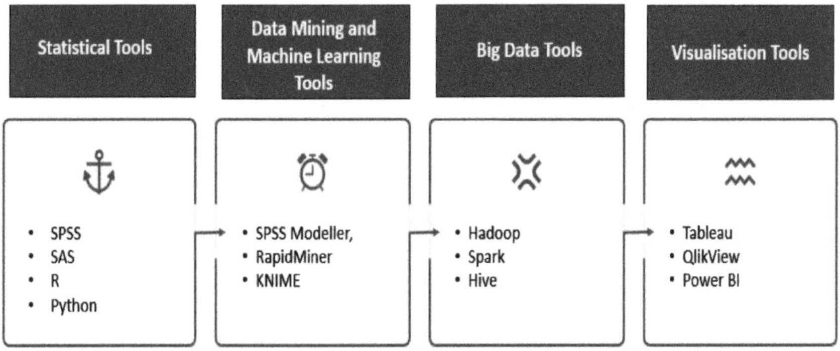

Fig. 7 Tools and Technologies for BA.
Source: Author's own

- **Data visualization tools:** These are the types of tools that will provide a visual representation of large and complex data which can then be communicated to the stakeholders in terms of insights and findings (Ansari et al., 2022) some popular visualization tools include Tableau, QlikView, and Power BI. Tableau is known for its user-friendly interface where you can generate interactive dashboards and visualizations, QlikView is yet another effective visualization tool which can accommodate large amounts of data making it highly suitable for data analysis and discovery and Power BI is also one of the popular one from Microsoft for interactive reporting and visualization (Toic et al., 2022).

7.1 Techniques used in Business Analytics

Some of the techniques used in BA are as follows

- **Regression Analysis:** Through this tool, it is possible to discover any relationship between two or more variables and predict the outcome (Vetter and Schober, 2018) where it models the relationship of the dependent variable with independent variables and is applicable in many business scenarios (Hoang and Watson, 2022). A marketing manager might use regression analysis, for instance, to understand the relationship between sales and changes made to advertising spending. The simplest form of regression analysis is "simple linear regression" – this is where a simple 'dependent-independent' relationship is formed and all the remaining forms of regression analysis are all-inclusive in 'multiple linear regression' forms. In simple linear regression, the connection between only two variables is analyzed (Montgomery et al., 2021).
- **Clustering:** This method aggregates similar data based on distance, similarity, or density and finds the corresponding patterns and forms in a large dataset (Ahmad and Khan, 2019) where it identifies different alternative solutions in business; for instance, it can enable the retailers to manage and develop marketing messages and promotions differently to all segments and make their campaigns more effective (Seyedghorban et al., 2020). The two major classification types of clustering are hierarchical clustering and k-means clustering (Anitha & Patil, 2022) where hierarchical clustering is defined as organising similar data points into a tree-like structure, while k-means clustering groups the data into fixed clusters based on distance (Govender and Sivakumar, 2020; Tabianan et al., 2022).
- **Classification:** The new prediction for a carrier relies on the characteristics of already classified observation data, aiming to assign a class to a new observation based on the traits of previously classified examples (Koziarski, 2020). This approach can generate predictions that assist in decision-making for future events, For instance, this classification method can be applied in various business contexts, such as a tool that categorizes applicants into risk levels based on their credit histories and other factors, helping banks determine whether to approve loan requests from applicants–an example of how classification processes are utilized in the banking sector. The fundamental types of classifications include binary and multivariate classification (Harangi, 2020) where in binary classification, there are only two classes that an observation can belong to, while multiclass classification allows a specific

observation to be associated with three or more category classifications (Bocken et al., 2022).

- **Time Series Analysis:** Using statistical techniques, time series analysis identifies trends, patterns, and anomalies in data over time. This method is utilized to model the behavior of a variable over time and make predictions based on past information (Wulff, 2017).

8. Application of Business Analytics

BA as a tool can equip firms with the insights and knowledge necessary to make better decisions, optimize operations, and achieve greater efficiency, some of the top BA applications in various industries are as follows:

- **Marketing Optimization:** Business analytics can examine consumer data and preferences to create well-targeted marketing strategies that enhance sales and boost customer engagement (Chahal et al., 2019). Properly directed information can significantly improve marketing efforts.
- **Sales Forecast:** By analyzing consumption behavior, businesses can utilize sales and demand forecasting models to enhance their inventory and resource planning (Kraus et al., 2020). Decisions made regarding future sales volumes are effectively based on historical data and predictive models.
- **Supply Chain Optimization:** Consequently, business analytics contributes to cost reduction and more efficient resource utilization. This leads to lower costs, shorter lead times, increased productivity, and ultimately, higher customer satisfaction (Araz et al., 2020).
- **Risk Management:** This enables managers to implement appropriate risk mitigation strategies, as it helps identify potential risks before they develop into significant issues (Leo et al., 2019). Additionally, it aids in reducing certain risks by allowing for their identification and prevention based on data analysis and trends.
- **Fraud Detection:** Financial analysis through business analytics can uncover unusual behavior patterns across various types of activities (Tang and Karim, 2019), addressing some of the most pressing needs in the financial sector and potentially leading to substantial cost savings for the industry.
- **Customer Relationship Management:** BA can enhance the client experience by analyzing client data and recognizing behavioural trends and patterns (Anshari et al., 2019), and having a robust CRM practice can help in improving efficiency and productivity.
- **Financial Planning and Analysis:** BA can examine financial data and optimize financial planning and analysis, allowing businesses to plan and assign their resources in a better way (Tong-On et al., 2021), and the advantage of BA is that it can look at huge volumes of historical data and can be extremely useful in financial planning and analysis.
- **Healthcare Analytics:** BA can be used to analyze healthcare data, reveal patient outcomes, predict disease patterns, and enhance clinical decision-

making (Wowak et al., 2023), and healthcare analytics, particularly BA in the healthcare industry, has emerged as a critical field of late with the rising prominence scientific-based decisions, and it has become imperative for healthcare organizations to leverage analytics to improve patient care, streamline operations, and optimize resource allocation.

- **Energy and Utilities:** BA can be used to monitor and manage the production, distribution, and consumption of energy in the energy and utilities sector. Prescriptive analytics can optimize energy consumption and waste minimization, and also forecast demand (Maran et al., 2022) and the integration of BA has become increasingly vital for optimizing operations, improving efficiency, and fostering sustainable practices.
- **Retail:** BA can help retailers better comprehend consumer behavior and preferences, optimize pricing strategies, and enhance inventory management. Prescriptive tools can optimize pricing and inventory levels, while predictive tools can estimate future demand (Hopkins, 2022) as one of the most dynamic and competitive sectors, the retail industry in India has witnessed a significant transformation with the integration of advanced analytics.
- **Government:** Government agencies can leverage business analytics (BA) by analyzing historical data to pinpoint areas for optimizing resource allocation, ultimately enhancing the effectiveness of public services (Mithas et al., 2013).
- **Manufacturing:** BA empowers factory owners to achieve significant improvements in production processes while reducing costs and enhancing product quality, predictive tools help forecast demand, and optimize production schedules, and prescriptive analytics assist in resource optimization and quality control (Sahoo, 2022).
- **Hospitality:** In the hospitality sector, businesses can utilize BA to refine pricing strategies, enhance customer experiences, and improve operational efficiencies, for instance, predictive analytics can forecast demand and optimize room rates, while prescriptive analytics can streamline labor management and boost customer satisfaction (Mariani, 2020).
- **Human Resources:** BA can enhance human resources by streamlining processes and aligning with business objectives (Harish et al., 2021) where it can specifically improve talent acquisition by predicting future hiring needs, assessing candidate qualifications, and optimizing recruitment strategies (Sousa et al., 2019; Rana et al., 2019).

9. Importance of Business Analytics

Numerous researchers have emphasized the significance of BA, and some of the most frequently cited reasons are as follows:

- *Better decision-making:* BA provides organizations with the data and information resources necessary for making informed, data-driven choices, for example, companies can uncover patterns or trends and identify relationships that might otherwise go unnoticed (Neto et al., 2023).

Table 3 Application of Business Analytics

Industry	*Descriptive Analytics*	*Diagnostic Analytics*	*Predictive Analytics*	*Prescriptive Analytics*
Retail	Sales analysis, inventory management, customer segmentation	Root cause analysis, customer churn analysis, A/B testing	Demand forecasting, price optimization, personalized recommendations	Supply chain optimization, dynamic pricing, product recommendations
Healthcare	Patient demographics, medical histories, clinical outcomes	Disease diagnosis, treatment efficacy analysis, readmission analysis	Disease progression modeling, patient risk stratification, clinical decision support	Personalized treatment plans, resource allocation optimization, clinical trial design
Finance	Transaction analysis, financial statements, credit scoring	Fraud detection, risk management, compliance monitoring	Portfolio optimization, credit risk modeling, market trend analysis	Automated trading, investment recommendations, loan approval
Manufacturing	Quality control, production output, supply chain efficiency	Equipment failure analysis, downtime analysis, yield analysis	Predictive maintenance, defect prediction, demand forecasting	Production planning optimization, supply chain risk management, vendor selection
Education	Student demographics, enrollment trends, graduation rates	Student performance analysis, course evaluations, retention analysis	Dropout prediction, student risk stratification, adaptive learning	Curriculum planning optimization, personalized learning, academic program evaluation
Marketing	Web traffic analysis, social media metrics, campaign performance	Brand sentiment analysis, competitor analysis, customer satisfaction analysis	Lead scoring, customer lifetime value modeling, market segmentation	Marketing mix optimization, campaign optimization, customer journey mapping
Sports	Game statistics, player performance, team records	Player injury analysis, opponent scouting, lineup optimization	Win probability modeling, player career trajectory analysis, fantasy sports analytics	Player salary optimization, game strategy optimization, team value assessment
Government	Public safety analysis, budget analysis, citizen satisfaction analysis	Crime trend analysis, emergency response analysis, infrastructure analysis	Resource allocation modeling, public health modeling, disaster response modeling	Policy optimization, regulatory compliance monitoring, citizen engagement optimization
E-commerce	Site traffic analysis, customer reviews, purchase history	Cart abandonment analysis, checkout optimization, pricing analysis	Cross-sell/up-sell modeling, purchase prediction, product recommendations	Supply chain optimization, customer service optimization, personalized marketing

- *Efficiency:* BA can enhance a firm's efficiency by streamlining operations and processes, eliminating waste, and improving overall productivity, and this includes conducting bottleneck analyses along production lines and optimizing supply chain management (Schnegg and Moller, 2022).
- *Competitive advantage:* BA should be integrated into the strategies and decision-making processes of businesses to gain a competitive edge and analytics empower organizations to spot new opportunities and trends that are crucial for adapting to the competitive landscape (Wang et al., 2019).
- *Insights from consumers:* By analyzing consumer data and behavior, businesses can gain a deeper understanding of their customers, allowing them to tailor products and services to better meet customer needs (Omar et al., 2019).
- *Risk management:* BA can be employed to identify and manage both external and internal business risks as analytics can help detect potential misconduct and ensure compliance with regulations (Araz et al., 2020), leading to more effective and proactive risk management.
- *Innovation:* Business Analysis helps organizations uncover new development and innovation opportunities by analyzing market trends and consumer behavior, businesses can identify unmet needs in the market and create new products and services to address those needs (Duan et al., 2020).

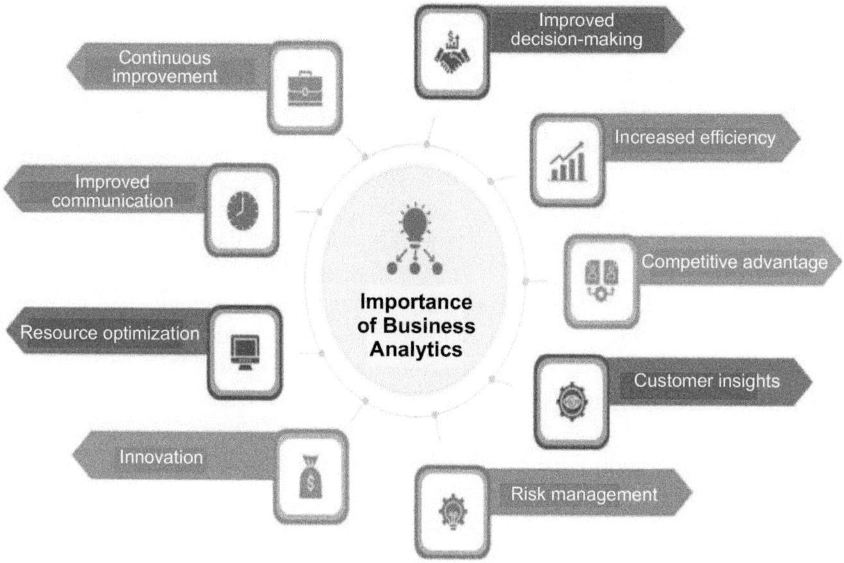

Fig. 8 Importance of Business Analytics.
Source: Author's own

- *Resource optimization:* Business Analytics is essential for maximizing an organization's resources by effectively managing staffing levels, innovative inventory practices, and more and this approach helps companies reduce costs and boost profits while maintaining efficiency (Handelsman, 2012).

- *Improved Communication:* Analytics can significantly enhance communication within an organization by providing clear insights into business performance which are easy to understand and act upon, aiding managers in making informed decisions and fostering better collaboration among team members (Larson, 2019).
- *Continuous Improvement:* Regular performance evaluations allow businesses to identify areas for improvement quickly, and this enables them to make data-driven decisions to address issues as they arise. BA supports organizations in remaining flexible and responsive to changes in the market (Shanks and Bekmamedova, 2012).

10. Future Directions in Business Analytics

Since its inception, BA has come a long way and continues to develop rapidly. Here are some emerging trends in BA:

- **Real-time analytics:** Real-time analytics will grow in importance as businesses seek to reply faster to varying conditions and client requirements, and businesses can make better decisions and analyze data in real-time (Kuznetsov et al., 2023).
- **Predictive modelling:** Predictive modelling is likely to play a vital role in BA as companies attempt to anticipate and respond to market changes (Singh, 2023), and by using ML algorithms and other analytics techniques, patterns and trends in data will be identified and future outcomes will be predicted.
- **Artificial intelligence and machine learning:** AI & ML will remain to improve BA, allowing businesses to automate processes, identify new opportunities, and make better decisions (Lampropoulos, 2023), and can distinguish patterns and insights in large data sets that are impossible for humans to detect.
- **Advanced data visualization:** Data visualization techniques will continue to evolve, making it less difficult for users to understand and present complex data sets and create interactive dashboards (Liu et al., 2023).
- **Cloud-based analytics:** As businesses look to reduce costs and increase flexibility, cloud-based analytics will continue to increase in popularity and the scalability option of the cloud enables industries to adapt to changing circumstances (Namjoshi, 2023).
- **Internet of Things (IoT):** IoT systems will enable firms to collect data and analyse from sensors, wearables, and other connected devices, and this information can be used to stimulate innovation and enhance business processes (Barani Sundaram et al., 2022).
- **Natural language processing:** NLP will increase in significance as businesses seek to obtain understanding from data, such as text and speech. NLP can be

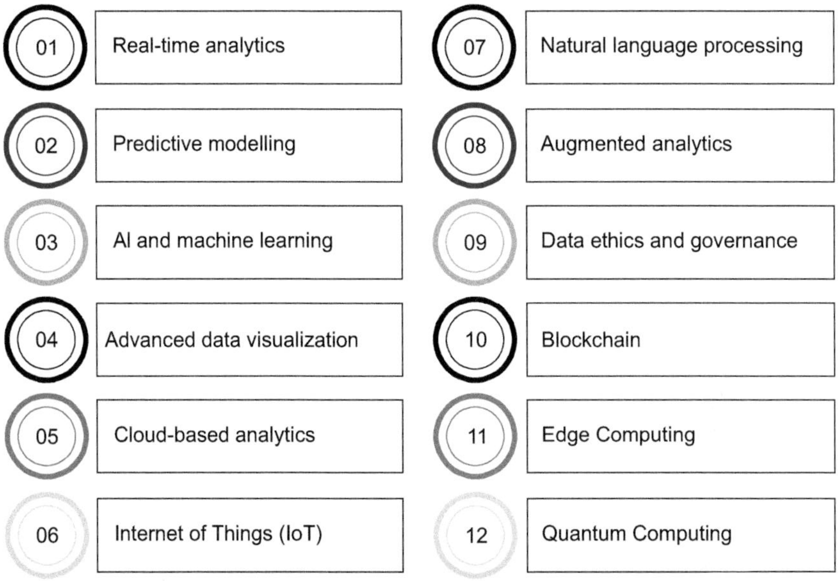

Fig. 9 Future Directions in BA.
Source: Authors's own

adopted to interpret client feedback, social media posts, and other sources of unstructured data (Gibson and Shibani, 2022).

- **Augmented analytics:** Augmented analytics will allow businesses to automate numerous data preparation, analysis, and interpretation processes. This will allow analysts and other professionals to focus on duties with greater value (Alghamdi and Al-Baity, 2022).
- **Data ethics and governance:** As organizations accumulate and analyze more data, the prominence of strong ethics and governance will increase. Businesses must ensure that they collect and utilise data in an ethical approach, while also adhering to all applicable regulations and standards (Yallop et al. 2023).
- **Blockchain:** Blockchain technology is gathering traction in BA due to its capacity to generate secure, tamper-proof transaction records. Blockchain can be applied to validate supply chain data, such as product origin and movement. This can increase accountability and transparency in industries with a high risk of fraud or counterfeiting (Deepa et al., 2022).
- **Edge Computing:** Instead of sending data to a centralized data centre, it deals with working with data nearer to its source. This can be especially useful in industries, such as manufacturing and transportation, that generate significant amounts of real-time data. Businesses can make better judgements in real-time by analyzing data at the edge (Liu et al., 2019).
- **Quantum Computing:** Although in its beginning stages, it has the likelihood to alter BA by processing enormous quantities of data at unprecedented rates (Gaurav et al., 2022). Complex optimisation problems, such as supply chain

optimisation and financial risk analysis, that are intractable using conventional computation methods can be solved using quantum computing.

With continued advancements in technology, data visualization, and automation, the future of BA is promising. Businesses will be able to become more agile, responsive, and competitive by collecting and analyzing in instantaneously, making predictions about future outcomes, and automating many of the processes involved in data preparation and analysis to maintain public trust and comply with applicable regulations, businesses will also need to address ethical and governance issues associated with data collection and use.

11. Challenges in Business Analytics

There are various significant obstacles that organizations face when implementing BA initiatives, some of them are:

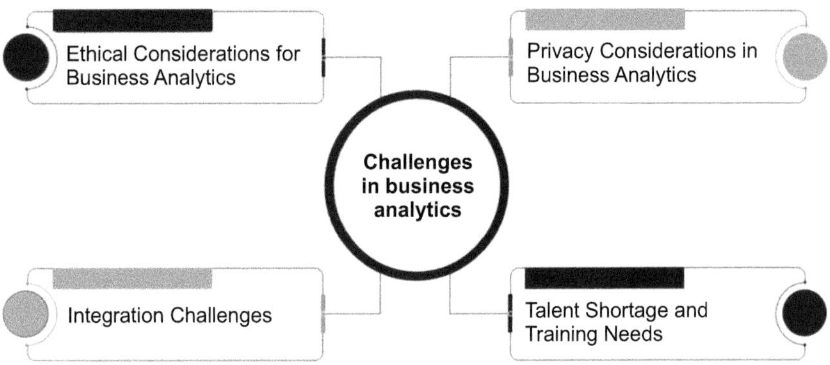

Fig. 10 Challenges in Business Analytics.
Source: Author's own

A few of the important challenges are highlighted in Fig. 10, the details description of each is as follows:

Ethical Considerations for Business Analytics

BA has evolved into an indispensable instrument for organizations as the use of data analytics increases, so do ethical concerns (Ogbuke et al., 2022) and these issues stem from the ethical implications of using data to make decisions due to the potential damage that can result from misusing or abusing data (Vidgen et al., 2020).

Additionally, it may result in the loss of employment which can occur when analytics reveal the potential for automation, resulting in job loss and organizations must ensure that they consider the potential impact on individuals and provide the necessary training and assistance can also raise ethical concerns and organizations must ensure that collected data is not used for manipulative or exploitative targeted marketing.

Privacy Considerations in Business Analytics

Privacy concerns are a major concern for businesses and raise concerns regarding the privacy of individuals and the potential for their personal information to be misused (Silaban, 2022). Organizations must ensure that they collect only the data required for their analytics, with the consent of the individuals whose information is collected. The use of third-party data is a noteworthy problem for the protection of privacy in BA. However, the use of third-party data raises concerns regarding data ownership and misuse potential, and to mitigate this issue, firms must draft agreements in place with third-party data providers (Chen, 2021).

Integration Challenges

Integrating BA into an organization's existing systems at times can be difficult as ensuring compatibility between various systems, being compatible (Khan et al., 2019) and using interoperable systems, cloud-based platforms, or employing application programming interfaces (APIs) can address this issue.

Talent Shortage and Training Needs

Data analysis, statistics, and data visualization are among the abilities required for BA however, it can be difficult to find employees with these skills, especially as demand for them continues to rise (Schmitt, 2023; Sharma et al., 2021b) and organisations must invest in training programmes, university partnerships, and/or the outsourcing of specific analytic duties.

Conclusion

This chapter examined the concept of BA and its implications in the current environment and the article has discussed numerous BA tools and techniques, such as statistical tools, data mining and ML tools, big data tools, and visualization tools the article also examines the different applications of BA, including customer, marketing, supply chain and financial analytics. In addition, the article has discussed the obstacles and opportunities in BA, such as ethical issues, privacy concerns, integration difficulties, talent shortages, and training requirements and the article has examined emerging trends in BA, such as AI and ML, IoT, natural language processing, predictive analytics, and data visualization and presented an exhaustive overview of the field of BA, emphasising its significance in the data-driven business environment of the present day. In conclusion, this paper has emphasised the significance of BA in the current data-driven business environment where we have investigated the various tools and techniques used in BA, the applications of BA, as well as the challenges and opportunities this field presents and the implications of emergent trends in BA for business practise. Finally, future research directions for this field have been identified, and it is evident that the future of BA is extremely promising, businesses must be prepared to adopt this technology and invest in the resources required to harness its potential.

References

Ahmad, A. and Khan, S.S. (2019). Survey of state-of-the-art mixed data clustering algorithms. *Ieee Access*, *7*, 31883–31902.

Agag, G., Shehawy, Y.M., Almoraish, A., Eid, R., Lababdi, H.C., Labben, T.G. and Abdo, S.S. (2023). Understanding the relationship between marketing analytics, customer agility, and customer satisfaction: A longitudinal perspective. *Journal of Retailing and Consumer Services*, *77*, 103663. https://doi.org/10.1016/j.jretconser.2023.103663.

Ajah, I.A. and Nweke, H.F. (2019). Big data and Business Analytics: Trends, platforms, success factors and applications. *Big Data and Cognitive Computing*, *3*(2), 32.

Alam, A. (2022). Cloud-Based E-Learning: Development of Conceptual Model for Adaptive E-Learning Ecosystem Based on Cloud Computing Infrastructure. In *Artificial Intelligence and Data Science: First International Conference, ICAIDS 2021, Hyderabad, India, December 17–18, 2021, Revised Selected Papers* (pp. 377–391). Cham: Springer Nature Switzerland.

Alghamdi, N.A. and Al-Baity, H.H. (2022). Augmented Analytics Driven by AI: A Digital Transformation beyond Business Intelligence. *Sensors*, *22*(20), 8071.

AL-Najjar, D., Al-Rousan, N. and AL-Najjar, H. (2022). Machine Learning to Develop Credit Card Customer Churn Prediction. *Journal of Theoretical and Applied Electronic Commerce Research*, *17*(4), 1529–1542.

Anitha, P. and Patil, M.M. (2022). RFM model for customer purchase behavior using K-Means algorithm. *Journal of King Saud University-Computer and Information Sciences*, *34*(5), 1785–1792.

Ansari, B., Barati, M. and Martin, E.G. (2022). Enhancing the usability and usefulness of open government data: A comprehensive review of the state of open government data visualization research. *Government Information Quarterly*, *39*(1), 101657.

Anshari, M., Almunawar, M.N., Lim, S.A. and Al-Mudimigh, A. (2019). Customer relationship management and big data enabled: Personalization & customization of services. *Applied Computing and Informatics*, *15*(2), 94–101.

Araz, O.M., Choi, T.M., Olson, D.L. and Salman, F.S. (2020). Role of analytics for operational risk management in the era of big data. *Decision Sciences*, *51*(6), 1320–1346.

Bag, S., Wood, L.C., Xu, L., Dhamija, P. and Kayikci, Y. (2020). Big data analytics as an operational excellence approach to enhance sustainable supply chain performance. *Resources, Conservation and Recycling*, *153*, 104559.

Bănărescu, A. (2015). Detecting and preventing fraud with data analytics. *Procedia economics and finance*, *32*, 1827–1836.

Banks, J. (1999). Introduction to simulation. In *Proceedings of the 31st conference on Winter simulation: Simulation—a bridge to the future-Volume 1* (pp. 7–13).

Barani Sundaram, B., Pandey, A., Abiko, A.T., Vijaykumar, J., Rastogi, U., Genale, A.H. and Karthika, P. (2022). Analysis of machine learning data security in the Internet of Things (IoT) circumstance. In *Expert Clouds and Applications: Proceedings of ICOECA 2021* (pp. 227–236). Springer Singapore.

Barga, R., Fontama, V., Tok, W.H. and Cabrera-Cordon, L. (2015). *Predictive analytics with Microsoft Azure machine learning* (pp. 221–241). Berkely, CA: Apress.

Batko, K. and Ślęzak, A. (2022). The use of Big Data Analytics in healthcare. *Journal of big Data*, *9*(1), 3.

Beaudry, J.S. and Miller, L. 2016. Research Literacy: A Primer for Understanding and Using Research. New York: The Guilford Press.

Bekar, E.T., Nyqvist, P. and Skoogh, A. (2020). An intelligent approach for data pre-processing and analysis in predictive maintenance with an industrial case study. *Advances in Mechanical Engineering*, *12*(5), 1687814020919207.

Berndt, J. (2020). *Scaling the ConceptCloud browser to very large semi-structured data sets: architecture and data completion* (Doctoral dissertation, Stellenbosch: Stellenbosch University).

Blashfield, R.K. and Aldenderfer, M.S. (1978). The literature on cluster analysis. *Multivariate behavioral research, 13*(3), 271–295.

Bocken, N.M., Harsch, A. and Weissbrod, I. (2022). Circular business models for the fastmoving consumer goods industry: Desirability, feasibility, and viability. Sustainable Production and Consumption, 30, 799–814.

Božič, K. and Dimovski, V. (2019). Business intelligence and analytics for value creation: The role of absorptive capacity. *International Journal of Information Management, 46*, 93–103.

Broby, D. (2022). The use of predictive analytics in finance. *The Journal of Finance and Data Science, 8*, 145–161. https://doi.org/10.1016/j.jfds.2022.05.003.

Burgess, S., Smith, G.D., Davies, N.M., Dudbridge, F., Gill, D., Glymour, M.M.,... and Theodoratou, E. (2019). Guidelines for performing Mendelian randomization investigations. *Wellcome Open Research, 4*.

Chahal, H., Jyoti, J. and Wirtz, J. (2019). Business Analytics: Concept and applications. *Understanding the Role of Business Analytics: Some Applications*, 1–8.

Charles, V., Garg, P., Gupta, N. and Agarwal, M. (2023). *Data Analytics and Business Intelligence: Computational Frameworks, Practices, and Applications*. CRC Press.

Chen, N., Chiang, N. and Storey, N. (2012). Business Intelligence and Analytics: From Big Data to Big Impact. *MIS Quarterly, 36*(4), 1165. https://doi.org/10.2307/41703503

Chen, R. (2021). Data breach on consumer behavior. In *Research Anthology on Privatizing and Securing Data* (pp. 1861–1879). IGI Global.

Chen, Y. and Lin, Z. (2021). Business intelligence capabilities and firm performance: A study in China. *International Journal of Information Management, 57*, 102232.

Chen, W., Milosevic, Z., Rabhi, F.A. and Berry, A. (2023). Real-Time Analytics: Concepts, Architectures, and ML/AI Considerations. *IEEE Access, 11*, 71634–71657. https://doi.org/10.1109/access.2023.3295694.

Cleveland, W.S. (2001). Data Science: an Action Plan for Expanding the Technical Areas of the Field of Statistics. *International Statistical Review, 69*(1), 21–26. https://doi.org/10.1111/j.1751-5823.2001.tb00477.x.

Codd, E.F. (1970). A relational model of data for large shared data banks. *Communications of the ACM, 13*(6), 377–387. https://doi.org/10.1145/362384.362685.

Das, S., Dey, A., Pal, A. and Roy, N. (2015). Applications of artificial intelligence in machine learning: review and prospect. *International Journal of Computer Applications, 115*(9).

Davenport, T.H. (2006). Competing on analytics. *PubMed, 84*(1), 98–107, 134. https://pubmed.ncbi.nlm.nih.gov/16447373.

Davenport, T.H., Harris, J.G. and Morison, R. (2009). *Analytics at Work: Smarter Decisions, Better Results*. https://ci.nii.ac.jp/ncid/BB01789501.

Davenport, T.H. (2018). From analytics to artificial intelligence. *Journal of Business Analytics, 1*(2), 73–80.

Deepa, N., Pham, Q.V., Nguyen, D.C., Bhattacharya, S., Prabadevi, B., Gadekallu, T.R. and Pathirana, P.N. (2022). A survey on blockchain for big data: approaches, opportunities, and future directions. *Future Generation Computer Systems*.

Di Lauro, S., Tursunbayeva, A., Antonelli, G. and Moschera, L. (2024). Disrupting human resource management with people analytics: a study of applications, value, enablers and barriers in Italy. *Personnel Review*. https://doi.org/10.1108/pr-11-2023-0927.

Dill, J., Earnshaw, R., Kasik, D., Vince, J. and Wong, P.C. (2012). Expanding the Frontiers of Visual Analytics and Visualization. In *Springer eBooks*. https://doi.org/10.1007/978-1-4471-2804-5.

Dixit, S. (2022). Artifical Intelligence and CRM: A Case of Telecom Industry. In *Adoption and Implementation of AI in Customer Relationship Management* (pp. 92–114). IGI Global.

Duan, Y., Cao, G. and Edwards, J.S. (2020). Understanding the impact of Business Analytics on innovation. *European Journal of Operational Research, 281*(3), 673–686.

Eckroth, J. (2018). A course on big data analytics. *Journal of Parallel and Distributed Computing, 118*, 166–176.

El-Haik, B. and Al-Aomar, R. (2006). *Simulation-based lean six-sigma and design for six-sigma*. John Wiley & Sons.

El Morr, C., Ali-Hassan, H., El Morr, C. and Ali-Hassan, H. (2019). Descriptive, predictive, and prescriptive analytics. *Analytics in healthcare: a practical introduction*, 31–55.

Farrugia, A. and Grima, S. 2021. A Model to Determine the Need to Modernise the Regulation of the Principle of Utmost Good Faith. *Journal of Financial Regulation and Compliance* 29: 454–473.

February, S.A.F. (2023). A Business Intelligence Effectiveness Model: Enhancing Organizational Decision-Making Capability. *International Journal of Strategic Decision Sciences (IJSDS), 14*(1), 1–23.

Ferraris, A., Mazzoleni, A., Devalle, A. and Couturier, J. (2019). Big data analytics capabilities and knowledge management: impact on firm performance. *Management Decision, 57*(8), 1923–1936.

Frazzetto, D., Nielsen, T.D., Pedersen, T.B. and Šikšnys, L. (2019). Prescriptive analytics: a survey of emerging trends and technologies. *The VLDB Journal, 28*, 575–595.

Gaurav, A., Chui, K.A. and Colace, F. (2022). Quantum Computing: A Tool in Big Data Analytics. *Cyber Security Insights Magazine, Insights2Techinfo, 3*, 10–14.

Gibson, A. and Shibani, A. (2022). Natural Language Processing-Writing Analytics. *by Charles Lang, George Siemens, Alyssa Friend Wise, Dragan Gaševic, and Agathe Merceron. 2nd ed. Vancouver, Canada: SoLAR*, 96–104.

Greasley, A. (2019). Simulating business processes for descriptive, predictive, and prescriptive analytics. In *Simulating Business Processes for Descriptive, Predictive, and Prescriptive Analytics*. de Gruyter. GmbH & Co KG, 2019.

Govender, P. and Sivakumar, V. (2020). Application of k-means and hierarchical clustering techniques for analysis of air pollution: A review (1980–2019). *Atmospheric Pollution Research, 11*(1), 40–56.

Gudivada, V.N., Irfan, M.T., Fathi, E. and Rao, D.L. (2016). Cognitive analytics: Going beyond big data analytics and machine learning. In *Handbook of Statistics* (Vol. 35, pp. 169–205). Elsevier.

Handelsman, D. (2012). Applying Business Analytics to optimize clinical research operations. In *Proceedings of the 2012 SAS Global Forum* (pp. 1–8).

Harangi, B., Baran, A. and Hajdu, A. (2020). Assisted deep learning framework for multi-class skin lesion classification considering a binary classification support. *Biomedical Signal Processing and Control, 62*, 102041.

Hariri, R.H., Fredericks, E.M. and Bowers, K.M. (2019). Uncertainty in big data analytics: survey, opportunities, and challenges. *Journal of Big Data, 6*(1), 1–16.

Harish, V., Krishnaveni, D. and Mansurali, A. (2021). Artificial Intelligence in Manufacturing. In *Reinventing Manufacturing and Business Processes Through Artificial Intelligence* (pp. 63–77). CRC Press.

Harish, V., Sharma, R., Rana, G., Nayyar, A., Parashar, B. (2023). Expanding the Scope of Learning Analytics in Blended Learning Environments. In: Ahuja, N.J., Kumar, A., Nayyar, A. (eds) Sustainable Blended Learning in STEM Education for Students with Additional Needs. Contributions to Environmental Sciences & Innovative Business Technology. Springer, Singapore. https://doi.org/10.1007/978-981-99-3497-3_14.

Hasan, R., Palaniappan, S., Mahmood, S., Abbas, A., Sarker, K.U. and Sattar, M.U. (2020). Predicting student performance in higher educational institutions using video learning analytics and data mining techniques. *Applied Sciences, 10*(11), 3894.

Hassan, N.R. (2019). The origins of business analytics and implications for the information systems field. *Journal of Business Analytics, 2*(2), 118–133. https://doi.org/10.1080/2573234x.2019.1693912.

Hoang, V.N. and Watson, J. (2022). Teaching binary logistic regression modeling in an introductory Business Analytics course. *Decision Sciences Journal of Innovative Education, 20*(4), 201–211.

Hopkins, E. (2022). Virtual Commerce in a Decentralized Blockchain-based Metaverse: Immersive Technologies, Computer Vision Algorithms, and Retail Business Analytics. Linguistic and Philosophical Investigations, (21), 203–218.

Jaggia, S., Kelly, A., Lertwachara, K. and Chen, L. (2020). Business Analytics. McGraw-Hill US Higher Ed USE.

Janiesch, C., Zschech, P. and Heinrich, K. (2021). Machine learning and deep learning. *Electronic Markets, 31*(3), 685–695.

Jantunen, A., Tarkiainen, A., Chari, S. and Oghazi, P. (2018). Dynamic capabilities, operational changes, and performance outcomes in the media industry. *Journal of Business Research, 89*, 251–257. https://doi.org/10.1016/j.jbusres.2018.01.037.

Kadaru, B.B. and UmaMaheswararao, M. (2017). An overview of general data mining tools. Int Res J EngTechnol, 4(9), 930–936.

Kakhki, M.D. and Sajadi, S.M. (2024). Business Analytics Affordances for Supply Chain Value Creation: A Technology-Organization-Environment Perspective. *International Journal of Production Economics, 276*, 109367. https://doi.org/10.1016/j.ijpe.2024.109367.

Kaur, M. and Kang, S. (2016). Market Basket Analysis: Identify the Changing Trends of Market Data Using Association Rule Mining. *Procedia Computer Science, 85*, 78–85. https://doi.org/10.1016/j.procs.2016.05.180.

Khan, S., Ali, S.A., Hasan, N., Shakil, K.A. and Alam, M. (2019). Big data scientific workflows in the cloud: Challenges and future prospects. *Cloud Computing for Geospatial Big Data Analytics: Intelligent Edge, Fog and Mist Computing*, 1–28.

Khan, M.Z., Zaman, F.U., Adnan, M., Imroz, A., Rauf, M.A. and Phul, Z. (2022). Comparative Case Study: An Evaluation of Performance Computation Between SQL And NoSQL Database. *Journal of Software Engineering, 1*(2), 14–23.

Klimberg, R. (2022). The evolution of Business Analytics and their impact on the service industry. In *Research Handbook on Services Management* (pp. 263–274). Edward Elgar Publishing.

Koziarski, M. (2020). Radial-based undersampling for imbalanced data classification. *Pattern Recognition, 102*, 107262.

Kraus, M., Feuerriegel, S. and Oztekin, A. (2020). Deep learning in Business Analytics and operations research: Models, applications and managerial implications. *European Journal of Operational Research, 281*(3), 628–641.

Kumar, S., Kar, A.K. and Ilavarasan, P.V. (2021). Applications of text mining in services management: A systematic literature review. *International Journal of Information Management Data Insights, 1*(1), 100008.

Kumar, A., Krishnamoorthy, B., Kamath, D. and D'Lima, C. (2022). Business Analytics adoption in firms. *International Journal of Business Information Systems, 40*(4), 464–492.

Kuznetsov, S.D., Velikhov, P.E. and Fu, Q. (2023). Real-Time Analytics: Benefits, Limitations, and Tradeoffs. *Programming and Computer Software, 49*(1), 1–25.

Lampropoulos, G. (2023). Artificial Intelligence, Big Data and Machine Learning in Industry 4.0. In *Encyclopedia of Data Science and Machine Learning* (pp. 2101–2109). IGI Global.

Larson, D. (2019). A review and future direction of Business Analytics project delivery. *Aligning Business Strategies and Analytics: Bridging Between Theory and Practice*, 95–114.

Leo, M., Sharma, S. and Maddulety, K. (2019). Machine learning in banking risk management: A literature review. *Risks, 7*(1), 29.

Lepenioti, K., Bousdekis, A., Apostolou, D. and Mentzas, G. (2019). Prescriptive analytics: Literature review and research challenges. *International Journal of Information Management, 50*, 57–70. https://doi.org/10.1016/j.ijinfomgt.2019.04.003.

Li, S., Gao, L., Han, C., Gupta, B., Alhalabi, W. and Almakdi, S. (2023). Exploring the effect of digital transformation on Firms' innovation performance. *Journal of Innovation & Knowledge*, *8*(1), 100317. https://doi.org/10.1016/j.jik.2023.100317.

Li, Q. and Li, Q. (2020). Overview of data visualization. *Embodying Data: Chinese Aesthetics, Interactive Visualization and Gaming Technologies*, 17–47.

Li, L. and Zhang, J. (2021). Research and analysis of an enterprise E-commerce marketing system under the big data environment. *Journal of Organizational and End User Computing (JOEUC)*, *33*(6), 1–19.

Liu, X., Alharbi, M.S., Chen, J., Diehl, A., Rees, D., Firat, E.E. and Laramee, R.S. (2023). Visualization Resources: A Survey. *Information Visualization*, *22*(1), 3–30.

Liu, Y., Soroka, A., Han, L., Jian, J. and Tang, M. (2020). Cloud-based big data analytics for customer insight-driven design innovation in SMEs. *International Journal of Information Management*, *51*, 102034.

Liu, D., Yan, Z., Ding, W. and Atiquzzaman, M. (2019). A survey on secure data analytics in edge computing. *IEEE Internet of Things Journal*, *6*(3), 4946–4967.

Maran, K., Senthilnathan, C.R., Usha, S. and Venkatesh, P. (2022, November). Business Analytics Contribution in the Growth of Indian Digital Business. In *2022 1st International Conference on Computational Science and Technology (ICCST)* (pp. 497–500). IEEE.

Mariani, M. (2020). Big data and analytics in tourism and hospitality: a perspective article. *Tourism Review*, *75*(1), 299–303.

Marzouk, M. and Enaba, M. (2019). Text analytics to analyze and monitor construction project contract and correspondence. *Automation in Construction*, *98*, 265–274.

Mattos, D.I., Bosch, J. and Olsson, H.H. (2021). Statistical models for the analysis of optimization algorithms with benchmark functions. *IEEE Transactions on Evolutionary Computation*, *25*(6), 1163–1177.

Mayer-Schnberger, V. (2013). *Big Data: A Revolution That Will Transform How We Live, Work and Think. Viktor Mayer-Schnberger and Kenneth Cukier*. http://dl.acm.org/citation.cfm?id=2588165.

Menezes, B.C., Kelly, J.D., Leal, A.G. and Le Roux, G.C. (2019). Predictive, prescriptive and detective analytics for smart manufacturing in the information age. *IFAC-PapersOnLine*, *52*(1), 568–573.

Mikalef, P., Pappas, I.O., Krogstie, J. and Pavlou, P.A. (2020). Big data and business analytics: A research agenda for realizing business value. *Information & Management*, 57(1), 103237.

Mirnaghi, M.S. and Haghighat, F. (2020). Fault detection and diagnosis of large-scale HVAC systems in buildings using data-driven methods: A comprehensive review. *Energy and Buildings*, *229*, 110492.

Mithas, S., Lee, M.R., Earley, S., Murugesan, S. and Djavanshir, R. (2013). Leveraging big data and Business Analytics [Guest editors' introduction]. *IT professional*, *15*(6), 18–20.

Mohammad, A.S. and Pradhan, M.R. (2021). Machine learning with big data analytics for cloud security. *Computers & Electrical Engineering*, *96*, 107527.

Montgomery, D.C., Peck, E.A. and Vining, G.G. (2021). *Introduction to Linear Regression Analysis*. John Wiley & Sons.

Muneeswaran, V., Nagaraj, P., Dhannushree, U., Ishwarya Lakshmi, S., Aishwarya, R. and Sunethra, B. (2021). A framework for data analytics-based healthcare systems. In *Innovative Data Communication Technologies and Application: Proceedings of ICIDCA 2020* (pp. 83–96). Springer Singapore.

Namjoshi, A.R. (2023). *Study of data analytics affecting SME's for business decisions* (Doctoral dissertation, Tilak Maharashtra Vidyapeeth).

Neto, L.C.M., Brewer, B.E. and Gray, A.W. (2023). Data on data: An analysis of data usage and analytics in the agricultural supply chain. *Applied Economic Perspectives and Policy*. 45(3), 1577–1591.

Ngiam, K.Y. and Khor, W. (2019). Big data and machine learning algorithms for health-care delivery. *The Lancet Oncology, 20*(5), e262–e273.

Ogbuke, N.J., Yusuf, Y.Y., Dharma, K. and Mercangoz, B.A. (2022). Big data supply chain analytics: ethical, privacy and security challenges posed to business, industries and society. *Production Planning & Control, 33*(2–3), 123–137.

Ohri, A. (2012). R for Business Analytics. In *Springer eBooks*. https://doi.org/10.1007/978-1-4614-4343-8.

Omar, Y.M., Minoufekr, M. and Plapper, P. (2019). Business Analytics in manufacturing: Current trends, challenges and pathway to market leadership. *Operations Research Perspectives, 6*, 100127.

Pape, T. (2016). Prioritising data items for business analytics: Framework and application to human resources. *European Journal of Operational Research, 252*(2), 687–698. https://doi.org/10.1016/j.ejor.2016.01.052.

Paradza, D. and Daramola, O. (2021). Business intelligence and business value in organizations: a systematic literature review. *Sustainability, 13*(20), 11382.

Patel, J.A. and Sharma, P. (2020). Online analytical processing for business intelligence in big data. *Big Data, 8*(6), 501–518.

Power, D.J., Heavin, C., McDermott, J. and Daly, M. (2018). Defining business analytics: an empirical approach. *Journal of Business Analytics, 1*(1), 40–53. https://doi.org/10.1080/2573234x.2018.1507605.

Pröllochs, N. and Feuerriegel, S. (2018). Business analytics for strategic management: Identifying and assessing corporate challenges via topic modeling. *Information & Management, 57*(1), 103070. https://doi.org/10.1016/j.im.2018.05.003.

Raghupathi, W.R.V. (2013). An Overview of Health Analytics. *Journal of Health & Medical Informatics, 04*(03). https://doi.org/10.4172/2157-7420.1000132.

Raghupathi, W. and Raghupathi, V. (2018). An Empirical Study of Chronic Diseases in the United States: A Visual Analytics Approach to Public Health. *International Journal of Environmental Research and Public Health, 15*(3), 431. https://doi.org/10.3390/ijerph15030431.

Raghupathi, W. and Raghupathi, V. (2021). Contemporary Business Analytics: An Overview. *Data, 6*(8), 86. https://doi.org/10.3390/data6080086.

Rana, G., Sharma, R. and Goel, A.K. (2019). Unraveling the power of talent analytics: Implications for enhancing business performance. *Business governance and society: Analyzing shifts, conflicts, and challenges*, 29–41.

Rana, G., Sharma, R. and Parashar, B. (2024, May). Application of Artificial Intelligence for the Success of Supply Chain Operations in the Age of Data Analytics. In *2023 International Conference on Smart Devices (ICSD)* (pp. 1–6). IEEE.

Rathore, B. (2023 A). Digital Transformation 4.0: Integration of Artificial Intelligence & Metaverse in Marketing. *Eduzone: International Peer Reviewed/Refereed Multidisciplinary Journal, 12*(1), 42–48.

Rathore, B. (2023 B). Integration of Artificial Intelligence & It's Practices in Apparel Industry. *International Journal of New Media Studies (IJNMS), 10*(1), 25–37.

Ridzuan, F. and Zainon, W.M.N.W. (2022). Diagnostic analysis for outlier detection in big data analytics. *Procedia Computer Science, 197*, 685–692. https://doi.org/10.1016/j.procs.2021.12.189.

Rodrigues, A.P., Fernandes, R., Bhandary, A., Shenoy, A.C., Shetty, A. and Anisha, M. (2021). Real-time Twitter trend analysis using big data analytics and machine learning techniques. *Wireless Communications and Mobile Computing, 2021*, 1–13.

Sahoo, S. (2022). Big data analytics in manufacturing: a bibliometric analysis of research in the field of business management. *International Journal of Production Research, 60*(22), 6793–6821.

Samuel, A.L. (1959). Some Studies in Machine Learning Using the Game of Checkers. *IBM Journal of Research and Development, 3*(3), 210–229. https://doi.org/10.1147/rd.33.0210.

Sarker, I.H. (2021). Data science and analytics: an overview from data-driven smart computing, decision-making and applications perspective. *SN Computer Science*, *2*(5), 377.

Schmitt, M. (2023). Deep learning in Business Analytics: a clash of expectations and reality. *International Journal of Information Management Data Insights*, *3*(1), 100146.

Schnegg, M. and Möller, K. (2022). Strategies for data analytics projects in business performance forecasting: a field study. *Journal of Management Control*, *33*(2), 241–271.

Schuff, D., Corral, K., St. Louis, R.D. and Schymik, G. (2018). Enabling self-service BI: A methodology and a case study for a model management warehouse. *Information Systems Frontiers*, *20*, 275–288.

Seddon, P.B., Constantinidis, D., Tamm, T. and Dod, H. (2016). How does business analytics contribute to business value? *Information Systems Journal*, *27*(3), 237–269. https://doi.org/10.1111/isj.12101.

Seyedghorban, Z., Tahernejad, H., Meriton, R. and Graham, G. (2020). Supply chain digitalization: past, present and future. *Production Planning & Control*, *31*(2-3), 96–114.

Shah, D. and Murthi, B.P.S. (2021). Marketing in a data-driven digital world: Implications for the role and scope of marketing. *Journal of Business Research*, *125*, 772–779.

Sharma, R., Singh, S.P. and Rana, G. (2019). Employer branding analytics and retention strategies for sustainable growth of organizations. *Understanding the role of business analytics: Some applications*, 189–205.

Sharma, R., Saini, A. and Rana, G. (2021). Big data analytics and businesses in industry 4.0. *Design Engineering*, 238–252.

Sharma, A.M. (2020), "Data Visualization", Kumari, S., Tripathy, K.K. and Kumbhar, V. (Ed.) *Data Science and Analytics*, Emerald Publishing Limited, Leeds, pp. 1–22.

Sharma, R., Rana, G., Sharma, R., Saini, R. and Prakash, C. (2021b). Revitalizing Talent Management Practices through Technology Integration in industry 4.0. *R. Sharma, R. Saini & C. Prakash, Role of Internet of Things (IoT) in Sustaining Disruptive Businesses (1st ed.).* New York USA: Nova Science Publishers.

Shanks, G. and Bekmamedova, N. (2012). Achieving benefits with Business Analytics systems: An evolutionary process perspective. *Journal of Decision Systems*, *21*(3), 231–244.

Sheng, J., Amankwah-Amoah, J., Khan, Z. and Wang, X. (2021). COVID-19 pandemic in the new era of big data analytics: Methodological innovations and future research directions. *British Journal of Management*, *32*(4), 1164–1183.

Silaban, D.P. (2022). Challenges in Business Analytics Implementation: A Comprehensive Review Using TOE Framework. *Milestone: Journal of Strategic Management*, *2*(2), 131–144.

Singh, U.P. (2023). Decision Making and Predictive Analysis for Real Time Data. *Advances in Data Science and Analytics: Concepts and Paradigms*, 21–38.

Sousa, M.J., Pesqueira, A.M., Lemos, C., Sousa, M. and Rocha, Á. (2019). Decision-making based on big data analytics for people management in healthcare organizations. *Journal of Medical Systems*, *43*, 1–10.

Sprague, R.H. (1980). A Framework for the Development of Decision Support Systems. *MIS Quarterly*, *4*(4), 1. https://doi.org/10.2307/248957.

Stadler, J.G., Donlon, K., Siewert, J.D., Franken, T. and Lewis, N.E. (2016). Improving the Efficiency and Ease of Healthcare Analysis Through Use of Data Visualization Dashboards. *Big Data*, *4*(2), 129–135. https://doi.org/10.1089/big.2015.0059.

Szukits, Á. and Móricz, P. (2023). Towards data-driven decision making: the role of analytical culture and centralization efforts. *Review of Managerial Science*, *18*(10), 2849–2887. https://doi.org/10.1007/s11846-023-00694-1.

Tabianan, K., Velu, S. and Ravi, V. (2022). K-means clustering approach for intelligent customer segmentation using customer purchase behavior data. *Sustainability*, *14*(12), 7243.

Tang, J. and Karim, K.E. (2019). Financial fraud detection and big data analytics–implications on auditors' use of fraud brainstorming session. *Managerial Auditing Journal, 34*(3), 324-337.

Toic, A., Poscic, P. and Jaksic, D. (2022). Analysis of Selected Business Intelligence Data Visualization Tools. In *Central European Conference on Information and Intelligent Systems* (pp. 25–32). Faculty of Organization and Informatics Varazdin.

Tong-On, P., Siripipatthanakul, S. and Phayaphrom, B. (2021). The implementation of business intelligence using data analytics and its effects towards on performance in the hotel industry in Thailand. *International Journal of Behavioral Analytics, 1*(2).

Troilo, M., Bouchet, A., Urban, T.L. and Sutton, W.A. (2015). Perception, reality, and the adoption of business analytics: Evidence from North American professional sport organizations. *Omega, 59*, 72–83. https://doi.org/10.1016/j.omega.2015.05.011.

Vidgen, R., Hindle, G. and Randolph, I. (2020). Exploring the ethical implications of Business Analytics with a business ethics canvas. *European Journal of Operational Research, 281*(3), 491–501.

Vetter, T.R. and Schober, P. (2018). Regression: the apple does not fall far from the tree. *Anesthesia & Analgesia, 127*(1), 277–283.

Wang, J., Yang, Y., Wang, T., Sherratt, R.S. and Zhang, J. (2020). Big data service architecture: a survey. *Journal of Internet Technology, 21*(2), 393–405.

Wang, S., Yeoh, W., Richards, G., Wong, S.F. and Chang, Y. (2019). Harnessing Business Analytics value through organizational absorptive capacity. *Information & Management, 56*(7), 103152.

Winter, R., Marjanovic, O. and Wixom, B.H. (2012). Introduction to the Business Analytics, Business Intelligence and Data Warehousing MiniTrack. *Hawaii International Conference on System Sciences*, 4128. https://doi.org/10.1109/hicss.2012.368.

Wixom, B.H., Yen, B. and Relich, M. (2013). Maximizing value from Business Analytics. *MIS Quarterly Executive, 12*(2).

Wolniak, R. and Grebski, W. (2023). The five stages of business analytics. *Scientific Papers of Silesian University of Technology Organization and Management Series, 2023*(178), 735–752. https://doi.org/10.29119/1641-3466.2023.178.41.

Wowak, K.D., Lalor, J.P., Somanchi, S. and Angst, C.M. (2023). Business Analytics in Healthcare: Past, Present, and Future Trends. *Manufacturing & Service Operations Management.*

Wulff, S.S. (2017). Time series analysis: Forecasting and control. *Journal of Quality Technology, 49*(4), 418.

Xia, B.S. and Gong, P. (2014). Review of business intelligence through data analysis. *Benchmarking: An International Journal, 21*(2), 300–311.

Yallop, A.C., Gică, O.A., Moisescu, O.I., Coroş, M.M. and Séraphin, H. (2023). The digital traveller: implications for data ethics and data governance in tourism and hospitality. *Journal of Consumer Marketing*, 40(2), 155–170.

Yoshikuni, A.C., Dwivedi, R., Zhou, D. and Wamba, S.F. (2023). Big data and business analytics enabled innovation and dynamic capabilities in organizations: Developing and validating scale. *International Journal of Information Management Data Insights, 3*(2), 100206. https://doi.org/10.1016/j.jjimei.2023.100206

Zaki, M. (2019). Digital transformation: harnessing digital technologies for the next generation of services. *Journal of Services Marketing. 33*(4), 429–435.

Zhang, Z., Xiao, Y., Fu, Z., Zhong, K. and Niu, H. (2022). A Study on Early Warnings of Financial Crisis of Chinese Listed Companies Based on DEA–SVM Model. *Mathematics, 10*(12), 2142. https://doi.org/10.3390/math10122142.

2 | Unraveling the Dimensions of HR Analytics and Sustainable HR Analytics: Paving to a New Paradigm

H.H.D.P.J. Opatha[1*] and N.W.K.D.K. Dayarathna[2]

In today's tough corporate environment, the adage 'do or die' rings true across industries, challenging firms to strive for survival in the face of relentless competition. In this context, the integration of HR analytics promising as a fundamental requirement for any forward-thinking firm. Despite its growing popularity in the business world, the practical use of HR analytics remains a concern, particularly in developing countries. One key factor is the HR analytics competency of HR professionals, which magnificently increases the application of HR analytics. In addition, on-time investments, supportive employees, top management support, ethical data governance, research culture, etc. are prerequisites for effective HR analytics adoption. This chapter sheds light on the critical role of HR analytics, promoting a better understanding of its relevance and applicability across diverse HR activities. Moreover, by highlighting the importance of HR analytics, this chapter hopes to raise collective awareness inside businesses, pressing them to leverage the potential of data-driven insights for strategic decision-making. By providing real-world examples and best practices, it is expected to close the knowledge gap between theory and practice and inspire businesses to embrace HR analytics as a potent instrument for resilience and long-term organizational success in the fast-paced market of today.

1. Introduction

HR analytics has emerged as a critical factor in enhancing the strategic value of Human Resource Management (HRM). Historically, HRM has been viewed as more subjective and theoretical than accounting, finance, marketing, and operations management, which rely heavily on quantitative data (Sharma et al., 2021). This perception has often presented challenges for HR professionals, who may struggle to implement their ideas and decisions without solid quantitative evidence (Patil and Priya, 2024).

In the dynamic, uncertain, complex, and ambiguous business environment of today, businesses are being asked to make decisions based on a growing amount

[1] 0000-0002-5243-0868

[2] 0000-0002-6875-6133

* Corresponding author: poojaopatha@sjp.ac.lk

DOI: 10.1201/9781003514282-2

of clear, trustworthy facts gathered from several sources. HR analytics represents a transformative tool that bridges the gap between subjective HRM practices and evidence-based decision-making. This gap-filling is of utmost importance as it enhances the strategic value of HRM (Jabir, Falih, and Rahmani, 2019). However, though HR analytics brings strategic value to the organization, it is being adopted by several organizations and the majority are not utilizing HR analytics to the maximum (Dooren, 2012; Rana et al., 2019). This may be because HR professionals do not possess the right attitude and competencies toward HR analytics.

Since sustainability has been a concern in every aspect, it is important to explore the role of HR analytics in the backdrop of sustainability. This chapter will explore the significance of HR analytics and its role in ensuring long-term sustainability in the organization. The chapter contains the definitions of HR analytics, the evolution of HR analytics, the dimensions of HR analytics, the importance of HR analytics, the process of HR analytics, HR metrics, and sustainable HR analytics (Sharma et al., 2019).

These sections of the chapter focus on the strategic usefulness of HR analytics and how it may transform traditional HR practices into evidence-based HR practices, ultimately contributing to sustainability. HR analytics makes the decision-making of the company more systematic and evidence-based, a competitive advantage for the organization and long-term survival (Kameswari, Palivela, Settur, and Solanki, 2023).

1.1 Objectives of the Study

[1] To explore the concept of HR Analytics
[2] To study the evolutionary Levels for HR Analytics
[3] To determine the significance of HR Analytics
[4] To explain the application of HR Analytics
[5] To identify HR metrics
[6] To explore the concept of sustainable HR Analytics

2. Background of HR Analytics

HR analytics has been subject to a variety of interpretations in the literature, each with a unique viewpoint on its essence and value. Table 1 contains a compilation of notable definitions proposed by experts in the area, providing a complete picture of the varied nature of human resource analytics. This anthology provides readers with rich insights, allowing for a deeper understanding of this critical topic within the larger context of Human Resource Management (HRM).

As we delve into these various definitions, we will engage in a voyage of investigation and discovery, uncovering the common features of HR analytics, i.e., application of analytical and statistical tools, analyzing people-related data and information, evidence-based decision-making, intending to prove the causal

Table 1 Definitions of HR Analytics.

Researcher	*Definition*
Boudreau and Ramstad (2004)	*"HR analytics is not just about* ***statistics and research design*** *as it goes beyond that and includes* ***establishing proper standards for HR analytics*** *and* ***improving HR analytical capabilities*** *across the organization".*
Kapoor and Sherif (2012)	*"HR analytics means using* ***business analytics models*** *to* ***analyze data and making appropriate decisions****".*
Dooren (2012)	*"A methodology that uses* ***statistical techniques and experimental aspects*** *to assess the causal relationship between HR functions and organizational performance and to provide trustworthy foundations for human capital decisions".*
Kirtane (2015)	*"HR analytics is an integrated process that helps to improve the quality of people-related choices,* ***which in turn improves individual and organizational performance****".*
Jain and Nagar (2015)	*"A* ***combination of quantitative and qualitative data*** *and information that provides significant insights for the organization's management to use when making decisions".*
Reddy and Lakshmikeerthi (2017)	*"Using data, analysis, and research to* ***understand the relationship between people management practices and business outcomes*** *like profitability, customer satisfaction, and quality".*
Bhattacharyya (2017)	*"****Analytical reasoning*** *applied to HRM functions".*
Marler and Boudreau (2017)	*"An* ***IT-enabled HR approach*** *that establishes business impact and facilitates data-driven decision-making by utilizing descriptive, visual, and statistical analyses of data about HR processes".*
Kiran, Sharma, and Brijmohan (2018)	*"By* ***combining statistical models to work-related data,*** *HR analytics enables business leaders to optimize human resource management by offering a* ***data-driven framework*** *for solving business problems".*
Jabir et al. (2019)	*"****Understanding how and why things happen****, creating ideas about the* ***best course of action****, and drawing conclusions about the best and worst outcomes based on data analysis are all part of HR analytics.".*
Bahuguna, Srivastava, and Tiwari (2023)	*"Considered an integrative approach, HRA uses* ***HR Metrics, HRIS, organizational data, big data, statistical techniques, and algorithms*** *to gather information, identify trends, and make* ***evidence-based decisions*** *about human resources".*

Source: Adapted from Opatha (2021).

relationship between HR functions and organizational performance, etc. HR analytics solely revolves around conducting research in HR-related matters. It encompasses a systematic process of gathering, analyzing, and interpreting data to inform various HR functions, ranging from recruitment and talent management to performance evaluation and organizational development.

According to the definitions provided earlier, HR analytics involves utilizing analytical methods and tools to examine HR data thoroughly. It aims to uncover solutions and enable the formulation of effective decisions concerning HR matters. These decisions are based on evidence derived from data analysis, all with the overarching goal of enhancing the strategic importance of HRM. Figure 1 illustrates the pertinent visual representation corresponding to the definition provided.

Fig. 1 HR Analytics Process.
Source: Author Developed

3. Four Evolutionary Levels for HR Analytics

HR decision-making differs significantly from decision-making in other management domains due to its focus on human management, a complex process influenced by the unique nature of human beings. Organizational failures often stem from erroneous HR decisions. For instance, selecting the wrong individual for a role can lead to diminished organizational performance and failure to meet stakeholder expectations, in the short and long term. To address this challenge, HR analytics has evolved through several stages. Smith (2013) outlines four evolutionary levels for HR analytics:

- Level 0 – Intuitive Decision Making: In the late 1980s and 1990s, professionals often relied on assumptions, opinions, and intuition to make business decisions. This reliance on gut feelings persisted in HR longer than in other functional areas.
- Level 1 – Reporting: With the proliferation of computers and software applications, organizations began to focus on data reporting. This stage involved data collection, governance, creation of data dictionaries, and implementation of self-service reporting systems, primarily addressing the question: "What happened?" based on reported data.
- Level 2 – Metrics and Dashboards: Advancements in IT technology facilitated the use of metrics (e.g., employee turnover rate, absenteeism rate) and dashboards, which visually present complex data using graphs and charts. This stage aimed to answer the question: "What is happening today?" in real-time.

- Level 3 – Strategic Analytics: The current stage of HR analytics incorporates advanced statistical tools and analysis. This stage enables HR professionals to make more comprehensive conclusions and decisions, moving into the realm of predictive analytics and beyond predictive analytics (prescriptive analytics) to address the questions: "What will happen in the future?" and "What should be done?"

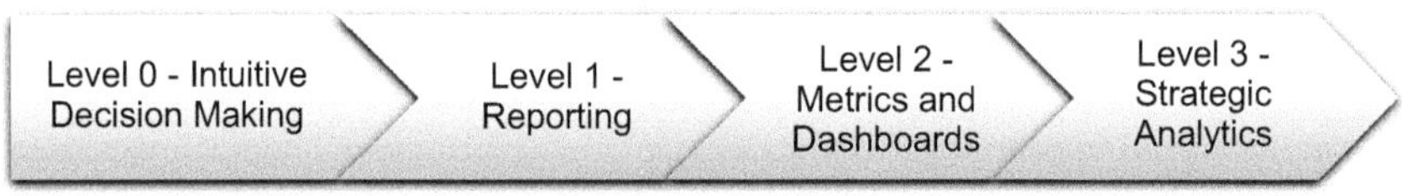

Fig. 2 Four Evolutionary Levels of HR Analytics.
Source: Author Developed

4. Three Dimensions of HR Analytics

HR analytics is of three types (Fitz-Enz and Mattox II, 2014; Fred and Kinange, 2015; Jabir et al., 2019). The illustration is shown in Fig. 3.

4.1 Descriptive HR Analytics

Descriptive analytics entails studying data to draw inferences about the current state of something. This analysis uses past data and statistical techniques like mean, median, variance, standard deviation, etc. For instance, analyzing the current absenteeism, turnover, satisfaction, etc., can be considered descriptive analytics. Through descriptive analytics, HR professionals can analyze past data and explain what is happening and what has happened in the organization.

4.2 Predictive HR Analytics

Predictive analytics means making predictions or projections about future HR-related aspects by analyzing past data. Statistical techniques such as correlation and regression analyses can be utilized to make such predictions in addition to other predictive models. For instance, organizations can predict the impact of employee satisfaction on organizational profitability using regression analysis. Moreover, finding the employee who is more likely to leave the organization in the future by conducting research is also predictive analytics. Hence, predictive analytics explains what will happen in the future.

4.3 Prescriptive HR Analytics

Prescriptive analytics means applying decision-making, management, and operations research methodologies to identify the best strategies to implement in alignment with the conclusions derived from descriptive and predictive analytics. For example, if predictive analytics shows that employee satisfaction has a positive impact on organizational profitability, what actions need to be implemented to enhance employee satisfaction? Prescriptive analytics enables organizations to make informed, evidence-based strategies and actions to enhance their HR performance and overall organizational performance.

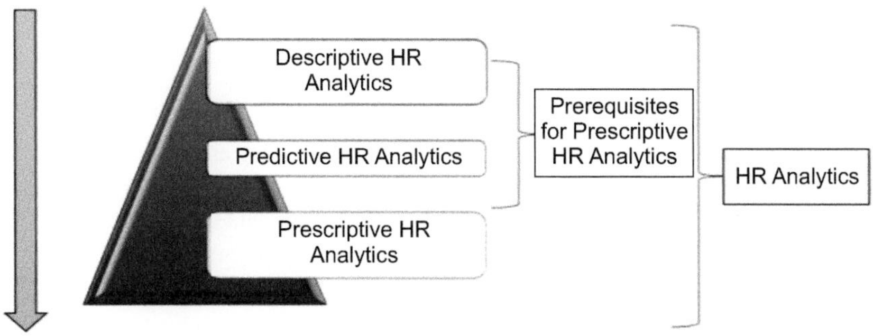

Fig. 3 Composition of HR Analytics.
Source: Opatha and Dyarathna (2024)

5. Importance of HR Analytics

As per Kirtane (2015), it is not a secret that HR analytics plays a pivotal role in enhancing effective and efficient decision-making to ensure sustainability while balancing the economic, social, and green aspects of the organization through comprehensive evaluation and feedback. Modern human resource data analytics and evidence-based practices optimize overall business performance (Patil and Priya, 2024; Rana and Sharma,2019). Various researchers list numerous advantages of HR analytics, including Kiran et al. (2018); Bhattacharyya (2017); Kirtane (2015); Reena, Ansari, and Jayakrishnan (2019); Reddy and Lakshmikeerthi (2017); and Fred and Kinange (2015).

Enhance Employee Performance

Employers can use HR analytics to assess workers' job performance and classify them based on their performance level. This paves the way for the organization to identify underperforming employees and initiate training programs targeting these employees' training needs to transform them into averagely performing or high-performing employees.

Maximize Return on Human Resource

Organizations can determine the return on investments using different HR metrics like the ROI of human capital and assess whether the return is sufficient to justify the cost of human resources.

Facilitate HR Professionals to Manage Career Aspirations of Employees

Retention rate of high potentials, career path ratio, time to promotion, internal promotion rate, etc., facilitate organizations in managing the career paths of their employees. HR managers can conduct research investigations to explore the career aspirations of employees and their impact on job performance.

Predict HR Requirements

HR analytics allows HR professionals to predict future employee needs using different analytics techniques like extrapolation, indexation, formal expert surveys, etc. Furthermore, HR analytics supports conducting HR audits to assess the current level of employee competencies and thereby develop replacement charts.

Identify Factors Affecting Employee Satisfaction and Productivity

It is easier for organizations to identify factors affecting employee satisfaction and productivity, and other employee-related matters, by conducting HR research. Surveys can be conducted separately, and the collected data can be analyzed with the help of statistical techniques using software like SPSS and SmartPLS.

Explore Reasons for Employee Turnover

HR analytics can predict the attrition rates of employees. Analyzing past trends and data relating to employee turnover sheds light on future prediction rates for employee attrition. Additionally, by conducting interviews and surveys among employees identified as having a high likelihood of leaving the organization, HR professionals can uncover the reasons behind their intention to leave.

Design and Execute Customized Training and Development Programs

HR analytics helps HR professionals develop effective training and development programs after carefully analyzing the training needs of employees identified through surveys. With the support of HR analytics, organizations can create and execute customized training programs that effectively address employees' skill gaps.

Empower HR Professionals to Quantify HR Contributions

Unlike other functional areas of the organization, it is challenging for HR to demonstrate the direct impact of HR functions on overall business performance and profitability. However, HR data-driven insights display the importance of HR for organizational success. As evidence-based HR measures HR's contribution through various HR metrics, the relationships and impacts among HR and organizational outcomes can be measured through statistical techniques. These empower HR professionals to quantify their contributions to organizational success.

HR analytics is promising as one of the important drivers of organizational sustainability in today's dynamic business landscape because of its immeasurable importance.

6. Process of HR Analytics

Bassi (2012) conceptualizes HR analytics as a cyclical process comprising several critical steps, each integral to deriving actionable insights and driving organizational effectiveness:

6.1 Determining Stakeholder Requirements

This initial step involves identifying the needs and expectations of key stakeholders within the organization. Understanding what information stakeholders require enables HR professionals to tailor their analytical efforts to address specific business challenges and opportunities.

6.2 Defining the HR Research Agenda

Once stakeholder requirements are determined, HR professionals define a clear research agenda outlining objectives, scope, and methodologies for their analytical endeavours.

6.3 Identifying Data and Information Sources

The third step involves gathering data from various sources, including employee interviews, surveys, social media, and market data. Moreover, it is critical to ensure the accuracy and recency of the data collected from the identified sources.

6.4 Gathering Data and Information

After identifying the sources, data need to be collected. When gathering data from different data sources, HR must align with data privacy policies and the relevancy of the data to the research questions and objectives.

6.5 Transforming Data and Information

Upon data collection, it must be converted into a structure that is easier for analysis. This may involve organizing the data appropriately, removing errors, and inconsistencies. Using different metrics and statistical techniques, raw data can be transformed into meaningful information directed toward effective decision-making.

6.6 Communication of Found Results

To ensure that the insights gained from analytics are turned into workable plans, the information and conclusions drawn from the data analysis must be shared with the pertinent company stakeholders.

6.7 Enabling Strategy Creation, Decision Making, Execution, and Learning

Finally, HR analytics enables HR professionals to strategize and execute these strategies appropriately. Moreover, evaluating these strategies once they have been implemented is also significant to ensure ongoing learning and continuous improvements.

7. Application of HR Analytics

Application of HR analytics might not be easier than said. It might take cost, effort, and time. Thus, the application of HR analytics is challenging unless the organization has the appropriate environment for it. HR analytics is not merely finding HR metrics but involves a rigorous process. HR analytics application involves logical research design, data collection, data analysis using statistical and other analytics models, and deriving evidence-based decisions and strategies. This process has to be facilitated by several supportive factors. Following the conceptual model proposed by Gurusinghe, Arachchige, and Dayarathna (2019), technology (technology competence), organizational (basic resources, analytical competencies of HR personnel, top management commitment, HR-business strategic alignment), and environmental (regulatory support, competitive pressure) factors significantly impact the successful adoption of HR analytics.

HR analytics finds extensive application in diverse domains, with turnover studies being a prominent example, as highlighted by Davenport, Harris, and Shapiro (2010). Beyond attrition, HR analytics is instrumental in selecting and monitoring organizational health indicators, identifying areas requiring attention, determining impactful actions on the bottom line, forecasting workforce levels, and adapting the workforce to changing business environments. Research exemplified by King (2016) demonstrates the effectiveness of HR analytics in predicting employee attrition, revealing insights such as the correlation between tenure in a position and the likelihood of departure. As a result, they could identify that the main reason for employee attrition was "time in position.". The longer an employee remained in a position without receiving a promotion, the more likely they were to leave the firm. This study, conducted in collaboration between a US-based company and a university, utilized R statistical software to develop a predictive model with an impressive 81% overall accuracy and an 84% recall accuracy. Such findings underscore the transformative potential of HR analytics in optimizing organizational outcomes.

According to Naula (2015), HR analytics is being used by HR professionals for the following purposes:

- Assessing employee engagement levels based on absence trends.
- Benchmarking HR performance and employee performance.
- Assessing recruitment and retention data.
- Talent management by tracking high performers and nurturing talent in-house.
- Linking compensation to overall business performance.
- Predict causes for attrition of employees.
- Assess future employee demand and supply etc."

Dooren (2012) researched major factors affecting the application of HR analytics and he interviewed several professionals from different business organizations. Some of the questions he asked in the interview were in what ways is your company achieving HR analytics goals?, what information is used, how is it used, and what measurements, statistics, and analytical tools are used? Explain the methodology used to produce HR analytics for each HR service area or subject, etc. The answers given to these questions during the interviews are mentioned below in Table 2.

Table 2 Interview Results.

Interview No.	*Answer*	*Application Area*
Interview 01	For basic headcount reporting, we report on hiring costs and turnaround times, measure customer satisfaction, and use SAP to manage the payroll system.	Customer and Labor Management Relations Recruitment and Selection Pay Management
Interview 02	Sex, seniority, age, and nationality data are also gathered for basic headcount reporting. Engagement measurement is currently being developed.	Labor Management Relations
Interview 03	In addition to conducting an engagement survey and using an HRIS system (PeopleSoft) and management information system (Cognos), we also report on headcounts, absence, and mobility (input/output from workers).	Labor Management Relations
Interview 04	Report on vacation days, absenteeism, etc., gather data on performance result differences that we can influence (high/low performers), and establish benchmarks for pay and incentives.	Labor Management Relations Performance Management Pay Management
Interview 05	We evaluate HR's operational efficacy. We solely utilize information from our own HRIS, including age, gender diversity, and the number of joiners and leavers.	Labor Management Relations Predictive and Prescriptive Analytics
Interview 06	Measure headcounts, absenteeism, etc.	Basic Descriptive Analytics
Interview 07	Performance reviews, potential and development ratings, Strategic Workforce Planning, etc.	Performance Management Training and Development Human Resource Planning
Interview 08	Measure absenteeism, mobility, employee satisfaction, PPM (Performance Potential Matrix) etc.	Labor Management Relations Performance Management
Interview 09	Measure the effectiveness of the HR department by gathering information on employee age, education, experience, skills, performance, and potential.	Labor Management Relations Performance Management
Interview 10	We look at how many vacancies are available, how many people have applied, what kinds of positions people have applied for, how many individuals have reported sick leave, and how well people are performing.	Recruitment and Selection Performance Management Labor Management Relations

Contd.

Table 2 *Contd.*

Interview No.	*Answer*	*Application Area*
Interview 11	Personal data (race, gender, age, function, level, etc.).	Basic Descriptive Analytics
Interview 12	Employee engagement surveys, succession planning, and employee mobility.	Labor Management Relations HR Planning

Source: Author Developed

According to Jain and Nagar (2015) and the authors' perspectives, the requirements listed in Table 3 pertain to various HRM functions and fields, all of which can be effectively addressed using HR analytics.

Table 3 Possible Areas to be Covered by HR Analytics in HR Functions.

HR function/field	*Why HR Analytics?*
Employee Planning and Staffing	(a) Job Analysis – Exploring job and job holder data and developing Job Descriptions (JD) and Job Specifications (JS). (b) To identify how many jobs have JDs and JSs. (c) To assess the competencies of internal employees and develop replacement charts. (d) To assess how many applicants possess appropriate Knowledge, Skills, and Attitudes (KSAs) that match the job specification. (e) To distinguish the best recruitment, selection, and induction method. (f) To develop selection schemes and initiate profile matching.
Training and Development	(a) To identify training needs by conducting surveys. (b) To investigate which training methods would have the maximum impact on employees' job effectiveness. (c) To measure the level of transfer of training of the employees. (d) To measure the return on investment of training programs.
Remuneration	(a) To explore the determinants of compensation. (b) Job evaluation - Measure the relative worth of the job. (c) To highlight the importance of remuneration programs on employees' satisfaction and morale. (d) To conduct pay surveys considering market data. (e) To research to see whether the organizational remuneration program attracts talented employees from the industry or not.
Performance Appraisal	(a) To measure the current performance level of the employees by developing performance schemes. (b) To identify high-performing and underperforming employees. (c) To explore the relationship between employee job performance and organizational outcomes.

Contd.

Table 3 *Contd.*

HR function/field	*Why HR Analytics?*
Health and Safety Management	(a) To specify which employee categories are more prone to workplace hazards. (b) To measure the accident ratio of the organization. (c) To research the relationship between the sound health and safety of employees and employee performance. (d) To explore safety issues in the organization (safety audits).
Grievances Handling	(a) To recognize the most common employee grievances. (b) To find the impact of employee grievances on organizational performance. (c) To ascertain the best methods of grievance handling. (d) To measure how many grievances have been solved.
Labour Management Relationships (LMR)	(a) To discover factors contributing to positive labour relations. (b) To pinpoint the relationship between healthy LMRs and organizational performance. (c) To detect which union actions are more detrimental to business performance. (d) To recognize reasons behind trade union actions through surveys.

Source: Author Developed.

8. HR Metrics

HR metrics serve as vital tools for evaluating HR practices, encompassing both operational and strategic areas within an organization. These metrics provide quantitative data about various HR processes, offering descriptive insights into aspects such as employee performance, turnover rates, training effectiveness, and recruitment efficiency.

Distinctions exist between HR metrics and HR analytics. While HR metrics provide descriptive data, HR analytics involves a more comprehensive approach that leverages HR metrics to derive actionable insights for effective decision-making. In essence, HR analytics entails the interpretation and synthesis of HR metrics to facilitate efficient HR decision-making processes.

Concerning Bucknall and Wei (2006), the following formulas and examples are stipulated under HR Metrics.

8.1 Absenteeism Rate

$$\text{Absenteeism rate} = 100$$

Number of working days lost for the period is the sum of the number of staff who failed to come to work over a specific period. Total working days are the total number of staff multiplied by the number of working days for a specific period.

The absenteeism rate can be calculated for specific departments or job categories within an organization. This allows the organization to identify which departments or job roles experience higher rates of absenteeism. By analyzing absenteeism rates across different segments of the workforce, the organization can pinpoint potential causes and develop targeted solutions to address them.

8.2 Accident Cost

Accident cost = Direct cost + Indirect cost

Direct Cost – This type of cost can be directly measured in terms of money.

- Medical injury and illness cost
- Compensation cost
- Building damage
- Tool and equipment damage
- Product and material damage
- Legal expenses
- Expenditure on emergency supplies
- Interim equipment rental

Indirect Cost – This type of cost cannot be measured directly in terms of money. Thus, managers need to apply several indirect ways to estimate the indirect cost.

- Investigation time (Time/number of hours multiplied by the hourly rate given for investigation).
- Wages paid for lost time – some employees may assist the injured person, and this is a lost time for the production process. (Lost time/hours multiplied by employee's rate of pay.)
- Cost of hiring and training replacements (Cost of advertisements to recruit a new employee + interview cost + induction cost + cost of training programs, etc.)
- Overtime payments (Number of hours of overtime multiplied by the hourly rate of overtime.)
- Lost production profits ([(unit price* number of damaged products) – (unit cost*number of damaged products)])
- Extra supervisory time (The amount of time spent completing accident forms, helping the injured person to take medical treatments, analyzing the causes of the accident, follow-up visits etc., multiplied by the supervisor's rate of pay).
- Decreased output of injured worker upon return to work (Decreased output multiplied by unit price).
- Loss of business and goodwill (Number of customers lost compared to before the accident multiplied by average customer's value).
- Fines, increased insurance premiums or other charges imposed against the organization as a result of the accident.

Most of the time, it is hard for the organization to collect data relating to indirect costs for a specific period. Therefore, it is recommended to estimate the direct cost and indirect cost for each type of accident (minor/major) once they occur and to calculate a ratio of indirect cost to direct cost. This would be helpful to estimate the indirect cost for a specific type of accident (minor/major) once they occur again in the future. However, the ratio needs to be updated as time passes.

Accident cost can be analyzed against several benchmarks such as historical trends within the organization, business units, competitive organizations, etc. If the organization's accident cost is increasing compared to historical data, it means that the organization has to take more preventive measures to reduce accidents in the workplace.

8.3 Average Time for Dispute Resolution

$$\begin{array}{c}\text{Average time for} \\ \text{dispute resolution}\end{array} = \frac{\text{No. of days taken to resolved disputes}}{\text{Total no. of disputes}} \times 100$$

If the dispute resolution time is increasing, this provides a signal to the organization that it has a problem with the process of dispute resolution and needs to take immediate action to redesign the process to reduce the cost.

8.4 HR Budget as a Percentage of Sales

$$\text{HR budget as a percentage of sales} = \frac{\text{Total HR department budget}}{\text{Total sales of the company}} \times 100$$

The higher increase in the HR budget as a percentage of sales indicates that the department is overstaffed, or a need to review HR work practices, or to investigate better technology or outsourcing options to reduce expenses. A decrease in ratio simply reflects how well the organization is leveraging its HR resources across its workforce, or it may reflect that there is a need for improvements in HR practices and the department is lacking staff.

8.5 Profit Per Employee

$$\text{Profit per employee} = \frac{\text{Operating profit}}{\text{No. of full time equivalet employees}}$$

Full-time equivalent employees refer to the total employees working 45 hours per week. A higher profit per employee indicates that the company can operate on low overhead costs and can operate the business with fewer employees.

8.6 Turnover Rate

$$\text{Turnover rate} = \frac{\begin{array}{c}\text{Number of employees left involuntaraly and} \\ \text{voluntaraly during the period}\end{array}}{\text{Number of total employees exist during the period}} \times 100$$

A high turnover rate means there is a need for the HR department to analyze the causes of turnover by conducting exit interviews, online questionnaires etc., and cost-effectively address those issues.

8.7 Retention Rate of Key Employees

$$\text{Retention rate of key employees} = \frac{\text{Number of employees remaining at the end of the period}}{\text{Total number of employees exist at the begining of the period}} \times 100$$

A higher retention rate compared with the industry norms indicates that employees' job satisfaction, involvement, commitment, and engagement are to a high degree. If the rate is low, the organization needs to take HR strategies to retain key employees.

8.8 Recruiting Efficiency Ratio

This metric is prepared to know the talented employees recruited compared to the total number of employees recruited during the period. If the ratio is high, that means the talented employees recruited compared to the total number of employees recruited during the period is high.

$$\text{Recruiting Efficiency Ratio} = \frac{\text{Talented people recruited during the period}}{\text{Total no. of employees recruited during the period}} \times 100$$

8.9 New Hire Performance Ratio

This ratio helps in recruitment validation and can be measured in terms of the performance results of recruits after onboarding and job placement. This will be further useful to find an effective way of recruiting people. A particular recruitment process is considered to have high validity if there are a lot of top performers hired from it relative to the total number of recruiters from the same procedure.

$$\text{New hire performance ratio} = \frac{\text{Number of high performers recruited from a specific recruitment process}}{\text{Total no. of employees recruited from a specific recruitment process}} \times 100$$

8.10 Recruitment Success Ratio

This represents the percentage of applicants who successfully finished the interview process concerning the total number of job hopefuls interviewed. If the rate is high, it indicates that a significant proportion of job candidates were successful in making it through the interview process. However, if the rate is more than the expected rate, it indicates that the organization has to reevaluate its interview schemes.

$$\text{Recruitment success ratio} = \frac{\text{Number of candidates succeful at the interview process}}{\text{Total no. of candidates interviewed at the interview process}} \times 100$$

8.11 *Yield Ratio*

Number of applicants who completed the stage/Total number of applicants who entered the stage. For example:

6:5 (30 applicants applied; after screening, 25 candidates were invited for the employment test)
5:4 (After the test, 20 candidates were called for the selection interview)
4:3 (Following the interview, 15 applicants' backgrounds were checked)
3:2 (After the background investigation, 10 applicants were selected for the medical test)
2:1 (Following the medical test, 5 applicants were selected)

8.12 *Cost Per Hire*

$$\text{Cost per hire} = \frac{\text{Total hiring cost}}{\text{Number of new recruits}}$$

Total hiring cost = Advertising costs + Agency fees + Employee referral cost + Travel expenses + Relocation expenses + Internal recruiter's costs + Administrative costs

An increase in cost per hire implies that the organization needs to move toward more cost-efficient recruitment methods rather than using existing methods continuously.

8.13 *Headcount Trained Rate*

$$\text{Headcount trained rate} = \frac{\text{Number of employees trained}}{\text{Total no of employees}}$$

8.14 *ROI of Training*

$$\text{ROI of Training} = \frac{\text{(Training Benefit-Training Cost)}}{\text{Training Cost}}$$

The return on investment is one important statistic that aids stakeholders in managerial decision-making. Analyzing HR analytics from an ROI-based perspective can increase the chances of the practical adoption of HR analytics (Ben-Gal, 2018).

Return on investment of training is useful to determine the success of implementing a training initiative, and also helpful to decide whether to invest money on training further is worth it or not. If the ROI of training is high, that means that the training effort is effective and efficient.

8.15 *Cost Per Trainee Hour*

$$\text{Cost per trainee hour} = \frac{\text{Total training cost}}{\text{Total no. of trainees} \times \text{total no. of training hours}}$$

Cost per trainee hour helps determine the cost-effectiveness of current training delivery methods and whether the system needs to be redesigned.

8.16 Training Transfer Ratio

A high ratio indicates that employees are effectively implementing the skills and insights acquired during training into their daily tasks and processes.

$$\text{Training transfer ratio} = \frac{\begin{array}{c}\text{Number of trained employees actualy}\\ \text{apply the trained practices}\end{array}}{\text{Total no. of trained employees}} \times 100\%$$

Organizations have the option to measure subjective HR concepts such as employee engagement, job satisfaction, commitment, and attitudes by employing various indexes developed through the creation of questionnaires. For instance, Opatha (2019) introduced a questionnaire designed to assess employee job involvement, utilizing a 5-point Likert scale (Table 4). By utilizing such questionnaires, organizations can construct indexes to quantify employee job involvement effectively. Similarly, organizations can design tailored questionnaires to gauge other subjective concepts, offering a quantitative means to indirectly measure these aspects.

Table 4 Questionnaire to Measure Job Involvement.

Statement	*Strongly Disagree*	*Disagree*	*Average*	*Agree*	*Strongly Agree*
My job is a central part of my overall life.	1	2	3	4	5
Much of my energy has been devoted to my job.	1	2	3	4	5
I feel I am strongly and emotionally attached to my job duties.	1	2	3	4	5
I am not a person without my job.	1	2	3	4	5
My values match my job values highly.	1	2	3	4	5
I mention my job first when I introduce myself.	1	2	3	4	5
My job is so attached to my life that I live, eat, and breathe on the job.	1	2	3	4	5
My major satisfaction comes from my job.	1	2	3	4	5
I am indeed physically immersed in my job.	1	2	3	4	5
I have a strong willingness to participate in decision-making.	1	2	3	4	5

Source: Opatha (2019)

9. Sustainable HR Analytics: A New Paradigm

Sustainable Human Resource Management (HRM) is the cornerstone of managing people in a way that balances all three aspects of sustainability i.e., economy, social responsibility, and greenery (Piwowar-Sulej, 2021; Aust (Ehnert), Matthews, and Muller-Camen, 2020). Renwick, Redman and Maguire (2013) have mentioned that sustainability means consuming natural resources and other resources sustainably without limiting the usage of these resources for future generations.

Though prevailing literature covers substantial research findings in terms of sustainable HRM, there remains a notable gap in the literature regarding the concept of sustainable HR analytics.

Therefore, this chapter hopes to bridge this literature gap by introducing a novel paradigm named "Sustainable HR Analytics." To determine the true image of sustainable HR analytics, it is essential to comprehend the extent of sustainable HRM. Sustainable HRM entails the following as per Kainzbauer, Rungruang and Hallinger (2021).

- HRM is viewed as a system that consists of HR strategies, HR policies, HR functions, etc.
- Organizational goals are diverse in terms of financial, social, and environmental.
- HRM is not only for shareholders but also for a wide range of stakeholders, such as CEOs, middle and line managers, HR professionals, employees, communities, society, and the environment.
- HR impacts can be either positive or negative, e.g., increasing employability, guaranteeing a harmonious work-life balance, and minimizing negative impacts.
- Long-term orientation.
- Inside-out perspective to outside-in perspective.

According to Dyllick and Muff (2015), as a company progresses towards higher levels of business sustainability, it undergoes three significant shifts:

- Shift 01: The organization's primary concern shifts from focusing solely on economic profits to encompassing a broader perspective that emphasizes the importance of social, environmental, and economic sustainability-related issues.
- Shift 02: The concept of business value evolves from being solely linked to shareholder value to embracing a more extensive value proposition that encompasses all three dimensions of the triple bottom line, namely people, planet, and profit.
- Shift 03: Instead of adopting an internal perspective, organizations adopt an external outlook. This is where people start to think of the economic, social, and green scopes of the organization and initiate new strategies and actions to enhance the financial, social, and environmental status of the organization.

In accordance with the above shifts, Aust et al. (2020) have identified four categories of sustainable Human Resource Management.

1. *Socially Responsible HRM:* This entails the HR actions taken by the organization to ensure the long-term well-being of the stakeholders, including the employees.
2. *Green HRM:* This entails the HR actions taken by the organization to ensure environmental sustainability that aligns with ecological values.
3. *Triple Bottom Line HRM:* This entails the HR actions taken by the organization to ensure the economic stability of the organization by making profits.
4. *Common Good HRM:* This entails the HR actions taken by the organization to ensure the societal well-being at large.

Opatha (2019) and Kramar (2022) have presented comprehensive definitions for sustainable HRM covering its three scopes i.e., planet, society, and profit. Sustainable HRM entails efforts to improve the environmental, social, and economic performance of the organization. Based on the above categorizations in the literature, the researchers of this paper categorize sustainable HRM as,

1. *Green HRM:* Focuses on the formulation of people management strategies that safeguard the environment.
2. *Socially Responsible HRM:* Concentrates on crafting people management practices that enhance the well-being and quality of life of individuals.
3. *Economic HRM:* Emphasizes the creation of people management strategies aimed at boosting financial profitability.

Sustainable HR analytics involves the integration of HRM sustainability principles into HR analytics processes. In essence, sustainable HR analytics embodies evidence-based HRM that addresses three key facets of sustainable HRM: environmental responsibility, social accountability, and financial profitability. Consequently, sustainable HR analytics is categorized into three distinct domains:

1. *Green HR Analytics:* This represents evidence-based HRM focused on the adoption of environmentally conscious people management practices.
2. *Socially Responsible HR Analytics:* This encompasses evidence-based HRM efforts dedicated to the advancement of people management practices that elevate individuals' quality of life.
3. *Economic HR Analytics:* This pertains to evidence-based HRM that concentrates on the development of people management strategies geared towards enhancing financial profitability.

Considering the above domains, *sustainable HR analytics can be defined as one of the emerging areas in the HR field that strives to help HR professionals integrate sustainability goals into their plans and operations which assess the evidence-based impact of human resources activities on environmental, social, and economic sustainability through which find possibilities to improve sustainability outcomes while retaining organizational performance.*

Traditional HRM is largely concerned with meeting the organization's immediate needs through short-term goals such as people resourcing, managing performance, managing rewards, training and development, and employee relations (Wilton, 2013). Its goal is to ensure that the staff is effectively and efficiently managed to meet the company's goals (Armstrong, 2009). In contrast, sustainable HRM focuses on the long-term well-being of the firm and its personnel. It seeks to incorporate environmental, social, and economic sustainability concepts into HR processes thereby achieving sustainable development goals (Kramar, 2022). This involves developing employee well-being, promoting diversity and inclusion, and addressing environmental problems such as decreasing the carbon footprint and encouraging eco-friendly practices (Mohiuddin, Hosseini, Faradonbeh, and Sabokro, 2022). Based on this differentiation, sustainable HR analytics and traditional HR analytics can be distinguished. Traditional HR Analytics focuses on evaluating data for common HR indicators such as attrition rates, recruitment efficacy, training outcomes, and performance evaluations. The focus is on optimizing HR processes and increasing organizational efficiency (Davenport and Harris, 2017). Sustainable HR Analytics goes beyond typical HR metrics to include data on environmental, social, and economic sustainability. The emphasis is on incorporating sustainability ideas into HR processes and increasing long-term organizational resilience. Figure 4 shows the aspects of sustainable HR analytics.

Integrating the above aspects into HR analytics can help organizations achieve their sustainability goals while enhancing overall HR effectiveness and contributing positively to the workforce and society. Several real-world case studies illustrate how organizations implement HR analytics to ensure long-term operational success while addressing environmental, social, and economic considerations. Best Buy has found that a 0.1% increase in employee engagement can boost annual operating income by over $100,000 per store. Seeing how significant this impact is, the company decided to conduct employee engagement surveys quarterly instead of just once a year (Davenport et al., 2010). Microsoft provides detailed diversity and inclusion statistics in its annual reports, focusing on representation across different levels of the organization and tracking progress toward its diversity goals. Google has developed a mathematical algorithm that proactively predicts which employees are most likely to leave the organization. This enables management to take timely actions to retain those employees effectively (Reddy and Lakshmikeerthi, 2017). Unilever's workforce analytics team demonstrated that analytics can be highly effective during times of crisis. They analyzed organizational networks and developed models to identify potential cost reductions, helping the company navigate challenges and optimize expenses (AIHR Academy). A noteworthy study conducted in Zimbabwe focused on improving hiring assessments to address road traffic accidents (RTAs) among drivers and identified some predictive factors that could be used as selection criteria for hiring new drivers, thereby enhancing safety and reducing the associated costs. Research done on Starbucks Coffee in Manado City has found that Green HRM indicators such as green recruitment, green development, and green reward significantly impact employee performance, and they do evaluations on these using analytics (Samola, 2022). These case studies underscore the importance of HR analytics in promoting organizational sustainability. They demonstrate how HR analytics plays a critical role in tracking progress toward sustainability by providing evidence-based insights into various

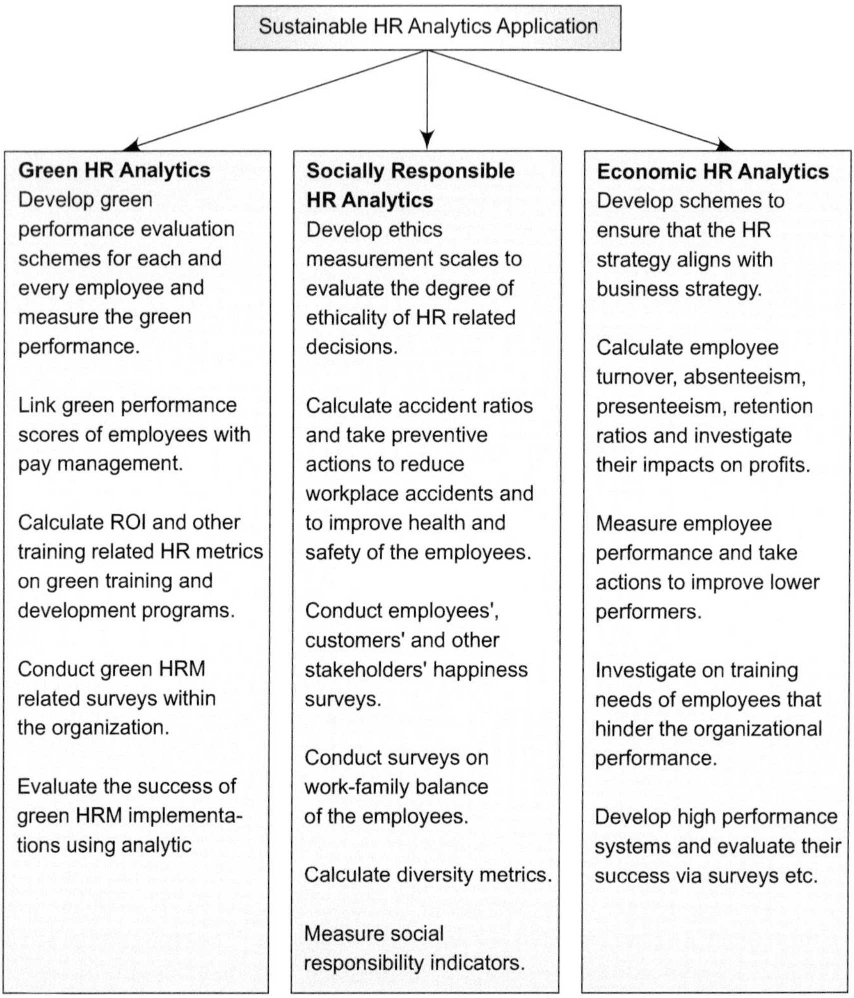

Fig. 4 Aspects of Sustainable HR Analytics
Source: Author Developed

aspects of human resource management. Through the use of data and analytics, organizations can effectively manage and improve their sustainability efforts, ensuring long-term success and alignment with environmental, social, and economic goals.

Conclusion

Every organization must take an innovative step forward that opens the door to competitive advantage in a competitive environment. In the light of the resource-based concept, human resource and human resource practices can derive the competitive advantage for the organization. The main challenge for organizations is to maintain and secure this advantage for the long term. That's why sustainability is pivotal to securing that expected competitive advantage over time. A powerful instrument that can support sustainability is HR analytics. Not using HR analytics can prevent organisations from reaching their pinnacles. Therefore, it is indispensable for organizations to place significant importance on applying HR analytics to accomplish competitive advantage through sustainability while evaluating the HR initiatives in terms of economy, social responsibility, and environment. In the future, HR analytics will mostly involve integrating Artificial Intelligence (AI) into HR procedures for the goals of implementation and assessment. This will enable organizations to make more accurate predictions than ever before. The organization must, however, ensure transparency and equity while using AI fairly and equitably. As a consequence of successful HR analytics, organizations can bring data-driven insights to the decision-making process. Further, by combining intuition with evidence, decision-makers can take the best actions to ensure sustainability.

References

AIHR Academy. (n.d.). *People analytics: Case study collection*. Retrieved from https://www.aihr.com/resources/AIHR_HR_Analytics_Case_Study_Collection.pdf.

Armstrong, M. (2009). *Armstrong's handbook of human resource management practice* (11th ed.). Kogan Page Limited.

Aust (Ehnert), I., Matthews, B. and Muller-Camen, M. (2020). Common good HRM: A paradigm shift in sustainable HRM? *Human Resource Management Review, 30*(1), 1–11.

Bahuguna, P.C., Srivastava, R. and Tiwari, S. (2024). Human resources analytics: Where do we go from here? *Benchmarking: An International Journal, 31*(2), 640–668.

Bassi, L. (2012). *HR analytics handbook*. Reed Business.

Ben-Gal, H.C. (2018). An ROI-based review of HR analytics: Practical implementation tools. *Personnel Review*.

Bhattacharyya, D.K. (2017). *HR analytics: Understanding theories and applications*. SAGE Publications.

Boudreau, J.W. and Ramstad, P.M. (2004). Talentship and human resource measurement and analysis: From ROI to strategic organizational change. *University of Southern California*.

Bucknall, H. and Wel, Z. (2006). *Magic numbers for human resource management: Basic measures to achieve better results*. John Wiley & Sons.

Davenport, T. and Harris, J. (2017). *Competing on analytics: Updated, with a new introduction: The new science of winning*. Harvard Business Review Press.

Davenport, T.H., Harris, J.G. and Shapiro, J. (2010). Competing on talent analytics. *Harvard Business Review*. Retrieved from https://hbr.org/2010/10/competing-on-talent-analytics

Dooren, J.V. (2012). HR analytics in practice (Master's thesis).

Dyllick, T. and Muff, K. (2015). Clarifying the meaning of sustainable business: Introducing a typology from business-as-usual to true business sustainability. *Organization & Environment, 29*(2), 156–174.

Fitz-Enz, J. and John Mattox, I.I. (2014). *Predictive analytics for human resources*. John Wiley & Sons.

Fred, M.O. and Kinange, U.M. (2015). Overview of HR analytics to maximize human capital investment.

Gurusinghe, R.N., Arachchige, B.J.H. and Dayarathna, D. (2021). Predictive HR analytics and talent management: A conceptual framework. *Journal of Management Analytics, 8*(2), 195–221.

Jabir, B., Falih, N. and Rahmani, K. (2019). HR analytics: A roadmap for decision making (case study). *Indonesian Journal of Electrical Engineering and Computer Science, 15*(2), 979–990.

Jain, A. and Nagar, N. (2015). An emerging trend in human resource management. *SS International Journal of Economics and Management, 5*(1), 1–10.

Kainzbauer, A., Rungruang, P. and Hallinger, P. (2021). How does research on sustainable human resource management contribute to corporate sustainability? A document co-citation analysis, 1982–2021. *Sustainability, 13*(1), 1–21.

Kameswari, J., Palivela, H., Settur, S. and Solanki, P. (2023). Identification, assessment, and optimization of key impact variables in people analytics using AI. In Tyagi, P., Chilamkurti, N., Grima, S., Sood, K. & Balusamy, B. (Eds.), *The adoption and effect of artificial intelligence on human resources management, Part A* (pp. 245–282). Emerald Publishing Limited.

Kapoor, B. and Sherif, J. (2012). Human resources in an enriched environment of business intelligence. *Kybernetes, 41*(10), 1625–1637.

King, K.G. (2016). Data analytics in human resources: A case study and critical review. *Human Resource Development Review, 15*(4), 487–495.

Kiran, K.S., Sharma, N. and Brijmohan, D.R. (2018). HR analytics: Transactional to transformational HR approach. *International Journal of Advance and Innovative Research, 5*(3), 1–11.

Kirtane, A. (2015). Corporate sustainable HR analytical practices. *Journal of Management & Administration Tomorrow, 4*(1), 33–40.

Kramar, R. (2022). Sustainable human resource management: Six defining characteristics. *Asia Pacific Journal of Human Resource, 60*, 146–170.

Marler, J.H. and Boudreau, J.W. (2017). An evidence-based review of HR analytics. *The International Journal of Human Resource Management, 28*(1), 3–26.

Mohiuddin, M., Hosseini, E., Faradonbeh, S.B. and Sabokro, M. (2022). Achieving human resource management sustainability in universities. *International Journal of Environmental Research and Public Health, 19*(2), 928.

Naula, S. (2015). HR analytics: Its use, techniques, and impact. *International Journal of Research in Commerce & Management, 8*(1), 47–52.

Opatha, H.H.D.N.P. (2019). *Sustainable human resource management*. Department of Human Resource Management, University of Sri Jayewardenepura.

Opatha, H.H.D.P.J. (2021). *HR analytics: Quantitative side of human resource*. Feather Print Services.

Patil, B.S. and Priya, M.R.S.R. (2024). HR data analytics and evidence-based practice as a strategic business partner. *Vilakshan: XIMB Journal of Management, 21*(1), 114–125.

Piwowar-Sulej, K. (2021). Human resources development as an element of sustainable HRM with the focus on production engineers. *Journal of Cleaner Production, 278*, 1–14.

Rana, G. and Sharma, R. (2019). Emerging human resource management practices in Industry 4.0. *Strategic HR Review, 18*(4), 176–181.

Rana, G., Sharma, R. and Goel, A.K. (2019). Unraveling the power of talent analytics: Implications for enhancing business performance. *Business governance and society: Analyzing shifts, conflicts, and challenges*, 29–41.

Reddy, P.R. and Lakshmikeerthi, P. (2017). HR analytics: An effective evidence-based HRM tool. *International Journal of Business and Management Invention, 6*(7), 23–34.

Reena, R., Ansari, M.M.K. and Jayakrishnan, S.S. (2019). Emerging trends in human resource analytics in the upcoming decade. *International Journal of Engineering Applied Sciences and Technology, 4*(8), 260–264.

Renwick, D.W., Redman, T. and Maguire, S. (2013). Green human resource management: A review and research agenda. *International Journal of Management Reviews, 15*(1), 1–14.

Samola, W.N.E. (2022). The impact of green human resource management on employee performance. *Jurnal EMBA: Jurnal Riset Ekonomi, Manajemen, Bisnis dan Akuntansi, 10*(2), 429–443.

Sharma, R., Singh, S.P. and Rana, G. (2019). Employer branding analytics and retention strategies for sustainable growth of organizations. *Understanding the role of business analytics: Some applications*, 189–205.

Sharma, R., Saini, A.K. and Rana, G. (2021). Big data analytics and businesses in industry 4.0. *Design Engineering*, 238–252.

Smith, T. (2013). *HR analytics: The what, why, and how*. Author.

Wilton, N. (2013). *An introduction to human resource management* (2nd ed.). SAGE.

3 Data Science for FinTech Innovations: Trends, Tools, and Applications

V. Harish[1*], V. Srividya[2] and M. Kalai Arasi[3]

Financial Technology (FinTech) revolutionizes traditional financial services through digital intrusion and innovations, enhancing efficiency and ubiquity. Data science is a field of study that augments the potential of data, using scientific methods to uncover patterns, get meaningful insights, and drive informed decisions for business. By leveraging advanced technologies like big data analytics, deep learning, and machine learning, data science plays a pivotal role in Fintech by enabling predictive insights and personalized services. This paper aims to investigate the core principles of Fintech in Data Science, Financial data types, Role of algorithms behind data science in FinTech while examining key components that facilitate these applications, Financial Data Ecosystem in FinTech, Tools and Technologies Driving Data Science in FinTech, Applications of Data Science in FinTech, Challenges in Implementing Data Science in FinTech. The findings highlight three prominent applications: Fraud Detection and Prevention, Personalized Financial Services, and Risk Management, which demonstrate significant impact on operational efficiency, enhanced customer loyalty, and accuracy. The top challenges identified are data privacy and security concerns, Integration with Legacy Financial Systems, and Managing Regulatory and Ethical Constraints. Addressing these challenges will give huge potential to firms in the form of various benefits as stated.

1. Introduction

The dawn of financial technology, known as FinTech, has pivotally transformed the global landscape of finance, it has evolved as a niche sector over the two decades in transforming traditional banking and its services like payments, lending, and investment to a digital format (Gust, 2021). This strong revolution is further anchored by the integration of data science which leverages statistical methods, advanced algorithms, machine learning, and deep learning techniques through data mining from big data to derive and extract actionable insights. This integration of FinTech and data science acts as an enabler that has opened opportunities for

[1] 0000-0002-3913-9419
[2] 0000-0001-7845-7076
[3] 0009-0002-3240-7369
* Corresponding author: harish@psgim.ac.in

DOI: 10.1201/9781003514282-3

innovation, resulting in efficiency enhancement, hyper-personalizing customer experiences, and ensuring optimal decision-making (Cao et al., 2021).

Data science plays a critical role in solving some of the most crucial challenges in the financial industry, like identifying fraud patterns in real-time and forecasting market trends. The application of data science has enabled and improved the agility of financial institutions, to adapt to the highly varying dynamics of the modern economy (Hasan et al., 2020; Sharma et al., 2021a) and its impact is most obvious and reflected in FinTech, where agility and adaptability are the core principles. Established financial institutions including startups are leveraging data science to disrupt traditional models by integrating advanced technologies with legacy systems to offer cheaper, faster, and more inclusive services.

The emergence of FinTech through its technological advancements represents a paradigm transition from its traditional conceptual financial services. Traditional banking is a time-consuming process that presumed and followed a one-size-fits-all approach, has been replaced by data analytics, hyper-personalization (Gunewardana and Jayamaha, 2022; Harish et al., 2025). The predictive and prescriptive analytics support optimal decision-making in a short duration, the intelligent virtual agents modelled and trained with deep learning and machine learning algorithms are revolutionizing financial investment planning and wealth management by offering data-driven investment decisions catered to individual financial needs and goals.

The four V's (Volume, Variety, Veracity, and Velocity) of Big data generated in the digital era by the finance sector are crucial for FinTech's growth (Awotunde et al. 2021). Every digital activity generates data such as online transactions, social media interaction, clickstream data, IoT-enabled financial activity, and so on, generating vast amounts that can be analyzed to gain deeper, valuable insights into market trends and customer behavior. Though big data is available, the potential to harness this data effectively determines the success of FinTech firms among rivals, which is attainable through data science, its tools, and methodologies that can transform raw data into meaningful insights, enabling financial institutions to make the best optimal decisions (Andronie et al., 2023).

Cloud computing, grid computing, quantum computing, and artificial intelligence have boosted the integration of data science into FinTech enabling its high-performance computing and processing of big data in real time (Zhu et al., 2024). These radical technological enablers promote the competency of small-scale FinTech start-ups to compete with established legacy institutions through their access to global financial services. In addition to the existing advancements, the introduction and integration of blockchain technology with data science are transforming sectors such as decentralized finance, secure transactions, and fraud detection. Despite these notable advancements, the application of data science in FinTech is still challenging due to issues such as cybersecurity risks, data privacy, system vulnerability, and data and algorithmic biases, which possess significant hurdles that are prioritized to be addressed to ensure sustainable growth (Chaudhary et al., 2024). In the financial sector, where decisions often have high-cost implications, it is crucial to ensure fairness and transparency in data and algorithms. Similarly, securing confidential financial data against breaches is critical, since the systems are highly vulnerable to the updated hacking methodologies and cyber threats have become casual.

Another primary concern is the regulatory framework surrounding FinTech innovations due to the rapid change of technologies and their obsolescence, the ability of older regulatory frameworks to accommodate FinTech innovations is limited, leading to uncertainties and compliance challenges for businesses (Gurrea-Martínez and Remolina, 2020).

Chapter Organization

The book chapter will be organized into the following chapters: Core Principles of Data Science in FinTech, Types of Financial Data, The Role of Algorithms in FinTech, Key Components of Data Science in FinTech, Financial Data Ecosystem in FinTech, Tools and Technologies Driving Data Science in FinTech, Applications of Data Science in FinTech, Challenges in Implementing Data Science in FinTech, Future Directions and Opportunities, and Conclusion.

In conclusion, integrated data science and FinTech indicate a rapidly emerging field with a huge potential to reengineer financial services. FinTech companies can deliver hyper-personalized and secure services to their customers by leveraging emerging technologies and advanced analytical tools and techniques. To realize its potential, it has to overcome significant challenges, especially in regulation, workforce readiness, and ethics. This chapter explores the concepts and themes, providing insights into the role of data science in shaping the future of FinTech and its implications for businesses, consumers, government, and society at large.

2. Objectives

The objectives of the chapter are as follows:

1. To understand the core principles of data science in FinTech
2. To identify the key components of Data Science in FinTech
3. To understand the tools and technologies driving Data Science in FinTech
4. To consolidate the applications of Data Science in FinTech
5. To list the various challenges in Implementing Data Science in FinTech

3. Methodology

The methodology that the chapter adopts is a systematic analysis of academic research papers, industry reports, whitepapers, and case studies and to consolidate relevant information, a range of sources, including academic databases like Google Scholar, JSTOR, Web of Science, Scopus, and Dimensions databases were looked into and filtered based on multiple parameters. The collected data were analyzed using a combination of thematic, critical, and comparative analysis. Thematic analysis involves identifying key themes and subthemes within the literature, while critical analysis evaluates the strengths, weaknesses, and limitations of existing research, identifying gaps in the literature, and comparative analysis involves comparing and contrasting different approaches, methodologies, and findings, identifying best practices and lessons learned from successful implementations. A combination of all three methods was adopted to meet the objectives of the chapter.

4. Analysis

4.1 Objective 1

4.1.1 Core Principles of Data Science in FinTech

Data science operates on key principles that make it a powerful tool in the FinTech industry. Illustrated in Fig. 1 and also includes:

1. Real-time Data-Driven Decision-Making

Like all the sectors, financial sectors increasingly rely on data in decision-making that apply to risk assessment, customer segmentation, and investment strategies. The historical and real-time data collected through multiple channels and omnichannel are collected through which organizations identify anomalies, predict trends, and make informed optimal choices to mitigate risks and enhance performance (Aro, 2024). This approach also empowers businesses to identify market growth opportunities and formulate or adjust strategies dynamically to be competitive in a rapidly changing market landscape (Solanki, 2024).

2. Deep Learning, Machine Learning, and Predictive Analytics

Deep Learning and Machine learning algorithms are integrated through FinTech applications revealing the potential capabilities of the power of data and analytics to predict customer behaviour, fraud patterns, and market movements (Paramesha et al., 2024). Predictive models help to determine loan eligibility, forecast credit defaults, and improve the accuracy of financial projections. Since predictive modelling is attained through analysing big data, it provides deeper insights that help firms identify potential threats, adapt to emerging risks, and capitalize on trending market opportunities more effectively (Bello, 2023).

3. Intelligent Automation, Reengineering and Efficiency

Data science enables Robotic Process Automation and Intelligent Process Automation through which the monotonous mundane tasks such as customer support, credit scoring, and regulatory compliance are automated through intelligent agents (Thangam et al., 2025). This decreases human error, and time-lapse in operations, and ensures consistency in effective service delivery. Further, the reengineering of business processes attained through automation allows financial institutions to streamline and scale their operations seamlessly in storing and handling larger volumes of transactions, utilize resources effectively for higher-value tasks, and foster innovation and growth.

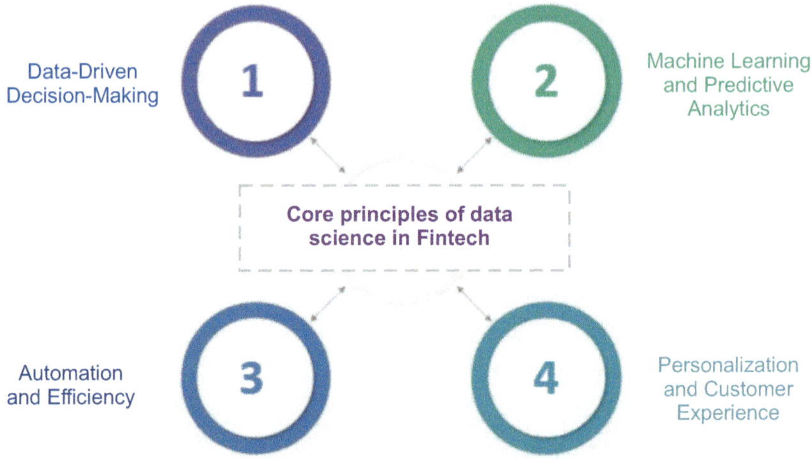

Fig. 1 Core Principles of Data Science in Fintech.

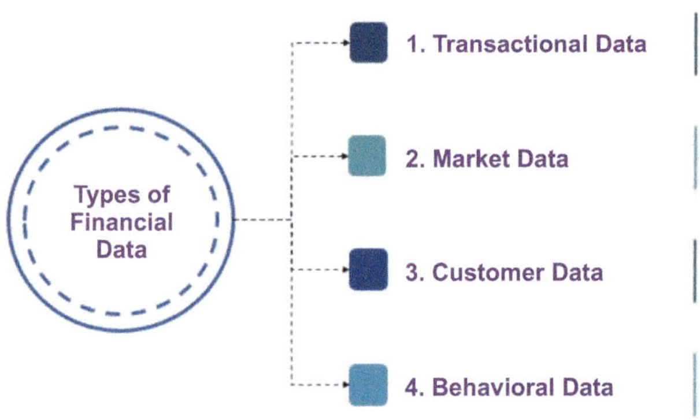

Fig. 2 Types of Financial Data.

4. Personalization, Recommendation, and Customer Experience

FinTech organizations use data sciences to tailor and recommend services that meet the demands of each unique customer. By collecting and analyzing spending habits, investment behaviour, credit history, and other personal data, these companies deliver personalized financial products and services that enhance customer satisfaction and loyalty (Saxena and Muneeb, 2024; Rana and Sharma,2019). This personalized approach promotes stronger customer relationships, trust, long-term stickiness, and loyalty, which helps the companies to avail the competitive advantage of the market and position themselves as leaders in customer-centric financial models.

5. Types of Financial Data

The application of data science in FinTech depends on the availability of various data types. Understanding these data types is crucial in collecting, processing, and analyzing the data science and its impact on financial operations shown in Fig. 2.

1. Transactional Data

This includes records of purchases, payments, and other financial transactions which are vital for tracking and understanding customer behavior and detecting outliers indicative of fraud (Assefa et al., 2020). FinTech companies gather and analyze transactional and behavioral data to gain insights into spending patterns and propensities. This helps them provide tailored financial solutions that match and satisfy their clients' changing needs.

2. Market Data

Real-time and historical data, including interest rates, stock prices, and economic indicators, are collected and analyzed to recommend investment decisions and trading strategies. The collected data is then processed and analyzed by financial institutions to identify patterns with economic and other indicators, forecast market trends, and execute precision trading, giving them a first-mover advantage and competitive edge in a highly volatile financial environment (Fataliyev et al., 2021).

3. Customer Data

Personal information, preferences, investment, and credit histories of customers are the key to understanding customer needs and designing personalized financial products for recommendations. Leveraging this data is an enabler for recommender systems and helps evaluate creditworthiness, mitigating and managing risks associated with loans and other financial services (Boiński, et al., 2022).

4. Behavioral Data

Data related to customer interactions in omnichannel platforms and their clickstream data such as browsing history and app usage patterns, attitudes, and intentions towards investment behavior helps FinTech firms optimize the usefulness of user interfaces and enhance effective engagement (Zhang et al., 2022). Analyzing behavioral data results in predictive models that equip the companies to anticipate user needs, take proactive measures to enhance loyalty, and deliver a seamless integrated customer experience.

5. The Role of Algorithms in FinTech

Algorithms are the lifeline of data science applications in FinTech. The raw data are processed with mathematical models, statistical tools, and computational procedures to extract valuable insights. Commonly used algorithms shown in Fig. 3 include:

1. Classification Algorithms

These are predominantly used in building predictive models for credit scoring to identify defaulters based on their financial history, loan payment history, and behavior. By labeling data as a predefined one, classification algorithms like decision trees, logistic regression, support vector machine, Naïve Bayes classifier and so on can be applied to build predictive models that predict the defaulters (Tian et al., 2021), ensuring the lenders assess risk more accurately and support reliable decision-making. They also play a key role in fraud detection by identifying outliers and suspicious transactions that diverge from normal customer behavior.

2. Clustering Algorithms

These help segment customers into groups based on similar patterns and characteristics, enabling targeted marketing, enhanced service, and product development. Clustering algorithms like K-Means, DBScan, Hierarchical

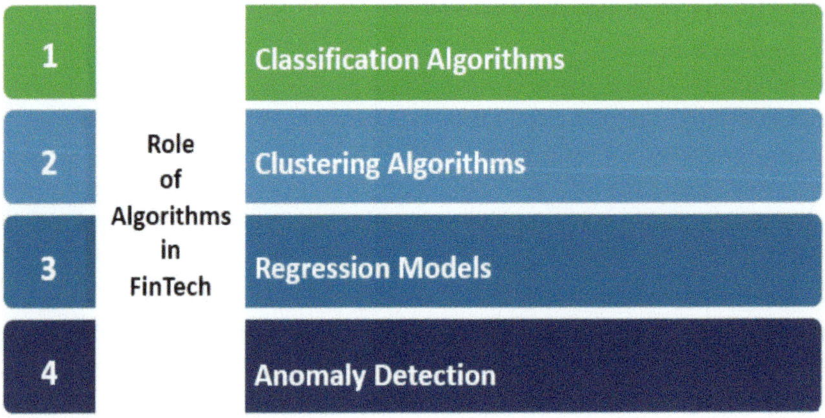

Fig. 3 Role of Algorithms in Fintech

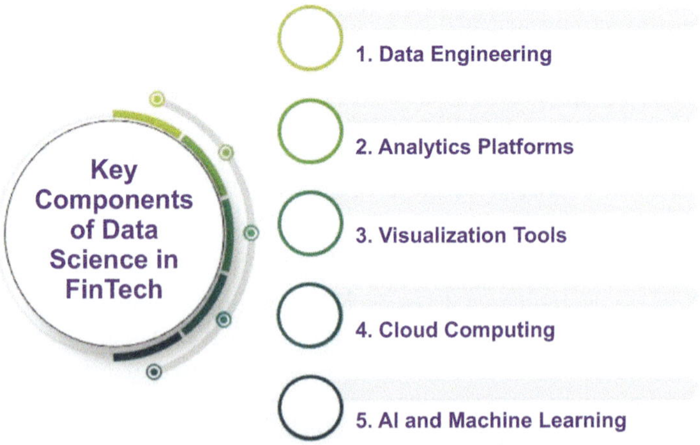

Fig. 4 Key Components of Data Science in Fintech

clustering, and so on allow FinTech companies to better understand customer preferences and behaviour and cater their offerings for every segment through which it can enhance user experiences. For instance, valuable customers can be identified and offered premium services, while at-risk customers can be targeted with retention strategies, while other segments can receive financial solutions as per their preference and position in the journey (Liu, 2024).

3. Regression Models

These models use algorithms like linear regression, support vector regression, decision trees, and ensemble models to forecast financial metrics such as stock prices, revenue, and loan default probabilities (Efendi et al., 2023). The relationship between variables is analyzed to build regression models that help businesses predict future outcomes with greater accuracy, supporting strategic decision-making. These models are also valuable for scenario analysis, which allows businesses to realistically recreate different economic scenarios, forecast, and plan accordingly.

4. Anomaly Detection

Algorithms like KNN, Random forest, Bayesian Networks, and ensemble methods are used for anomaly detection. This is critical for identifying outliers and irregularities in regular transaction patterns, which could predict and identify fraud or system errors. These models analyze big transactional data in real-time, flagging unusual transactions or activities before they result in significant issues (Panda, 2024). This capability enhances security, building customer trust by safeguarding their financial assets.

5. Objective 2

5.1 Key Components of Data Science in FinTech

The successful integration of data science into FinTech relies on a combination of technological and analytical components shown in Fig. 4.

1. Data Engineering

The primary step in a data science project involves collecting, storing, and organizing data. FinTech companies use advanced data techniques and databases to store and manage structured, semi-structured, and unstructured data efficiently. Proper data engineering ensures that raw data is cleansed, pre-processed, accessible, and ready for analysis, which is the base for building accurate predictive models and deriving actionable insights (Verma, 2024).

2. Analytics Platforms

Tools such as Hadoop and Spark enable the processing of big datasets and these platforms support real-time analytics, which is crucial for applications like fraud detection and high-frequency trading (Lee et al., 2022). By leveraging these platforms, FinTech firms possess the potential to process massive volumes of transactions with high speed and accuracy, enabling them to stay competitive in the dynamic financial market.

3. Visualization Tools

FinTech companies use visualization tools like Tableau and Power BI to project complex insights in a user-friendly visualized format. These technologies assist decision-makers in understanding trends and patterns by offering business intelligence insights based on previous data. Visualization tools simplify the interpretation of data through storytelling by creating interactive visualizations of dynamic dashboards supporting stakeholders to make optimal decisions (Petare et al., 2024).

4. Cloud Computing

FinTech companies' scalability is managed and supported by the adoption of cloud services like Google Cloud and AWS for storage and servers, which offers cost-effective services to the companies and the potential to handle big data (Lăzăroiu et al., 2023). Cloud computing enhances business models through streamlining, attained through radical innovation and continuous integration of tools and services while ensuring data security and compliance.

5. Artificial Intelligence and Machine Learning

Data-oriented models offer many FinTech applications like Intelligent agents and Robo-advisors in which Deep Learning and Machine Learning act as enablers to understand, learn, and analyze patterns and trends from historical data to improve their performance over time. These technologies create real-time predictive and prescriptive models for fraud detection and personalized financial services that make AI and ML the vital components of modern FinTech solutions (Pattnaik et al., 2024).

5.2 Financial Data Ecosystems In Fintech

The financial data ecosystem is an interconnected network of data sources, processes, and users. It encompasses a variety of stakeholders, including banks, payment interfaces and processors, insurance companies, and customers.

1. Data Sources

Financial data is collected from various sources with the historical data of investment and transactions, social media, market feeds, and IoT devices that contribute to the variety, richness, and quality of the data used in FinTech applications (Allioui and Mourdi, 2023). Companies can acquire significant insights into consumer behavior, risk concerns, and market trends by tapping and aggregating data from these various sources, enhancing their ability to deliver personalized services and positioning themselves as market leaders.

2. Data Processors

The collected raw data is then cleansed, normalized, and pre-processed using advanced analytics tools to ensure consistency and accuracy. Data pre-processing is critical for eliminating errors and biases in datasets and preparing them for analysis. It also ensures compliance with regulations, which is vital in the highly regulated financial sector (Nand et al., 2023).

3. End Users

Businesses use processed data for analytical applications that help them decide on essential business activities such as loan approval and personal, investment, and financial management. Companies depend on these analytical insights to develop innovative personalized financial products, while consumers benefit from advanced technological tools in optimizing investments, tracking their spending behavior, and achieving financial goals (Cong et al., 2021). This highlights the pivotal role of data in driving financial innovation, strategies, and decisions.

5.3 Interplay Between Data Science and FinTech Innovation

The interrelationship between data science and FinTech is remarkable through its automation and revolutionizing innovation. For instance, NLP (Natural Language Processing) advancements have transformed rule-based chatbots into Intelligent Agents to provide personalized financial advice in vernacular languages. Similarly, integrated blockchain technology with data analytics is revolutionizing the tamper-proof transparent transactions that are validated and recorded. The foundations of data science in FinTech spotlight its transformative impact on the financial industry. By leveraging analytical techniques, radical tools, and varied data types, FinTech firms can create value for customers, addressing complex challenges and driving responsible and sustainable growth. The subsequent sections of this chapter will explore deeper into the emerging trends, tools, and real-world applications that illustrate the dynamic relationship between data science and FinTech.

5.4 Objective 3: Tools and Technologies Driving Data Science in Fintech

The rapid growth of FinTech is driven by the advanced tools and technologies that empower the seamless integration of data science into financial services. These tools enable the effective collection, storage, processing, and analysis of data through which business reengineering happens because of continuous improvement in innovation. Figure 5 explores the key tools and technologies that underpin data science in FinTech, emphasizing their functionality and impact on the industry.

1. Artificial Intelligence (AI), Machine Learning (ML) and Deep Learning (DL)

AI, ML, and DL are the lifelines of most FinTech applications, offering the potential capabilities to enhance process automation and decision-making and improve customer experiences. With the transformation of rule-based chatbots to Intelligent agents capable of providing instant customer support and fraud detection systems that monitor transactions in real-time, predictive modeling for defaulters with real-time predictive analytics (Selvaraj et al., 2024), Fintech has become essential for financial institutions. Through continuous adaptive learning and training models through real-time and augmented data, AI tools enable operational efficiency through automation, help businesses to be safe from fraudsters, and deliver highly personalized services that enhance customer satisfaction and loyalty (Pattnaik et al., 2024).

Key Tools: TensorFlow, Scikit-learn, and PyTorch provide robust frameworks for building and deploying AI and ML models that solve critical financial challenges.

2. Big Data Platforms

Big data tools and technologies are vital to handle both structured and unstructured big data, generated through multiple channels in the financial sector (Sharma et al., 2021b; Awotunde et al., 2021). They allow FinTech companies to collect, store, process, and analyze data efficiently, through which they interpret and extract effective insights that support business decision-making. With real-time and augmented analytics support, big data empowers firms to stay proactive and agile to market changes and customer needs, providing a competitive advantage (Simpson et al., 2024).

Key Tools: Apache Spark, Hadoop, and AWS Big Data Solutions are predominantly adopted for managing and analyzing big data effectively.

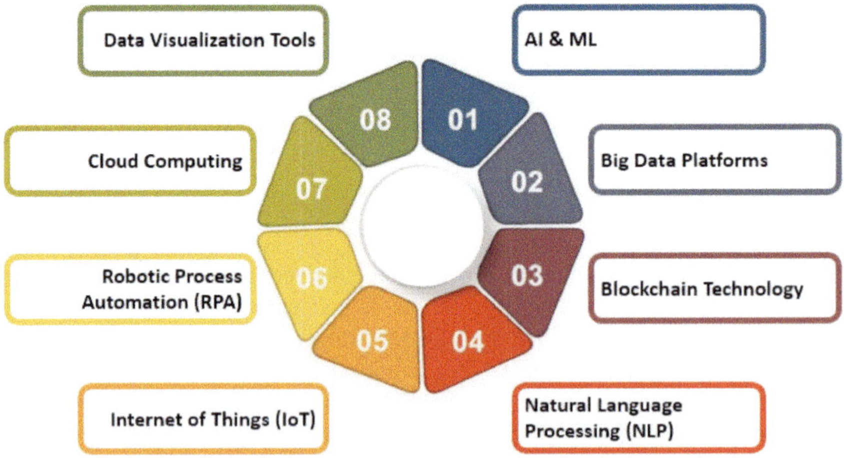

Fig. 5 Tools and Technologies Driving Data Science in FinTech.

3. Blockchain Technology

Blockchain offers a decentralized, transparent, tamper-proof, and secure framework for recording transactions which is a transformative technology in FinTech (Roy and Tinny, 2024). Its transparent and absolute nature has revolutionized payments, asset management, investments, and compliance. Blockchain has become the backbone for stakeholders' loyalty because of its tamper-proof secured framework, which also enables the creation of responsible financial ecosystems (Müller, 2021).

Key Tools: Ethereum, Hyperledger, and Chainalysis are popular platforms that power blockchain-based solutions in FinTech.

4. Natural Language Processing (NLP)

NLP enables smart human-machine interactions attained through NLU (Natural Language Understanding) which empowers the interpretation and responses in vernacular languages (Park et al., 2024). In FinTech, it is utilized to analyze textual and vocal data and provide user-friendly conversational interfaces. This technology reduces the complexity of human-machine interaction, ensuring seamless customer service which could be attained through actionable insights derived from unstructured text data like emails, reviews, or social media posts (Rane et al., 2024).

Key Tools: NLTK, SpaCy, and Google Dialogflow are essential for building NLP-driven applications in FinTech.

5. Internet of Things (IoT)

IoT devices like wearable payment systems and smart ATMs are predominantly available in FinTech (Bhat et al., 2023). By collecting real-time data through sensors, these devices boost operational efficiency conveniently. The IoT integration FinTech has promoted dynamic financial solutions like activity-based and risk-averse calculated insurance and investments, and real-time asset tracking, which transformed traditional practices (Chen and Liao, 2021).

Key Tools: IoT platforms like ThingWorx, AWS IoT Core, and Google Cloud IoT facilitate the seamless integration of IoT into FinTech ecosystems.

6. Robotic Process Automation (RPA)

RPA automates redundant mundane and rule-based tasks, permitting the FinTech workforce to focus on complex and crucial processes (Faiz et al., 2023). Through manual intervention elimination in processes like account reconciliation and compliance reporting, RPA ensures error reduction and faster turnaround times, which eventually boosts operational efficiency (Mah et al., 2023).

Key Tools: Blue Prism, UiPath, and Automation Anywhere lead the market in enabling automation across diverse FinTech operations.

7. Cloud Computing

Cloud computing supports cost-effective scalable solutions for storing and processing financial data. It acts as an enabler for firms to access advanced computing technologies without high-cost infrastructure investments. The scalability and flexibility of cloud computing solutions ensure that businesses can quickly become scalable and adapt to market demands while sustaining robust security and compliance standards (Prakash et al., 2021; Lăzăroiu et al., 2023).

Key Tools: Microsoft Azure, AWS, and Google Cloud Platform are the primary tools adopted for cloud-based FinTech solutions.

8. Data Visualization Tools

Data visualization tools support easy understanding and interpretation of complex datasets by transforming them into comprehensible visual representations that aid in stakeholder communication and optimal decision-making (Kharakhash, 2023). Through their interactive visualization and dashboards, these tools enable firms to interpret easily and communicate insights effectively for technical and non-technical stakeholders to derive multi-criterion data-driven decisions with clarity and confidence (Petare et al., 2024).

Key Tools: Power BI, Tableau, and D3.js are most prevalently used to create real-time interactive and informative data visualizations in FinTech.

The various array of tools and technologies discussed above highlights the innovative infrastructure and framework that enables FinTech companies to harness the power of data science effectively. These technologies provide effective solutions for the unique and complex challenges faced by the financial industry.

5.5 Objective 4: Applications of Data Science in FinTech

Data science through its radical innovation has become a basis in FinTech which helps companies to take proactive measures by addressing critical challenges while creating transformative opportunities. From fraud prevention to personalized services, data science is revolutionizing financial institutions' operations in delivering value. Its applications enable smarter decisions, reduce risks efficiently, and enhance customer satisfaction. The following sections detail the key applications of data science in FinTech along with examples and real-world impacts.

1. Fraud Detection and Prevention

Fraud detection is one of the most significant applications of data science in the financial industry. Advanced deep learning and machine learning algorithms analyze historical transaction patterns and investor behaviors to identify anomalies, thus ensuring real-time fraud prevention. These systems secure consumers and financial institutions from financial losses (Stojanović et al., 2021, Tatineni and Mustyala, 2024).

Table 1 Fraud Detection and Prevention.

Applications	*Impact*	*Case Study*
Real-Time Monitoring: Models continuously track and monitor transactions to flag anomalies such as high-value transactions and transactions to unknown accounts.	Higher accuracy in detection aids in proactive fraud prevention measures.	PayPal integrated data science to detect fraud in real-time, reducing paybacks, and millions are saved annually.
Behavioral Biometrics: Clickstream analytics analyses every click, typing speed, mouse movement, navigation, and device usage to authenticate users.	Enhanced customer convenience with high security and reduced risks.	Behavioral tools enable advanced user authentication in omnichannel platforms.

2. Personalized Financial Services

Personalization has become an essential feature for customer relationships in FinTech by the adoption of data science. Businesses tailor their financial products and services to meet the needs of specific customers by evaluating past customer data on behavior and transactions. This increases customer satisfaction and promotes sustainable financial inclusion (Rana et al., 2024; Sahu, 2020, Aderemi et al., 2024).

Table 2 Personalized Financial Services.

Applications	*Impact*	*Case Study*
Robo-Advisors: Deep Learning and Machine learning assess the risk tolerance and financial goals of individual customers and offer personalized investment portfolios.	Boosts effective customer satisfaction and trust in financial planning.	Platforms like Betterment offer personalized portfolio services which enhances customer loyalty and retention.
Credit Scoring: Integrates other source data like utility payments and social media interactions for precise risk assessment.	Expands socially sustainable credit access for all.	Enhanced credit scoring qualifies more individual customers for loans.

3. Risk Management

Data science facilitates proactive strategies in mitigating risks involved in a business and supports better risk management for maintaining financial stability. Predictive modelling helps businesses identify and analyze vulnerabilities, enabling companies to act proactively to prevent crises (Chowdhury and Kulkarni, 2023, Liu, 2024).

Table 3 Risk Management

Applications	*Impact*	*Case Study*
Loan Risk Assessment: Predictive models build models for defaulter identification through which it predicts the probability of default occurrence based on historical data.	Reduces loan defaults and protects financial stability.	JPMorgan Chase through its predictive models predicts and monitors credit risk globally.
Market Risk Analysis: Analyzes market trends based on geopolitical events to predict potential disruptions.	Enables firms to take preventive measures to safeguard assets.	Analytics-driven strategies formulation helps companies to navigate highly volatile market conditions.

4. Customer Relationship Management (CRM)

Data science promotes CRM by analyzing customer behavior, feedback, and reviews, allowing businesses to enhance stronger relationships. Predictive models help firms retain customers and improve customer satisfaction and loyalty levels (Sanodia, 2024, Rahman et al., 2024).

Table 4 Customer Relationship Management.

Applications	*Impact*	*Case Study*
Churn Prediction: Through churn models built by advanced predictive analytics, firms identify the at-risk customers enabling personalized targeted retention strategies.	Higher retention rates and improved customer loyalty.	CRM tools support retention strategy formulation in competitive markets and increasing revenue.
Sentiment Analysis: Uses NLP (Natural Language Processing) and other text analytics to analyze feedback and identify pain points of customers.	Provides optimal actionable insights for service improvement.	NLP tools allow FinTech firms to propose innovative offerings based on customer sentiment.

5. Automated Trading and Investment Strategies

Algorithmic automated trading and investment interfaces are widely implemented among the advanced applications of data science in FinTech. By leveraging big data, these systems execute trades and support efficient portfolio management (Cao et al., 2021, Andronie et al., 2023).

Table 5 Automated Trading and Investment Strategies.

Applications	*Impact*	*Case Study*
High-Frequency Trading (HFT): Executes trades in nanoseconds based on real-time data analysis.	Maximizes profit in highly volatile markets.	Hedge funds like Renaissance Technologies achieve high returns consistently.
Portfolio Optimization: Recommends portfolio adjustments based on asset performance and current market conditions.	Improves portfolio returns and minimizes risks.	Tailored financial goals are attainable through quantitative platforms to help investor maximize their profit.

6. Regulatory Compliance and Reporting

Regulatory compliance is one of the major challenges for financial institutions in which data science streamlines the process through automated tools to ensure adherence to regulations and reduce costs (Grassi and Lanfranchi, 2022, Yıldırım, 2023).

Table 6 Regulatory Compliance and Reporting.

Applications	*Impact*	*Case Study*
Anti-Money Laundering (AML): Analyze and detect money laundering activities and patterns.	Improves regulatory compliance and reduces risks.	Deep Learning and Machine learning ensure compliance and abidance with global anti-money laundering laws.

Contd.

Table 6 *Contd.*

Applications	*Impact*	*Case Study*
Regulatory Reporting: Automates report generation, ensuring accuracy and adherence to standards.	Reduces manual human errors and enhances operational efficiency.	Intelligent automated systems streamline reporting for Basel III and GDPR compliance.

7. Enhancing Financial Inclusion

Data science improves financial availability and literacy, which makes it easier for everyone, including marginalized groups, to have sustainable access, particularly in emerging economies. Radical innovations like mobile banking and microfinance promote financial inclusion and economic growth (Adeoye et al., 2024, Bello, 2023).

Table 7 Enhancing Financial Inclusion.

Applications	*Impact*	*Case Study*
Microfinance: Predictive analytics determines the creditworthiness of every individual customer without credit histories.	Provides sustainable financial access to all.	M-Pesa leverages data science to serve millions of unbanked individuals in society.
Mobile Banking: Data-driven analytical platforms offer affordable financial services through mobile devices.	Enhances convenience through a user-friendly interface and promotes extensive adoption.	Mobile solutions transform rural and low-income communities through digital literacy and inclusive services.

8. Blockchain Analytics and Cryptocurrency Management

Blockchain analytics supported by data science is essential in fintech due to the adoption of cryptocurrencies and digital asset management. These tools ensure transparency and security in cryptocurrency transactions (Lăzăroiu et al., 2023, Renduchintala et al., 2022).

Table 8 Blockchain Analytics and Cryptocurrency Management.

Applications	*Impact*	*Case Study*
Transaction Analysis: Monitors blockchain transactions to detect outliers.	Builds trust and ensures regulatory compliance.	Chainalysis tools identify suspicious transactions and enhance security.
Market Prediction: Analyzes price patterns, trends, and sentiment to forecast cryptocurrency movements.	Enables optimal trading and investment decisions.	Predictive models provide insights for cryptocurrency investors.

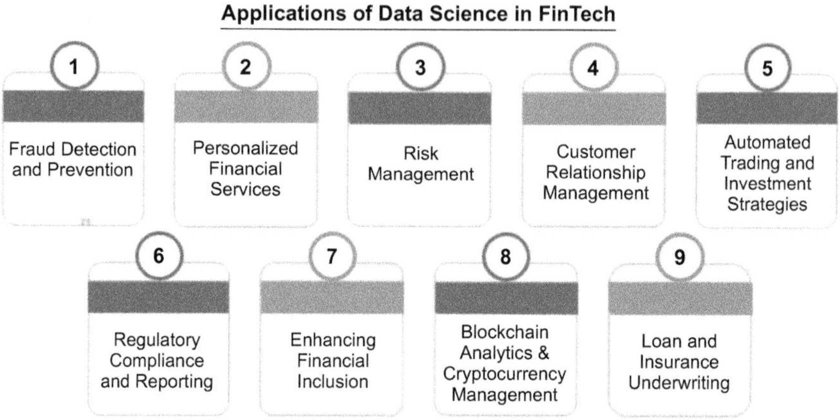

Fig. 6 Applications of Data Science in Fintech.

9. Loan and Insurance Underwriting

Data science has transformed better, faster, and more accurate underwriting processes using data-driven decisions adopted in predictive models (Crouhy et al., 2021).

Table 9 Loan and Insurance Underwriting.

Applications	*Impact*	*Case Study*
Loan Underwriting: Evaluates risk profiles using traditional data sources and historical data.	Reduces default rates and improves loan approval accuracy.	Companies like Lemonade use AI for instant underwriting and claims processing.
Insurance Premium Calculation: Offer personalized premiums by assessing risk factors.	Builds fair and customer-centric pricing models.	Deep Learning and Machine learning tools enhance customer satisfaction in insurance services.

5.6 *Objective 5: Challenges in Implementing Data Science in FinTech*

Implementing data science in the FinTech sector can revolutionize and reengineer the present business models, but the implementation has several challenges. These obstacles range from technological, workforce, and regulatory domains, highlighting the difficulty of integrating cutting-edge emerging technologies into a highly regulated and traditional industry. This section explores four major challenges that FinTech companies face: data privacy and security concerns, integration with legacy systems, managing regulatory and ethical constraints, and addressing workforce skill gaps.

1. Data Privacy and Security Concerns

Data is the lifeline of FinTech which aids in driving radical innovations in areas such as fraud detection, personalized financial products, and credit scoring. However, firms are exposed to serious privacy and security vulnerabilities when they gather, store, process, and analyze vast volumes of sensitive customer data. These risks are directly proportional to the growing cyberattack threats, which can lead to massive financial losses, damage to reputations, and disasters to companies, highlighting the need for the adoption of a proactive approach to data protection (Aldboush and Ferdous, 2023). Transparency about rules and guidelines on the utility of data usage and allowing customers to possess control of their data utility can alleviate concerns and foster trust. Companies must ensure clear and transparent communication about privacy policies to encourage their committed users to data security (Mehrban et al., 2020).

2. Integration with Legacy Financial Systems

Financial institutions mostly depend on legacy IT infrastructure that has been integrated into their traditional processes for decades. These systems are often designed for traditional banking operations and cannot handle the demands of the contemporary market (Mustyala, 2023). The integration process is not just technical but also involves cultural and psychological traits, pushing organizations to shift their mindset and adapt to a rapidly changing digital ambiance. The technological shift from legacy systems is quite expensive, which can be resolved through the adoption of hybrid solutions and cloud-based integrations that can provide a cost-effective way to modernize without disrupting existing operations (Patel, 2023).

3. Managing Regulatory and Ethical Constraints

The regulatory environment for FinTech is complex and varies significantly depending on geographical locations. Ethical concerns such as transparency, fairness, and accountability further complicate the adoption and utilization (Boppiniti, 2023). As FinTech innovations outperform the regulatory framework, companies must equate compliance with emerging technologies and changing customer needs to remain agile and innovative. For global FinTech firms, navigating varying regulations across geographical locations demands a coordinated approach. Centralized compliance teams are supported with intelligently automated tools that help reduce the burden of adhering to multiple legal frameworks concerning geographical locations (Rizinski et al., 2022).

4. Workforce Upskilling and skill management

The rapid evolution and innovation of data science technologies have created knowledge and skill gaps in the FinTech industry. Organizations need cross-disciplinary professionals possessing expertise in technical and financial domains (Wu and Kao, 2022). The challenge is to recruit talent who are interdisciplinary and capable of merging data science and finance concepts to derive valuable insights while promoting a culture of continuous learning within the workforce.

Developing a skilled workforce proficient in advanced technical tools and essential interpersonal skills ensures teams can effectively implement data-driven solutions and adaptively communicate their impact to diverse stakeholders (Doherty and Stephens, 2023).

6. Future Directions and Opportunities

The FinTech sector continues to evolve at an amazing pace, with data science serving as its driving factor and future advancements aiming to revolutionize the industry further, unlocking new and greater opportunities and overcoming challenges. A few key areas poised for significant progress include quantum computing, decentralized finance (DeFi), emerging market innovations, and sustainability-driven green finance analytics.

Quantum computing is set to redefine FinTech analytics by introducing unparalleled computational power which will enable rapid optimization of complex financial models, offering dynamic solutions for portfolio management and risk assessment. Enhanced cryptographic security, a critical requirement for secure transactions, enhancing customer trust is another promising benefit. Industry-Academia partnership is essential to overcome these barriers, paving the way for real-time and augmented analytics.

Decentralized finance (DeFi) attained with the help of Blockchain provides ubiquitous access to financial services and the field of data science plays a major role by enhancing the reliability of smart contracts, detecting fraud, and designing financial literacy and inclusive financial services for all.

Sustainability and green finance analytics are emerging since FinTech attempts to support and attain global environmental goals. Advanced data science methods aid in measuring carbon emissions, evaluating the environmental impact of investments, and promoting responsible investing through consumer awareness campaigns. Predictive and prescriptive models help firms forecast risks and align financial activities with sustainability goals, ensuring that green finance is the basis of the industry's future. The Data science integration in FinTech helps to attain the three pillars of sustainability (Environment, Economy, and Social) while contributing to a greener economy by delivering value-driven financial solutions.

Conclusion

Incorporating data science with FinTech enables revolutionary transformation through radical innovation, resulting in expanded opportunities. While data science has revolutionized the financial sector with higher benefits like personalized customer services, enhanced risk management, and streamlined automated operations efficiency, it still faces a lot of challenges like data privacy concerns, legacy system integration, and workforce knowledge and skill gaps to unlock its full potential. As the FinTech industry advances, emerging technologies like quantum computing, decentralized finance (DeFi), and green finance analytics enhance further innovation and sustainability (Sharma et al., 2024).

FinTech is evolving with the rising societal needs and data science continues to steer this evolution by adopting advanced tools abiding by ethical and regulatory compliance and helping FinTech firms to expand their capabilities and public trust. In conclusion, integrating data science and FinTech represents a transformative evolution and exponential growth by embracing innovation, attained through stakeholder collaboration, can lead FinTech toward a future where present market needs and future global requirements are satisfied.

Reference

Adeoye, O.B., Addy, W. A., Ajayi-Nifise, A. O., Odeyemi, O., Okoye, C. C. and Ofodile, O. C. (2024). Leveraging AI and data analytics for enhancing financial inclusion in developing economies. *Finance and Accounting Research Journal*, 6(3), 288–303.

Aderemi, S., Olutimehin, D.O., Nnaomah, U.I., Orieno, O.H., Edunjobi, T.E. and Babatunde, S.O. (2024). Big data analytics in the financial services industry: Trends, challenges, and future prospects: A review. *International Journal of Science and Technology Research Archive*, 6(1), 147–166.

Aldboush, H.H. and Ferdous, M. (2023). Building trust in fintech: an analysis of ethical and privacy considerations in the intersection of big data, AI, and customer trust. *International Journal of Financial Studies*, 11(3), 90.

Allioui, H. and Mourdi, Y. (2023). Exploring the full potentials of IoT for better financial growth and stability: A comprehensive survey. *Sensors*, 23(19), 8015.

Andronie, M., Iatagan, M., Uță, C., Hurloiu, I., Dijmărescu, A. and Dijmărescu, I. (2023). Big data management algorithms in artificial Internet of Things-based fintech. *Oeconomia Copernicana*, 14(3), 769–793.

Aro, O.E. (2024). Data analytics as a driver of digital transformation in financial institutions. *World Journal of Advanced Research and Reviews*, 24(1).

Assefa, S.A., Dervovic, D., Mahfouz, M., Tillman, R.E., Reddy, P. and Veloso, M. (2020, October). Generating synthetic data in finance: opportunities, challenges and pitfalls. *In Proceedings of the First ACM International Conference on AI in Finance* (pp. 1–8).

Awotunde, J.B., Adeniyi, E.A., Ogundokun, R.O. and Ayo, F.E. (2021). Application of big data with fintech in financial services. *In Fintech with artificial intelligence, big data, and blockchain* (pp. 107–132). Singapore: Springer Singapore.

Bello, O.A. (2023). Machine learning algorithms for credit risk assessment: an economic and financial analysis. *International Journal of Management*, 10(1), 109–133.

Bhat, J.R., AlQahtani, S.A. and Nekovee, M. (2023). FinTech enablers, use cases, and role of future internet of things. *Journal of King Saud University-Computer and Information Sciences*, 35(1), 87–101.

Boiński, P., Sienkiewicz, M., Bębel, B., Wrembel, R., Gałęzowski, D. and Graniszewski, W. (2022). On customer data deduplication: Lessons learned from a R&D project in the financial sector. *In CEUR Workshop Proceedings* (Vol. 3135).

Boppiniti, S.T. (2023). Data Ethics in AI: Addressing Challenges in Machine Learning and Data Governance for Responsible Data Science. *International Scientific Journal for Research*, 5(5).

Cao, L., Yang, Q. and Yu, P.S. (2021). Data science and AI in FinTech: An overview. *International Journal of Data Science and Analytics*, 12(2), 81–99.

Chaudhary, G., Manna, F., Khalane, M.V. P. and Muthukumar, E. (2024). Cybersecurity Challenges In Fintech: Assessing Threats and Mitigation Strategies For Financial Institutions. *Educational Administration: Theory and Practice*, 30(5), 1063–1071.

Chowdhury, D. and Kulkarni, P. (2023, March). Application of Data Analytics in Risk Management of Fintech Companies. *In 2023 International Conference on Innovative Data Communication Technologies and Application (ICIDCA)* (pp. 384–389). IEEE.

Chen, C.C. and Liao, C.C. (2021, September). Research on the development of Fintech combined with AIoT. *In 2021 IEEE international conference on consumer electronics-Taiwan (ICCE-TW)* (pp. 1–2). IEEE.

Cong, L.W., Li, B. and Zhang, Q.T. (2021). Alternative data in fintech and business intelligence. *The Palgrave handbook of FinTech and blockchain*, 217–242.

Crouhy, M., Galai, D. and Wiener, Z. (2021). The impact of fintechs on financial intermediation: A functional approach. *The Journal of FinTech*, 1(01), 2031001.

Doherty, O. and Stephens, S. (2023). Hard and soft skill needs: higher education and the Fintech sector. *Journal of Education and Work*, 36(3), 186–201.

Efendi, S., Nasution, M.K. and Herman, M. (2023, February). The Role of Detection Rate in MAPE to Improve Measurement Accuracy for Predicting FinTech Data in Various Regressions. *In 2023 International Conference on Computer Science, Information Technology and Engineering (ICCoSITE)* (pp. 874–879). IEEE.

Faiz, S.M., Nasreen, A. and Chaithanya, R. (2023). Streamlining and Optimizing End to End Fintech Industry Process through RPA. *International Journal of Automation and Smart Technology*, 13(1), 1–8.

Fataliyev, K., Chivukula, A., Prasad, M. and Liu, W. (2021). *Stock market analysis with text data: A review*. arXiv preprint arXiv:2106.12985.

Grassi, L. and Lanfranchi, D. (2022). RegTech in public and private sectors: the nexus between data, technology, and regulation. *Journal of Industrial and Business Economics*, 49(3), 441–479.

Gunewardana, G. and Jayamaha, A. (2022). Peril or Promise—How the SME Banking Sector Will Perform in 2022 and beyond? Do We See an Overreliance on Recovery? *Technology and Investment*, 13(3), 83–91.

Gurrea-Martínez, A. and Remolina, N. (2020). Global challenges and regulatory strategies to fintech. Banking & Finance Law Review (Forthcoming, 2020, Issue 36.1), SMU Centre for AI & Data Governance Research Paper, (2020/01).

Gust, M. (2021). Traditional Banking and Disruption Fintech. Evolutions, Causes, Typologies. *Revista Economia Contemporana*, 6(1), 57–72.

Harish, V., Sharma, R. and Rana, G. (2025). Unveiling the Digitalization Dilemma: A Study of Business Analytics Adoption in Small and Mid-Sized Manufacturing Firms Post-COVID-19. In: Singh, R., Gehlot, A. (eds) Business Data Analytics. ICBDA 2023. *Communications in Computer and Information Science*, vol 2358. Springer, Cham.

Hasan, M.M., Popp, J. and Oláh, J. (2020). Current landscape and influence of big data on finance. *Journal of Big Data*, 7(1), 21.

Kharakhash, O. (2023). Data visualization: transforming complex data into actionable insights. *Automation of technological and business processes*, 15(2), 4–12.

Lăzăroiu, G., Bogdan, M., Geamănu, M., Hurloiu, L., Luminița, L. and Ștefănescu, R. (2023). Artificial intelligence algorithms and cloud computing technologies in blockchain-based fintech management. *Oeconomia Copernicana*, 14(3), 707–730.

Lee, D.K.C., Lim, J., Phoon, K.F. and Wang, Y. (Eds.). (2022). Applications and Trends in Fintech II: Cloud Computing, *Compliance, and Global Fintech Trends (Vol. 5)*. World Scientific.

Liu, Y. (2024). Discussion on the Enterprise Financial Risk Management Framework Based on AI Fintech. *Decision Making: Applications in Management and Engineering*, 7(1), 254–269.

Mah, P.M., Muzam, J., Pełech-Pilichowski, T., Mbuch, D.T. and Ako, E. (2023). AI and Robotic Process Automation in Fintech: Analyzing the Shift towards Digitized Customer Services and Operational Efficiency. *Zeszyty Naukowe. Organizacja i Zarządzanie/Politechnika Śląska*, 187, 399–417.

Mehrban, S., Nadeem, M.W., Hussain, M., Ahmed, M.M., Hakeem, O., Saqib, S., ... and Khan, M.A. (2020). Towards secure FinTech: A survey, taxonomy, and open research challenges. *Ieee Access*, 8, 23391–23406.

Müller, S. (2021). The new ecosystem of the digital age: impact of blockchain technology on the accounting environment and financial statement fraud detection (Master's thesis, ISCTE-Instituto Universitario de Lisboa (Portugal)).

Mustyala. A. (2023). Migrating Legacy Systems to Cloud-Native Architectures for Enhanced Fraud Detection in Fintech. *EPH-International Journal of Science And Engineering*, 9(1), 16–26.

Nand, P., Jain, V., Le, D.N., Chatterjee, J.M., Kannan, R. and Verma, A.S. (Eds.). (2023). *Deep Learning for Healthcare Services IoT and Big Data Analytics*. Bentham Science Publishers.

Panda, K.C. (2024). Fraud Detection in Banking Applications: Machine Learning Approach. *Journal of Technological Innovations*, 5(1).

Paramesha, M., Rane, N.L. and Rane, J. (2024). Artificial intelligence, machine learning, deep learning, and blockchain in financial and banking services: A comprehensive review. *Partners Universal Multidisciplinary Research Journal*, 1(2), 51–67.

Park, S., Wang, X., Menassa, C.C., Kamat, V.R. and Chai, J.Y. (2024). Natural language instructions for intuitive human interaction with robotic assistants in field construction work. *Automation in Construction*, 161, 105345.

Pattnaik, D., Ray, S. and Raman, R. (2024). Applications of artificial intelligence and machine learning in the financial services industry: A bibliometric review. Heliyon.

Patel, K. (2023). Big Data in Finance: An Architectural Overview. *International Journal of Computer Trends and Technology*, 71(10), 61–68.

Petare, P.A., Bdair, M., Singh, M.N., Ateeq, K. and Akila, R. (2024). *Big Data analytics in Fintech: Revolutionizing risk management and decision making, ACTA Scientiae*, 7(1), 605–617.

Prakash, C., Saini, R. and Sharma, R. (2021). Role of internet of things (IoT) in sustaining disruptive businesses. *Internet of things and businesses in a disruptive economy*.

Rahman, M.H., Das, A.C., Shak, M.S., Uddin, M.K., Alam, M.I., Anjum, N., ... and Alam, M. (2024). Transforming customer retention in Fintech industry through predictive analytics and machine learning. *The American Journal of Engineering and Technology*, 6(10), 150–163.

Rana, G. and Sharma, R. (2019). Assessing Impact of Employer Branding on Job Engagement: A Study of Banking Sector. Emerging Economy Studies, 5(1), 7–21. https://doi.org/10.1177/2394901519825543.

Rana, G., Sharma, R. and Sharma, B. (2024). Impact of Financial Inclusion for Achieving Sustainable Development for Removing Poverty in all Forms: A Review Study from India. In: Mansour, N., Baral, S., Garg, V. (eds) E-Financial Strategies for Advancing Sustainable Development. *Sustainable Finance*. Springer, Cham. https://doi.org/10.1007/978-3-031-67523-2_20.

Rane, N., Choudhary, S. and Rane, J. (2024). Artificial intelligence, machine learning, and deep learning for sentiment analysis in business to enhance customer experience, loyalty, and satisfaction. Available at SSRN 4846145.

Renduchintala, T., Alfauri, H., Yang, Z., Pietro, R.D. and Jain, R. (2022). A survey of blockchain applications in the fintech sector. *Journal of Open Innovation: Technology, Market, and Complexity*, 8(4), 185.

Rizinski, M., Peshov, H., Mishev, K., Chitkushev, L.T., Vodenska, I. and Trajanov, D. (2022). Ethically responsible machine learning in fintech. *IEEE Access*, 10, 97531–97554.

Roy, A. and Tinny, S.S. (2024). Cybersecurity and Blockchain for Secure Financial Transactions: Evaluating, Implementing, and Mitigating Risks of Digital Payments. *International Journal of Applied and Natural Sciences*, 2(1), 38–48.

Sahu, M.K. (2020). Machine Learning Algorithms for Personalized Financial Services and Customer Engagement: Techniques, Models, and Real-World Case Studies. *Distributed Learning and Broad Applications in Scientific Research*, 6, 272–313.

Sanodia, G. (2024). Enhancing CRM Systems with AI-Driven Data Analytics for Financial Services. *Turkish Journal of Computer and Mathematics Education*, 15(2), 247–265.

Saxena, A. and Muneeb, S.M. (2024). Transforming Financial Services Through Hyper-Personalization: The Role of Artificial Intelligence and Data Analytics in Enhancing Customer Experience. In *AI-Driven Decentralized Finance and the Future of Finance* (pp. 19–47). IGI Global.

Selvaraj, A., Krothapalli, B. and Shanmugam, L. (2024). AI and Machine Learning Techniques for Automated Test Data Generation in FinTech: Enhancing Accuracy and Efficiency. *Journal of Artificial Intelligence Research and Applications*, 4(1), 329–363.

Simpson, B.D., Johnson, E., Adeleke, G.S., Amajuoyi, C.P. and Seyi-Lande, O.B. (2024). Leveraging big data for agile transformation in technology firms: Implementation and best practices. *Engineering Science & Technology Journal*, 5(6), 1952–1968.

Sharma, R., Saini, A.K. and Rana, G. (2021a). Big data analytics and businesses in industry 4.0. *Design Engineering*, 238–252

Sharma, R., Saini, R., Prakash, C. and Prasad, V. (Eds.). (2021b). *Internet of things and businesses in a disruptive economy*. Nova Science Publishers.

Sharma, R., Rana, G. and Agarwal, S. (2024). Green innovations for industrial development and business sustainability: Models and Implementation Strategies. CRC Press.

Solanki, A. (2024). Unleashing Insights: Exploring the Power of Behavioral Real-Time Analytics Platform in FinTech. Journal Homepage: http://www. ijmra. us, 14(05).

Stojanović, B., Božić, J., Hofer-Schmitz, K., Nahrgang, K., Weber, A., Badii, A., ... and Runevic, J. (2021). Follow the trail: Machine learning for fraud detection in Fintech applications. *Sensors*, 21(5), 1594.

Tatineni, S. and Mustyala, A. (2024). Enhancing Financial Security: Data Science's Role in Risk Management and Fraud Detection. *ESP International Journal of Advancements in Computational Technology* (ESP-IJACT), 2(2), 94–105.

Thangam, D., Pavan, K.A., Patil, S., Park, J.Y., Kandasamy, R. and Chikkandar, R.J. (2025). Intelligent Process Automation and Its Relevance to Various Industries. *In Advancements in Intelligent Process Automation* (pp. 387–412). IGI Global.

Tian, X., He, J.S. and Han, M. (2021). Data-driven approaches in FinTech: a survey. *Information Discovery and Delivery*, 49(2), 123–135.

Verma, R. (2024). Harnessing Advanced Data Engineering for Enhanced Efficiency and Customer Satisfaction in Financial Services. *SSRG International Journal of Computer Science and Engineering*, 11(7), 22–27.

Wu, A.C. and Kao, D.D. (2022). Mapping the sustainable human-resource challenges in southeast Asia's FinTech sector. *Journal of Risk and Financial Management*, 15(7), 307.

Yıldırım, U. (2023). Exploring the trends, challenges and opportunities of regulatory technology (RegTech) in the financial industry: A systematic literature review.

Zhang, W., Yan, S., Li, J., Tian, X. and Yoshida, T. (2022). Credit risk prediction of SMEs in supply chain finance by fusing demographic and behavioral data. Transportation Research Part E: Logistics and Transportation Review, 158, 102611.

Zhu, J., Xu, T., Zhang, Y. and Fan, Z. (2024). Scalable Edge Computing Framework for Real-Time Data Processing in Fintech Applications. *International Journal of Advance in Applied Science Research*, 3, 85–92.

4 Marketing Analytics to Identify Success and Impact

Deepak Halan[1*] and Park Thaichon[2]

With the growing complexities of the external environment and the increasing competition, it is becoming more important for firms to measure their success and impact. Marketing analytics plays a key role in measuring and optimizing the success and impact of various marketing activities. This chapter explores the role of marketing analytics, from the initial data collection to the advanced predictive insights stage. Diverse techniques and applications have been discussed that organisations can use to evaluate the return on investment (ROI) apart from other key metrics. Marketing analytics enables businesses to decode very large quantities of data, optimize marketing efforts, and finally drive business growth. This chapter delves into the evolution of data mining and the different types of marketing data that can be data mined while emphasising big data. Several businesses are using big data in predicting customer lifetime value (LTV) amongst other applications. Customer touchpoints are rich sources of data and their management is vital to improve customer relationships. This chapter also discusses various types of customer touchpoints and their better management and optimization. With the growing importance of social media, it is vital to examine customer-centric impact using sentiment analysis.

1. Introduction

In the dynamic world of modern marketing, technology and data-driven strategies play a pivotal role in achieving success (Kotler, 2009). Firms operate in a hyper-connected setting where every virtual or physical customer transaction results in valuable data. This chapter explores the complex associations between marketing analytics, customer touchpoints, big data, and sentiment analysis, emphasizing their integrated role in developing successful marketing strategies (refer to Fig. 1). Marketing analytics is the central pillar that converts raw information into practical insights that enable marketing managers to make smart decisions throughout the customer journey. Brands can enhance their tactics to meet dynamic customer expectations by analyzing the data 24*7 and examining emerging trends. This

[1] 0000-0002-0528-0844

[2] 0000-0001-7512-7362

* Corresponding author: deepak.halan@jaipuria.ac.in

DOI: 10.1201/9781003514282-4

analysis is possible only because of customer touchpoints—the diverse ways, such as websites, social media platforms, and physical stores via which prospects and customers come in contact with the company or brand (Rana et al., 2021). These interactions serve as important data collection opportunities that provide deep insights into consumer behavior, choices, and sentiments. The wealth of information generated builds up big data which consists of a massive and varied set of data derived from online transactions, offline transactions, social media communication, sensors, and more (Gandomi and Haider, 2015; Sharma et al., 2021). Big Data fuels marketing analytics on a day-to-day basis. Marketing analytics enables businesses to decode very large quantities of data, optimize marketing efforts, and drive business growth. This chapter delves into the evolution of data mining and the different types of marketing data that can be mined. Several businesses are leveraging big data in forecasting customer lifetime value (LTV) amongst other applications. Predictive analysis is helping several brands determine strategies to retain customers, increase sales, and improve overall customer experience.

Sentiment analysis is important as it facilitates collecting valuable consumer insights, which include emotions. These insights are gathered from different social media sites by using techniques like natural language processing (NLP) and text mining. Sentiment analysis enables marketing teams to investigate the consumer sentiments as being more positive or negative. This polarisation is based on the attitude reflected by the online customer reviews, and social media comments. Sentiment analysis augments conventional analytics, thus providing the marketing team access to comprehensive knowledge of, consumer attitude, expectations, and brand perception. This chapter delves into how this know-how can be transformed into customer-centric marketing. This chapter delves into customer touchpoints that generate big data and data mining and sentiment analysis that enable marketers to create customer-centric campaigns that improve customer engagement and help retain customers, thus building brand and customer loyalty.

Businesses need to evaluate their success and impact the external environment, and the competitive landscape is becoming more complicated (Rana et al., 2019). In this age of data-driven decision-making, marketing analytics plays a key role in understanding, measuring, and optimizing the success and impact of various marketing activities. With the availability of sophisticated technologies to gather and store data, marketers can crunch huge quantities of both organised and unorganised data. The total quantity of this data is so high that advanced technologies like artificial intelligence and machine learning are required to mine useful insights from this data (Sharma et al., 2024). This chapter explores the role of marketing analytics, from the initial data collection to the advanced predictive insights stage with the help of examples. Diverse models, situations, and applications have been discussed, which organisations can use to evaluate the return on investment (ROI), the enhancement in customer experience, and improvement in brand health, apart from other key metrics.

1.1 Objective of The Study

This chapter explores the complex associations between marketing analytics, customer touchpoints, big data, and sentiment analysis, emphasizing their integrated role in developing successful marketing strategies. One of the objectives of this chapter is to examine the various types of customer touchpoints and their better management and optimization. It delves into the evolution of data mining and the different types of marketing data that can be data mined while emphasising big data. This chapter also explores the role of marketing analytics, from the initial data collection to the advanced predictive insights stage with the help of several examples. This chapter is organised as follows: Initially, relevant literature is reviewed to gain an in-depth preliminary understanding and the methodology of the study is briefly outlined. Thereafter, big data and data mining, customer touchpoints, and sentiment analysis have been discussed. Finally, the concluding thoughts and future scope are presented.

2. Methodology

This study is based on a detailed review of journal and news articles to derive actionable insights on the complex associations between marketing analytics, customer touchpoints, big data, and sentiment analysis. For academic articles, reputable databases like Taylor and Francis, Elsevier, Wiley, Springer, and Emerald Insight were used to identify papers from top-ranking journals indexed in Scopus and Web of Science. Examples of journals cited include Psychology & Marketing, Harvard Business Review, and Journal of Business & Industrial Marketing. Articles outside the marketing domain, such as those related to IT or data science, were excluded, as were non-peer-reviewed materials, theses, and non-English publications. Only full-text articles were analyzed to ensure a comprehensive understanding, yielding 36 news articles and journal papers that provided a wealth of valuable insights for this study.

3. Literature Review

Marketing analytics, a "technology-enabled and model-supported approach to harness customer and market data to enhance marketing decision making" (Lilien, 2011, p. 5) encompasses two kinds of applications. Some applications engage their users for decision support while others work in an automated manner and do not require any decision-making intervention (Germann, Lilien, and Rangaswamy, 2013). There is considerable stress on chief marketing officers to deliver a return on investment from a varied marketing programs portfolio and given the easy availability of data as well as marketing analytical tools, it is best to adopt an integrated analytics approach (Bhandari, Singer, and Van Der Scheer, 2014). Marketing analytics is essentially a marketing-sensing capability while customer relationship management (CRM) and brand management are customer-linking capabilities. Integrated, these capabilities are vital accelerators of marketing performance (Cao and Tian, 2020). Businesses can use marketing analytics to capture data-based information and insights for improving their marketing capabilities and corporate performance (Cao, Duan, and Banna, 2019). The significance of marketing analytics has considerably increased in this current age of big data and advanced technology.

Marketing analytics can integrate big data and provide relevant insights that cut across touchpoints, thus enabling businesses to predict consumer behavior. Predictive analytics allows the identification of early churn signals which could manifest as a deviation in buying behavior or a more negative sentiment analysis based on online customer reviews (Basu, Lim, Kumar, and Kumar, 2023). Marketers strive to maintain as uniform a brand image as possible, at all customer touchpoints (Davenport and Phillips, 2016). Customer sentiments are sources of customer experiences across various touchpoints and sentiment analysis yields rich insights on brand associations, brand assessments, word-of-mouth, and buying of customers (Mishra, 2022). Marketing analytics can yield actionable knowledge to enhance customer experiences, across the various touchpoints in a customer's journey. This study examines current data sources like. (Patel and Gahletia, 2020). Social media has emerged as one of the most effective sources since it facilitates communication by consumers and interaction amongst them, daily (Hemmati, Arzanagh, and Rahmani, 2024). Businesses need to conduct sentiment analysis and opinion mining on big data sourced from social media touchpoints like Facebook, Twitter, and Instagram every day, to constantly monitor customer satisfaction (Mach-Król and Hadasik, 2021).

"Big Data" refers to extensive datasets encompassing nearly all aspects of consumer behavior, enabling predictive analytics that forecast how consumers are likely to behave in the future. In contrast, "Small Big Data" involves multiple survey datasets that allow for deeper and more nuanced analysis (Warshaw, 2016). The optimal approach to answering complex research questions often lies in combining these two types of data. Big data excels at identifying patterns and providing insights into the "what" of consumer behavior, while surveys capture attitudes and opinions, shedding light on the "why." For instance, web analytics and transaction data—important sources of big data—might reveal that, on average, women spend less time on online shopping transactions compared to men on a particular site. By combining surveys and big data, businesses can triangulate the insights and provide "rich data" that offers a detailed understanding of consumer behavior and choices. Hence there is a shift to "rich data", highlighting the higher weightage given to substance and utility vis-à-vis quantity (Callegaro and Yang, 2018).

Customer DNA encompasses the complex integration of individual characteristics, choices, and behaviors typical of digital-age consumers. The concept of customer DNA offers a more comprehensive and holistic view of each consumer by building on the principles of Customer Lifetime Value (CLV), which emphasizes the value of long-term customer connections. This includes demographics, purchase history, digital footprints, social media interactions, and feedback. Marketing intelligence has continuously evolved—from measuring public sentiment to utilizing advanced tools like artificial intelligence (AI) to analyze extensive datasets. Integrating marketing intelligence with Customer DNA presents tremendous opportunities, enabling marketers to deliver highly personalized experiences, anticipate future trends and demands, and address critical ethical considerations (Rao, Reddy, Chakradhar, Deepak, and Inampudi, 2024). Advanced technologies like AI, machine learning (ML), virtual reality (VR), augmented reality (AR), blockchain, IoT combined sentiment analysis, predictive analytics, and cross-channel integration are leading to hyper-personalization, real-time marketing, and ethical considerations (Behare et al., 2024). Big Data's potential for identifying marketing trends, enhancing marketing decisions, creating new revenue strategies, and identifying marketing success and impact is undeniable However, its growing use raises privacy concerns, identity theft, security, and consumer manipulation issues (Bollier, 2010). Addressing these concerns is crucial as we advance with Big Data initiatives. Deploying technologies like AI and IoT with CRM systems is making it possible for businesses to be in touch with customers 24*7, track their movements, gather data to analyze their purchase and product usage patterns, and ultimately understand reasons behind churn (Gupta and Ravi Kumar, 2022). However, many of these technologies are heavily reliant on data and analytics, which frequently encounter challenges related to the reliability of data sources, the demands of data infrastructure, data security concerns, and the consistency and accuracy of data models and insights. Also, from an organizational standpoint, measuring the tangible impact of marketing analytics remains a significant challenge (Srivastava, Kishore, and Dhingra, 2021). Faced with an overload of data, organizations often grapple with its complexity and struggle to provide customers with a seamless, consistent experience. The key priority should be integrating all platforms and channels, ensuring equal focus on physical and digital touchpoints. By establishing a unified customer data repository, businesses can better personalize interactions and enhance the overall customer experience (Fenech and Perkins, 2015).

4. Big Data and Data Mining

4.1 Big Data

Imagine, you are at the mall, inside your favourite departmental store. After shopping, you steer your cart towards the billing desk. The employee at the billing counter asks for your customer ID and then scans the barcode of each item in your cart, one by one. Details of the items that you have shopped for, get digitally transmitted to the billing system. The system recognises you since on your first visit to this department store, you were issued a loyalty card with a customer ID after completing a brief form with personal details such as age, gender, income, and occupation. A few days later, an email arrives from the store with special offers on headphones, inspired by your purchase of an iPod in the previous visit. Let us consider a different customer touchpoint. You log in to your favorite e-commerce site, intending to purchase a smartphone. After adding it to your shopping cart, you ask ChatGPT to conduct a sentiment analysis of the 580 online customer reviews. Based on the results, you decide against buying it. Soon, you notice that advertisements for the same handset appear every time you are on Facebook and Instagram. Your favorite e-commerce site has assigned a virtual identification number to you, to track your activities and uses this data to purchase targeted advertisements for products they know you are interested in—a practice often referred to as "ghost marketing." With these transactions, you have contributed to the realm of big data. Data collection and analysis occur across various industries, with retail, telecommunications, and finance leading the way (PTI, 2013). Among these, the retail sector stands out due to its high level of interaction with a diverse customer base, thus driving retailers to understand their consumers better. The term "Big Data", now widely popular, refers to the use of massive datasets to enhance business efficiency and responsiveness to customer needs. Big Data and data mining play a pivotal role in modern marketing analytics, equipping businesses with the tools to derive actionable insights from extensive structured and unstructured datasets. These technologies enable marketers to evaluate the effectiveness of their campaigns, products, and strategies with a data-driven approach. In the marketing context, big data encompasses a wide range of information, including customer behaviors, preferences, demographics, and interactions across various touchpoints. Big data can be immensely useful for marketers as key benefits include: (i) comprehensive customer understanding: (ii) integration of data from multiple sources to build a 360-degree customer profile (iii) improved decision-making: delivery of detailed insights into customer trends, enabling well-informed, strategic decisions and (iv) real-time adaptability: facilitation of dynamic adjustments to campaigns and strategies based on real-time data.

Big Data has emerged as a transformative system of knowledge, reshaping the very objects of study while influencing how we understand human networks and communities (Boyd and Crawford, 2012). The sheer volume and complexity of big data require advanced technologies like Hadoop, HBase, and CouchDB to process and analyze information—tasks that would be nearly impossible using traditional methods or human intelligence alone. Unsurprisingly, the potential of big data goes beyond marketing and has captivated a wide range of disciplines, including computer science, physics, economics, mathematics, political science, bioinformatics, and sociology. Professionals across these fields are eager to access the vast amounts of information generated about people, objects, and their interactions (Boyd and Crawford, 2012). Organizations capitalize on big data via data warehouses, which consistently gather information from diverse sources 24*7 and systematically store it in virtual repositories. Data warehousing and online analytical processing (OLAP) are pivotal for decision support, necessitating unique database technologies compared to traditional transaction processing (Chaudhuri and Dayal, 1997). These data sources can include retail activities, market surveys, or customer transactions like those in banking. The process often begins with a comprehensive customer database containing demographic details such as age, family size, and income. This data must extend beyond simple mailing lists to effectively capture purchasing behaviors recorded through various means such as store scans, catalogue orders, and customer databases. These behaviors often reveal genuine preferences, sometimes contradicting stated preferences in surveys. Analyzing this data yields valuable actionable insights. For instance, 3M leverages customer feedback for product improvements, while data mining aids in market segmentation and customer satisfaction assessment, MCI Communications scrutinizes vast volumes of customer call data to devise new discount plans. Similarly, Land's End, a retail chain, can pinpoint which among its approximately 2 million customers would benefit from receiving personalized mailers featuring clothing items tailored to their specific wardrobe preferences. Tesco, the British supermarket giant, customizes its communications by notifying targeted customer groups, such as wine enthusiasts or cheese lovers, about special sales or promotions for these items. Telecom companies like Airtel and Vodafone utilize advanced tools to analyze data for insights into customer behavior, uncovering patterns that allow them to predict scenarios such as high-usage customers who may be unaware of certain services. These examples highlight how big data is reshaping business-consumer interactions, and enhancing experiences right from the purchase to the after-sales service stage.

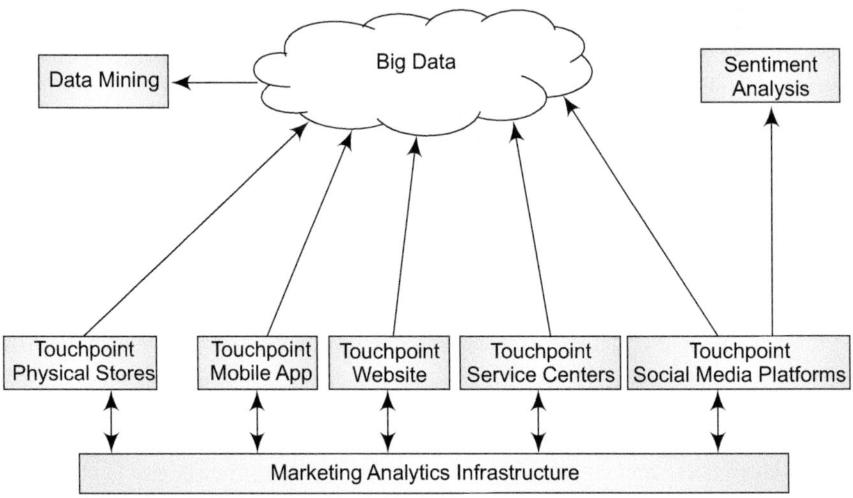

Fig. 1 The integrated role of customer touchpoints, big data, and sentiment analysis in marketing analytics.
Sources: Authors' own

4.2 Data Mining as the Forefront of Big Data

Marketing analytics is deeply rooted in data mining and statistical analysis (Chen, Chiang, Storey, 2012). However, integrating big data with marketing analytics for decision-making and operations presents significant challenges (Hanssens and Pauwels, 2016). Data mining involves examining data from warehouses to extract meaningful insights, identifying patterns such as frequently purchased products, testing hypotheses, or developing customer profiles based on consumption levels and satisfaction. This information guides marketing initiatives such as developing special promotions and loyalty programs, integral to CRM. These activities are assessed and refined over time. Traditionally, marketing data analysis involves analysts closely engaging with data and bridging it with user needs (Fayyad, Piatetsky-Shapiro, and Smyth, 1996). However, often, data remains siloed within organizations, impeding its full potential. Take, for example, the auto industry in India, where valuable data from car dealers, marketing insights, and data from customers during car servicing interactions, exist separately. Without integration, this data's true value is diminished. Marketing managers in organizations, across banks, telecom companies, and retail chains, effectively employ data mining. They capture data from large customer bases daily, although many companies have yet to harness big data's potential. Data mining can significantly enhance customer strategy, aiming for reduced costs and higher revenue and profits. It helps businesses attract new customers, retain existing ones, address complaints, and understand customer defections. Netflix depends on data mining to understand viewer choices. It mines data based on the watch history of viewers, the ratings given by them, and the way they search for content. Based on these insights, Netflix makes customised recommendations to viewers as per their individual preferences. The insights also support content creation. Big Data" has become a buzzword that has captured the attention of industry leaders worldwide. It differs from traditional data, not solely generated by internal transaction systems but also sourced externally from the internet, sensors, public data initiatives like the Human Genome Project, and multimedia repositories. Big Data has revolutionized analysis by shifting the focus from time-consuming data preparation to insightful analysis. Previously, business intelligence primarily reflected past events without offering predictions. Now, thanks to Big Data, nearly accurate forecasts are possible, as demonstrated by companies like Google, eBay, LinkedIn, and Amazon. Data mining today helps calculate a customer's Lifetime Value (LTV), which measures retention and predicts potential defections. LTV represents the net value a customer brings to a company. When a customer leaves, the lost profit equals the LTV. Data mining calculates LTV scores, which guide attention and perks to high-value customers, encouraging loyalty. Moreover, data mining predicts customer churn, assigning probabilities to potential defections. Organizations can categorize customers based on their likelihood to remain, using feedback to curb churn rates. Data mining reveals hidden trends and patterns, offering insights into product, pricing, distribution, and promotion strategies.

4.3 The Power of Big Data in Marketing

Big data encompasses large data sets, rapid data generation, complexity, authenticity, and quality in achieving intended outcomes. It offers unparalleled capacity to collect and analyze marketing data (Lazer et al., 2009). Big Data has been defined by Demchenko, Grosso, De Laat, and Membrey et al. (2013) through the five V's framework: Volume: Massive datasets from customer interactions, online behaviors, and sales records; Velocity: Real-time data streaming from platforms like social media, CRM systems, and e-commerce sites; Variety: Data in multiple forms, including structured data (e.g., transaction logs), unstructured data (e.g., social media posts), and semi-structured data (e.g., JSON files); Veracity: The accuracy of the data and information and Value: The value of the data to the business, which can come from insight discovery and pattern recognition. Beyond large sets and analysis tools, big data represents a computational shift in research (Burkholder, 1992). Companies are leveraging big data for marketing across the world. Amazon, for instance, uses data-driven recommendations to boost sales. Starbucks leverages big data from its mobile app, loyalty program, and customer purchase history to offer location-based promotions. For example, the app informs customers of nearby Starbucks outlets and offers deals customised to their favorite coffee drink or seasonal favorites. This amalgamation of big data augments customer experience and increases footfalls into Starbucks outlets. Traditional industries employ data-science teams to exploit data resources for competitive advantage, enhancing revenue and reducing costs (Provost and Fawcett, 2013). The expected value from these techniques is rising, enabling data-driven insights across organizational levels. Big data allows untraditional means of data analysis, such as GPS systems integrating traffic data for optimal routes (LaValle, Lesser, Shockley, Hopkins, and Kruschwitz, 2010). UPS, the world's leading courier company, gathers extensive data on the approximately 16.3 million packages it delivers daily and processes 39.5 million tracking requests each day. One of its latest sources of big data comes from telematics sensors installed in nearly 50,000 company trucks. These sensors monitor various metrics, including speed, direction, braking patterns, and more. Political marketing too has come a long way from conventional methods in today's digital age. Political parties across the globe are using data analytics extensively in marketing their party and candidates. The 2012 Obama Election Campaign achieved remarkable success by leveraging insights derived from data analytics. The campaign manager implemented an innovative, metrics-driven strategy and assembled an analytics team five times larger than the one that worked on Obama's 2008 campaign (Scherer, 2012). The Obama for America (OFA) team developed detailed, multi-variable profiles for 15 million "persuadable" voters, incorporating up to 1,000 variables per voter, including demographics, psychographics, and credit card transactions. Additionally, OFA had identified the names of approximately 69 million Americans who voted for Obama in 2008. The 2012 Obama campaign revolutionized political strategy by utilizing data analytics and experimental methods to build a winning coalition, one vote at a time. This approach disrupted the longstanding reliance on TV advertising in U.S. politics, introducing a new model: a national campaign with the precision and personalization of a local ward election, where individual voter interests were identified and addressed. A key tactic involved "experiment-informed programs"

(EIPs), developed in collaboration with the Analyst Institute, to evaluate the effectiveness of different messages in shaping public opinion. For instance, the campaign discovered that voters aged 45 to 65 were more responsive to Obama's Medicare arguments than those over 65, who were already eligible for the program. These insights informed a persuasion model that rated, on a 0-to-10 scale, the likelihood of a voter being influenced toward Obama after a single interaction with a campaign volunteer. This data-driven approach allowed the campaign to focus its efforts strategically, optimizing outreach and impact (Issenberg, 2013). Such comprehensive data management, powered by Big Data, played a pivotal role in securing Obama's victory.

5. Customer Touchpoints

Customers could get introduced to a brand through a social media platform, interact through website pages on an FAQs section where they can ask questions about products and services offered, and touch and feel a product by walking into a physical store. These are all instances where direct customer contact occurs and the social media platform, the website, and the physical stores are all examples of customer touchpoints. Touchpoints can be categorised as: face-to-face touchpoints – such as brick-and-mortar stores, service centers, and events; database-driven touchpoints – such as telephone, e-mail, SMS, and mass media touchpoints such as advertising and website. To enhance a brand's mindshare, it is crucial to engage the target audience across various touchpoints such as at home, work, educational institutions, during travel, and leisure activities like dining or shopping. These touchpoints collectively influence consumer perceptions and decision-making regarding a brand or organization. Understanding these touchpoints is essential for organizations as they drive customer experiences and form the foundation of the customer relationship. All successful brands collect and analyse data across touchpoints to capture valuable customer insights that help them to develop more targeted, effective, and impactful marketing strategies. Apple studies data from in-store customer transactions, such as product queries, service appointments, and participation in demonstrations. It then analyses this data to shortlist which of its products are most in demand and how customer experience at its retail outlets can be enhanced. This results in enhanced product showcases, superior training for employees, and improved store layouts. Coca-Cola utilises data from its IoT-enabled vending machines, which monitor purchase trends, machine locations, and inventory levels. Data from its smart vending machines touchpoint enables Coke to identify the optimal location plan for these machines, the area-wise popularity of its drinks, and the adaptation required in its marketing initiatives. Walmart consolidates data from three touchpoints: online shopping site, mobile app, and brick-and-mortar locations to produce a seamless shopping experience. It monitors online searches for groceries and associates this data with brick-and-mortar purchases to optimize inventory. This market analysis provides Walmart with valuable insights into product availability and personalized marketing enhancement. McDonald's monitors customer orders across three touchpoints: drive-thru lanes, self-service kiosks, and mobile app. This enables the fast-food giant to identify popular dishes, high-footfall hours, and regional likings. This marketing analytics initiative enables McDonald's to optimize menu offerings, fine-tune staffing levels, and run customised promotions. Post-purchase touchpoints, such as after-sales service, loyalty programs, and warranty activities, also can enhance customer experiences, fulfil brand promises, and boost brand loyalty and advocacy. Companies have developed innovative post-purchase strategies, including self-service channels and customer service hotlines, to ensure superior customer experiences compared to competitors and encourage repeat sales. For instance, Bajaj service centers for bikes, contact customers post-service to ensure satisfaction, offering free additional servicing if needed, which strengthens emotional connections and loyalty.

The rapid advancement of technology and changes in customer lifestyles have introduced new touchpoints, offering multiple ways for customers to interact with a brand. However, this also necessitates effective management of these diverse touchpoints. A well-balanced mix of touchpoints leads to improved customer experiences, fostering better relationships, higher profits, and increased customer loyalty. A comprehensive Touchpoint Management Plan, such as the one adopted by Philips (refer to Box), ensures consistent messaging across all channels. Mapping touchpoints allows organizations to monitor the impact of strategic initiatives, enabling the measurement of cause and effect at each touchpoint. This information, combined with an understanding of customer profitability, aids in optimizing marketing and sales budgets. For example, a company simplified its product database to streamline processes across touchpoints, saving customers significant time and boosting satisfaction.

Box: Touchpoint Management at Philips

Philips employs the Touchpoint Tool, a 6-point framework, to enhance customer interactions. This methodology identifies and prioritizes touchpoints, aligning them with brand positioning. The 'Touchpoint Wheel' maps each customer interaction, from pre-purchase to post-purchase. Significant touchpoints are prioritized and evaluated against Philips' three pillars: 'designed around you,' 'easy to experience,' and 'advanced.' This framework guides improvement programs, tested with real customer insights.

The Touchpoint Tool helps managers view business from a customer perspective, focusing on customer needs in marketing activities. For instance, Philips' Semiconductors Division improved product seminars by making them more interactive based on customer feedback. The tool is also valuable for new product launches or re-launches. For example, marketing strategies for Senseo, a popular gift item, leveraged word-of-mouth as a key touchpoint, saving on advertising costs. Philips' marketing managers are trained in the Touchpoint Tool, enabling them to enhance each customer interaction, leading to increased sales and stronger customer relationships.

6. Sentiment Analysis

Social media has become a powerful communication tool. Rather than seeking product suggestions from neighbours or family, people now rely on social sites and reviews. This has prompted companies to explore various online marketing strategies. Many businesses have built substantial followings on social media platforms and are now working to convert these followers into paying customers, as demonstrated by campaigns from Dell and Sony. Consumers today connect with their favorite brands online, beyond the control of manufacturers and retailers, seeking validation for their choices through social media. Unlike before, when consumer engagement just ended with a purchase, the digital age keeps consumers engaged with brands post-purchase via social media. Brands use social media to share content or reviews, launch new products, and gather feedback, resulting in comprehensive engagement at all brand levels. Research indicates that conventional media are more influential during the pre-purchase phase, while subsequent stages involve various online media. In the digital age, consumer engagement continues post-purchase through social media, unlike in the past. Smart marketers analyze this ongoing consumer behavior to inform their marketing strategies, spending, and overall brand decisions. While consumers still seek a clear brand promise and value, the main shift is in the touchpoints. Marketers need to aim to create a seamless customer experience by developing a plan based on their product offerings and customer segments. Apple, for example, has streamlined product descriptions and created explanatory videos for consistency across touchpoints. Today, digital touchpoints provide marketers with opportunities to gather and use customer information to enhance experiences. The speed, precision, and interactivity of these digital touchpoints shape the quality of brand interactions. Drury (2007) notes that social media combines professionally edited content with user-generated content, appealing to a fragmented media audience. For instance, Starbucks launched “My Starbucks Idea,” a platform for consumer feedback, evolving into the “Ideas in Action” blog to inform users about the impact of their suggestions. Social media influences consumers, making it crucial for studying consumer behavior and providing valuable insights. It offers a unique consumer panel different from traditional research, with insights gathered from interactions rather than formal questionnaires. This consumer-driven approach allows clients and researchers to initiate dialogues without controlling them. Consumers voluntarily join and choose their level of engagement. Dell’s “listening room” exemplifies a tool within its social-media strategy, monitoring thousands of daily Tweets in multiple languages to filter relevant business discussions. Insights from these interactions have influenced Dell’s product development and marketing communication, enhanced company-customer relations, and driven significant growth in community engagement (Veer et al., 2022). Gathering social media feedback is just one application area; it also offers opportunities for competitive analysis by listening to discussions about rival brands. This is particularly beneficial for late adopters, allowing them to learn from competitors’ mistakes. Crowdsourcing, a modern approach, taps into the collective intelligence of the target audience for product innovation, offering significant time and cost savings compared to traditional research.

Sentiment analysis, also called opinion mining, is an analysis of text analysis using natural language processing (NLP), machine learning, and data mining tools to investigate the emotional tonality. It classifies textual comments and reviews as positive, negative, or neutral. This enables marketers to capture customer sentiments about specific products, services, or brands. Sentiment analysis is extensively conducted on unstructured data, which is in the form of customer reviews, social media posts, survey responses, and service interactions. Amazon leverages sentiment analysis to mine customer reviews for knowledge of product performance. For example, online reviews that contain words like "poor quality" or "value for money" enable Amazon to identify the pain and gain points. Delta Airlines, a key airline in the United States and the oldest, conducts sentiment analysis on data collected from fliers on exiting the flight, using the survey method. The analysis enables Delta to identify problems like delays, in-flight service, or booking experience quality concerns. The airline then brainstorms for the best solutions for making specific service enhancements. Samsung leverages sentiment analysis to benchmark customer perception of its offerings vis-à-vis those of rivals such as Apple. Sentiment analysis enables Samsung to manage its brand reputation by fine-tuning aspects like screen resolution or pricing. Ultimately, this helps in marketing strategy refinement based on the pros and cons. H&M, a multinational fashion retailer known for its fast fashion business model, brings sentiment analysis into chatbot interactions to evaluate customer satisfaction on a real-time basis and achieve chatbot optimization. In the case of a dissatisfied customer, the bot highlights the interaction for some form of human intervention thus improving customer care. Pepsi taps sentiment analysis during major campaigns like the Super Bowl to gauge campaign effectiveness. It tracks comments, tweets, and posts to understand if the campaign struck the right chord with its target audience and utilises the insights gained to optimise future campaigns. Toyota does sentiment analysis to manage crises such as vehicle recalls. By tracking consumer sentiment on various social media and news platforms, Toyota addresses customer issues on a proactive basis and reinforces trust in its brand. Nike utilises communication on social media platforms like Instagram and Twitter for sentiment analysis and tracking of campaign effectiveness. For example, during key events like the launch of a new running shoe, Nike monitors hashtags and consumer comments to evaluate the level of audience engagement. This analysis enables Nike to alter its messaging and product offerings for better synchronisation with its target audience. Coca-Cola leverages big data to keep track of brand mentions and customer sentiment across various social media platforms. Data mining enables Coke to identify trending subjects, evaluate consumer opinions, and investigate the effectiveness of its marketing campaigns. For example, during key sporting events like the FIFA World Cup, Coca-Cola customises its advertising campaigns on a real-time basis depending on audience engagement and sentiment analysis.

Coca-Cola also uses social media to develop new products like vitamin water, which is a beverage that contains water, water-soluble vitamins, and minerals, and is frequently sweetened with sugar or other sweeteners. It could also contain electrolytes. For example, it engaged its audience in creating a new Vitamin water flavor through a "flavor creator lab" on Facebook, completing the process in months with minimal investment compared to traditional methods. Social media can also be leveraged for online marketing research and polls, as seen with Revlon India's Facebook campaign to understand Indian beauty habits. Within two weeks, 75,000 consumers participated, providing valuable insights that helped refine Revlon's advertising strategy, focusing on the ColorStay Whipped Crème – a type of foundation that is a part of Revlon's ColorStay line, known for its long-wearing and sturdy cosmetic offerings.

7. Conclusion and Future Scope

This chapter explores the complex associations between marketing analytics, customer touchpoints, big data, and sentiment analysis. Data mining began with professors and researchers applying their statistical expertise to create algorithms that underpin the data mining techniques being used today. These algorithms were initially implemented using programming languages like FORTRAN and were primarily academic applications. Data mining remains largely a specialized activity, conducted by dedicated teams with terminology rooted in academia, which can make it appear somewhat enigmatic and removed from everyday marketing requirements. Over time, vendors will continue to develop tools to make these algorithms more accessible and user-friendly. Today when a business problem arises, the IT team typically addresses the data warehousing aspects, while a separate data mining team, which is a part of the marketing team, provides analytical insights or recommendations to guide actions. Ignoring data mining could be detrimental to companies. Predictive analysis based on Big Data represents a valuable opportunity for gaining business intelligence and achieving competitive success. The future is likely to see a rise in hybrid data mining methods that combine diverse models for improved performance, functioning like trained artificial "experts." These models, akin to human expert consultations, can also be integrated with human expertise. In the future, these artificial experts may evolve into autonomous intelligent software agents (Kovalerchuk and Vityaev, 2000). Social media significantly impacts consumer behavior and offers a valuable medium for studying consumer interactions and gathering insights, as it serves as a consumer-driven environment. Beyond product and service feedback, crowdsourcing is emerging as an effective method for new product development. Creating a coherent customer experience strategy based on offerings and target segments is crucial in the future. Integrating traditional and digital marketing remains challenging, as does merging insights from both sources. Modern marketers are used to delivering messages, however, social media requires attentive listening and timely responses, necessitating dedicated resources. Hence this chapter concludes that the integrated role of customer touchpoints, big data, and sentiment analysis is vital in providing effective marketing analytics to identify impact and ultimately develop successful marketing strategies.

In the future, we can expect more seamless integration of data from various customer touchpoints while using marketing analytics to tap big data and run sentiment analysis to keep a finger on the emotional pulse of the consumers. This will improve brand agility, customer-centrism, and marketing strategy novelty, promising success in the form of continuous growth and competitive edge. The larger companies have more touchpoints, increasing the need for effective management, though touchpoint management is relevant for businesses of any size. Companies face intense competition and discerning customers, prompting top management to align brand strategies with customer needs. Every customer touchpoint needs to ultimately support business objectives. Companies aiming to outpace competitors will need to analyze their customer interactions and brand communication. Only brand investments identified by such analysis can produce significant results and success. With technology advancement, marketing will become smarter and more effective. For example, a marketer has developed a computer program that uses security cameras in physical stores to provide detailed insights into customer behavior. This technology can track the number of customers in the store at any given time, identify the areas they visit, and highlight the products that capture their attention. The software integrates this data with other variables such as staffing levels, weather conditions, product variety, and placement to determine the factors driving sales. Looking ahead, advancements in computer programs, combined with video surveillance, may allow retailers to analyze customers' gender, interpret facial expressions, and assess gestures. Such insights could help businesses understand the reasons behind a purchase decision, or lack thereof, enabling them to refine their strategies and better meet customer expectations.

As technology progresses, marketing analytics will become more essential to decision-making and customer engagement strategies. The amalgamation of customer touchpoints, big data, and sentiment analysis will shape the future of marketing by enabling more hyper-personalized, data-driven, and customer-centric approaches. Marketing analytics will build upon customer touchpoints across diverse channels such as brick-and-mortar retail outlets, online shopping sites, social media platforms, and mobile apps to provide customised experiences. We are likely to see the provision of more real-time tailor-made customer experiences, merging of IoT devices like wearables and smart shelves to collect touchpoint insights and enhance customer engagement, and predictive analytics powered by AI and ML to forestall customer needs proactively. Future marketing strategies will considerably bank upon amalgamated customer data pools (CDPs) that synchronise offline and online transactions. We can expect cutting-edge CDPs that merge data across touchpoints into a centralized platform. While marketing analytics to identify success and impact, offers high potential, marketers face many hurdles in effectively tapping customer touchpoints, big data, and sentiment analysis. These barriers are often due to technological, organizational, and ethical difficulties. For example, customer touchpoints cut across multiple channels, including websites, mobile apps, social media, physical stores, email, and call centers, and merging data from these diverse sources into one consolidated view is a considerable hurdle to be crossed. Every marketing team will need to overcome these barriers to realise the full potential of marketing analytics.

References

Basu, R., Lim, W.M., Kumar, A. and Kumar, S. (2023). Marketing analytics: The bridge between customer psychology and marketing decision-making. *Psychology & Marketing, 40*(12), 2588–2611.

Behare, N., Chaudhari, M., Sharma, S., Sane, A.C., Kharate, S., Waghulkar, S.,... and Pawar, P. (2024). Emerging Trends in Data-Driven Marketing. Data-Driven Marketing for Strategic Success, 323–358.

Bhandari, R., Singer, M. and Van Der Scheer, H. (2014). Using marketing analytics to drive superior growth. *Mckinsey Quarterly.*

Bollier, D. (2010). The promise and peril of Big Data. Washington DC: The Aspen Institute.

Boyd, D. and Crawford, K. (2012). Critical questions for Big Data: Provocations for a cultural, technological and scholarly phenomenon. *Information, Communication & Society*, vol.15, No. 5, pp. 662–679

Burkholder, L. (ed.) (1992) Philosophy and the Computer, Westview Press, Boulder, San Francisco, and Oxford.

Callegaro, M. and Yang, Y. (2018). The role of surveys in the era of "Big Data". *In The Palgrave handbook of survey research* (pp. 175–192). Palgrave Macmillan, Cham.

Cao, G. and Tian, N. (2020). Enhancing customer-linking marketing capabilities using marketing analytics. *Journal of Business & Industrial Marketing, 35*(7), 1289–1299.

Cao, G., Duan, Y. and El Banna, A. (2019), "A dynamic capability view of marketing analytics: evidence from UK firms", *Industrial Marketing Management*, vol. 76, pp. 72–83

Chaudhuri, S. and Dayal, U. (1997). An overview of data warehousing and OLAP technology. *ACM Sigmod record*, 26(1), 65–74.

Chen, H., Chiang, R.L.H. and Storey, V.C. (2012). Business intelligence and analytics: From Big Data to big impact. *MIS Quarterly*, vol. 36, No. 4, pp. 1165–1188.

Davenport, T.H. and Phillips, J. (2016). The future of marketing automation. *Applied Marketing Analytics, 2*(3), 213–224.

Demchenko, Y., Grosso, P., De Laat, C. and Membrey, P. (2013, May). Addressing big data issues in scientific data infrastructure. *In 2013 International conference on collaboration technologies and systems* (CTS) (pp. 48–55). IEEE.

Drury, G. (2007). 'Opinion Piece: Social Media: Should Marketers Engage and How Can it be Done Effectively', *Journal of Direct, Data and Digital Marketing Practice*, vol. 9, no. 3, pp. 274–277.

Fayyad, U., Piatetsky-Shapiro, G. and Smyth, P. (1996). From data mining to knowledge discovery in databases. *AI magazine*, 17(3), 37.

Fenech C. and Perkins B. (2015), Made-to-order: the rise of mass personalization, *The Deloitte Consumer Review*, Deloitte.

Gandomi, A. and Haider, M. (2015). Beyond the hype: Big data concepts, methods, and analytics. *International journal of information management*, 35(2), 137–144.

Germann, F., Lilien, G.L. and Rangaswamy, A. (2013). Performance implications of deploying marketing analytics. *International Journal of Research in Marketing, 30*(2), 114–128.

Gupta, C.P. and Ravi Kumar, V.V. "Artificial Intelligence and Internet of Things: Revolutionizing the implementation of Customer Relationship Management," 2022 ASU International Conference in Emerging Technologies for Sustainability and Intelligent Systems (ICETSIS), Manama, Bahrain, 2022, pp. 60–66, doi: 10.1109/ICETSIS55481.2022.9888821.

Hanssens, D.M. and Pauwels, K.H. 2016. Demonstrating the value of marketing. *Journal of Marketing* 80 (6): 173–190.

Hemmati, A., Arzanagh, H.M. and Rahmani, A.M. (2024). A taxonomy and survey of big data in social media. *Concurrency and Computation: Practice and Experience*, *36*(1), e7875.

Issenberg, S. (2013). How President Obama's campaign used big data to rally individual voters. *Technology Review*, *116*(1), 38–49.

Kotler, P. (2009). Marketing management: A south Asian perspective. Pearson Education India.

Kovalerchuk, B. and Vityaev, E. (2000). Data mining in finance: Advances in relational and hybrid methods. Springer Science & Business Media.

LaValle, S., Lesser, E., Shockley, R., Hopkins, M.S. and Kruschwitz, N. (2010). Big data, analytics and the path from insights to value. MIT Sloan Management Review.

Lazer, D., Pentland, A., Adamic, L., Aral, S., Barabási, A., Brewer, D., Christakis, N., Contractor, N., Fowler, J., Gutmann, M., Jebara, T., King, G., Macy, M., Roy, D. and Van Alstyne, M. (2009). Computational social science. *Science*, vol. 323, No. 5915, pp. 721–723.

Lilien, G.L. (2011). Bridging the academic–practitioner divide in marketing decision models. *Journal of Marketing*, 75(4), 196–210.

Mach-Król, M. and Hadasik, B. (2021). On a certain research gap in big data mining for customer insights. *Applied Sciences*, *11*(15), 6993.

Mishra, M. (2022). Customer experience: Extracting topics from tweets. *International Journal of Market Research*, 64(3), 334–353.

Patel, P. and Gahletia, S. (2020). Deriving Actionable Insights from Big Data to Enhance Customer Experiences Across the Consumer Journey. *International Journal of Responsible Artificial Intelligence*, *10*(8), 1–9.

Provost, F. and Fawcett, T. (2013). Data science and its relationship to big data and data-driven decision making. *Big data*, 1(1), 51–59.

Rana, G., Agarwal, S. and Sharma, R. (Eds.). (2021). *Employer branding for competitive advantage: Models and implementation strategies*. CRC Press.

Rana, G., Sharma, R. and Goel, A.K. (2019). Unraveling the power of talent analytics: Implications for enhancing business performance. *Business governance and society: Analyzing shifts, conflicts, and challenges*, 29–41.

Rao, P.B., Reddy, M.V.B., Chakradhar, P., Deepak, D. and Inampudi, P. (2024). Decoding Customer DNA: Leveraging Marketing Intelligence for Future Engagement. In Improving Service Quality and Customer Engagement With Marketing Intelligence (pp. 1–37). IGI Global.

Sharma, R., Saini, R., Prakash, C. and Prasad, V. (Eds.). (2021). *Internet of things and businesses in a disruptive economy*. Nova Science Publishers.

Sharma, R., Rana, G. and Agarwal, S. (2021). Techno innovative tools for employer branding in industry 4.0. In *Employer branding for competitive advantage* (pp. 171–180). CRC Press.

Sharma, R., Harish, V. and Rana, G. (2024). Navigating Risk In the Age of Artificial Intelligence: Assessing and Identifying Risks with AI Strategies. *Educational Administration: Theory and Practice*, *30*(4), 2130–2140.

Scherer, M. (2012). Inside the secret world of the data crunchers who helped Obama win. *Time Magazine*, 180(21), 56.

Srivastava, V., Kishore, S. and Dhingra, D. (2021). Technology and the future of customer experience. In *Crafting customer experience strategy* (pp. 91–116). Emerald Publishing Limited.

Veer, C., Kumar, P. and Sharma, R. (2022). Green Entrepreneurship: an avenue for innovative and sustainable product development and performance. In *Entrepreneurial Innovations, Models, and Implementation Strategies for Industry 4.0* (pp. 63–77). CRC Press.

Warshaw, C. (2016). The Application of Big Data in Surveys to the Study of Elections, Public Opinion, and Representation.

5 | Harnessing Data for Sustainability: Applications and Challenges in Environmental Analytics

Ashutosh Kumar Choudhary[1*] and Deepak Srivastava[2]

Environmental analytics is decisive in attaining the Sustainable Development Goals (SDGs) by offering valuable insights into complex ecological and social systems. This data-centric method enables organizations, policymakers, and communities to synchronize their efforts with global sustainability objectives, promoting a fair and balanced approach to development while protecting the planet for future generations. It utilizes a variety of analytical techniques and tools to manage, interpret, and present complex environmental data effectively. This chapter covers the concept of environmental analytics, its core elements, applications, and the significance of environmental data analysis for decision-making to promote sustainable development. It summarizes the techniques, tools, and models used in environmental data analysis and explores the challenges and opportunities.

1. Introduction

Environmental analytics is essential for achieving the Sustainable Development Goals (SDGs) by providing actionable insights into complex environmental systems. It integrates data-driven insights with scientific modeling to analyse and interpret complex environmental data. This empowers policymakers, scientists, and industries to make informed decisions that promote sustainability and environmental resilience. It is the application of advanced data science techniques, statistical methods, and computational tools to analyze and interpret environmental data. It enables the examination of complex ecological systems, resource management, and the impacts of human activity on the environment. By integrating diverse datasets, such as satellite imagery, sensor readings, and field observations, environmental analytics provides actionable insights for addressing challenges like climate change, pollution, and biodiversity loss. By enabling specific tracking of indicators such as air quality, water availability, and biodiversity, environmental analytics supports targeted interventions to mitigate climate change, conservation

[1] 0000-0002-5368-7366
[2] 0000-0002-7440-8311
* Corresponding author: akchoudhary@srhu.edu.in

DOI: 10.1201/9781003514282-5

of ecosystems, and promote sustainable resource use. This data-driven approach empowers organizations, policymakers, and communities to align their actions with global sustainability targets, ensuring a sustainable approach to development while safeguarding the planet for future generations. Environment statistics is an emerging statistical field in most countries and evidence-based policies and decision-making are needed to support sustainable development (Environment Statistics, 2024). At the regional and global level, scientists, policy analysts, and decision-makers are dependent on consistent, reliable, up-to-date information for the research community, preparing assessments, developing appropriate approaches for sustainable development, facilitating policy action, allocation of funds, and keeping informed on the state and trend of the environment (WESR, 2024). Environmental decision-making, including the selection of actions impacting the environment, occurs at various scales. From individuals choosing eco-friendly products to policymakers legislating national guidelines, effective decision-making hinges on a clear understanding of the environmental challenge (Ciobanu et al., 2022). Environmental policies at different levels depend on the timely and reliable availability of useful and integrated geospatial, earth observation and remote sensing, and real-time satellite imagery data. The United Nations Secretary-General raises, at the highest level, in his 2018 report on progress towards the sustainable development goals, "… The availability of high-quality, timely and disaggregated data is vital for evidence-based decision making and to ensure accountability for implementation of the 2030 Agenda..." (WESR, 2024). The provision of Big Data and Big Science creates the foundation for green solutions, with impact on billions of people, on the ground, effective solutions for big challenges, and priorities on the environment, from pollution to biodiversity and climate change. Big Data improves the lives of billions of people around the world and impacts countries and regions towards the achievement of sustainable development (WESR, 2024). Monitoring and analysis of environmental data contribute to resource management by providing critical information on various factors that enable sustainable resource distribution and utilization. In agriculture, for instance, monitoring weather patterns and soil conditions helps optimize crop yields and minimize environmental impact, which further contributes to sustainable food production (Ali and Khattak, 2022). Recent trends in environmental analytics reflect a merging of advanced technologies and innovative methodologies to address global ecological challenges. Big data analytics and machine learning (ML) enable the analysis of complex datasets from field studies, IoT sensors, and satellites. Real-time monitoring using IoT devices is another transformative development. Sensors now provide instant data on air and water quality, supporting immediate responses to environmental hazards. Additionally, collaborative platforms and open data initiatives foster international cooperation, making global environmental data more accessible. These innovations are paving the way for more precise and sustainable solutions, enhancing our ability to monitor and mitigate environmental issues effectively (Chalmeta and Barqueros-Muñoz, 2021).

2. Core Elements of Environmental Analytics

Environmental analytics plays a decisive role in advancing sustainability efforts by leveraging data science to improve environmental performance. The objectives for data collection are centred on understanding key environmental parameters, such as air quality, water usage, and energy consumption. Modern technology for data collection, including IoT sensors and remote sensing, enables the real-time gathering of large volumes of environmental data. The methodology of data collection is often multi-faceted, combining sensor networks, satellite imagery, and field data inputs to ensure comprehensive coverage. Once the data is collected, advanced data analysis techniques, powered by artificial intelligence (AI) and machine learning, help to identify patterns and insights. These tools support decision-making by optimizing process parameters and predicting future environmental impacts. Machine Learning (ML) and Deep Learning (DL) are the sub-fields of AI that train the models to learn from data without being programmed. When ML and DL are integrated they show high potential for process optimization, information-centric decision-making, and scientific innovation (Maganathan et al., 2020). Machine learning models, in particular, enable the optimization of resource management and energy efficiency, driving sustainable practices across industries. By integrating AI-driven analysis, companies can make informed decisions that enhance environmental performance while minimizing waste and energy consumption, ultimately contributing to broader sustainability goals. Fig. 1 shows the core elements of environmental analytics and its correlation with environmental performance and sustainability.

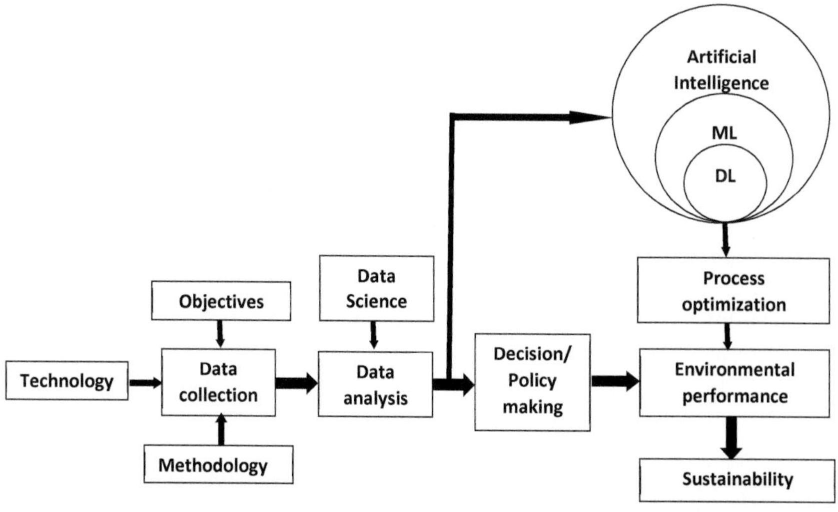

Fig. 1 Core elements of Environmental Analytics and Sustainability.

2.1 Data Sources and Collection Methods

Data sources and collection methods enable the acquisition of accurate, diverse, and comprehensive datasets for analysis. The key data sources include satellite remote sensing, which provides global coverage for monitoring land use, vegetation, and atmospheric conditions; *in situ* sensors for real-time measurements of air and water quality; and crowd-sourced data from citizen science initiatives. The collection methods such as environmental sampling, automated IoT-based monitoring systems, and drone technology enhance data precision and spatial resolution. Satellite imagery plays a pivotal role in environmental monitoring by providing comprehensive, high-resolution data on ecological and atmospheric changes over time and across large spatial scales. It enables the assessment of biodiversity loss, deforestation, urbanization, and climate change impacts with high accuracy. Advanced remote sensing technologies, such as multispectral and hyperspectral imaging, capture detailed information on vegetation health, water quality, and land-use patterns. Wulder (2008) and Drusch (2012) published the applications of NASA's Landsat and ESA's Sentinel missions for tracking forest cover changes and carbon sequestration. Additionally, thermal imaging from satellites aids in monitoring sea surface temperatures and urban heat islands, which are critical for climate studies. By integrating satellite data with Geographic Information Systems (GIS) and ML, researchers can detect pollution hotspots, predict ecological trends, and evaluate the effectiveness of conservation policies. These technologies are significant for global environmental initiatives to promote sustainable development.

Satellite imagery is a basis of environmental monitoring, offering significant insights into large-scale and long-term ecological and atmospheric changes. It provides high-resolution data on phenomena such as deforestation, desertification, urban expansion, and climate change impacts. Remote sensing technologies like multispectral, hyperspectral, and radar imaging enable the detailed observation of land use, vegetation health, and water bodies (Wulder et al., 2008). For example, NASA's Landsat program and the European Space Agency's Sentinel satellites have been instrumental in tracking deforestation in the Amazon and monitoring global crop health. Thermal imaging from satellites also aids in assessing urban heat islands and sea surface temperatures, which are vital for understanding climate dynamics. Satellite imagery thus plays a pivotal role in supporting global sustainability goals and environmental decision-making (Drusch et al., 2012). The integration of Internet of Things (IoT) data into environmental analytics platforms has revolutionized real-time monitoring and decision-making for environmental management. IoT-enabled sensors collect high-frequency data on air and water quality, temperature, humidity, and soil conditions, providing insights into environmental changes. This data is transmitted to cloud-based analytics platforms, where ML algorithms and big data tools process it to identify trends, predict risks, and optimize resource usage. By combining IoT data with Geographic Information Systems (GIS) and remote sensing, these platforms enable comprehensive environmental assessments, supporting global efforts like the UN Sustainable Development Goals (SDGs).

Big data and open data sources are significant in advancing environmental analytics, offering vast, diverse, and high-resolution datasets to understand and address complex environmental challenges. Big data refers to large and complex datasets generated from multiple sources, like IoT devices, remote sensing, weather stations, and social media (Nativi, 2015). These datasets provide detailed insights into phenomena such as biodiversity loss, environmental pollution, and climate change. For example, NASA's Earth Observing System (EOS) generates petabytes of data from satellites like Landsat and MODIS, which are extensively used for environmental modeling and monitoring. Big data analytics tools like Apache Hadoop and Spark enable efficient processing and visualization of these datasets, allowing for trend analysis and predictive modeling (NASA, 2023). Open data sources ensure that critical environmental data is freely available to researchers, policymakers, and the public. Some key sources are Copernicus Open Access Hub (COAH) which provides data from the European Space Agency's sentinel satellites, used for monitoring land, oceans, and atmospheric changes (ESA, 2023a); NOAA National Centers for Environmental Information (NCEI) which offers global weather, climate, and oceanic data (NOAA, 2023); Global Biodiversity Information Facility (GBIF) is an open-access repository for biodiversity records worldwide and World Bank Open Data (WBOD) which hosts datasets on environmental indicators, such as deforestation rates and carbon emissions (GBIF, 2024). These data sources are instrumental in creating transparency, enabling cross-disciplinary research, and supporting global SDGs initiatives. The integration of big and open data in environmental analytics has improved the accuracy of predictive models, facilitated real-time environmental monitoring, and supported the development of mitigation and adaptation strategies. For example, open data from Sentinel-2 satellites has been widely used to monitor agricultural practices and deforestation (Drusch et al., 2012).

Crowd-sourced data is increasingly significant in environmental analytics and decision-making, enabling real-time, localized, and large-scale data collection through contributions from individuals, communities, and citizen scientists. Leveraging mobile devices, online platforms, and sensor networks, crowdsourcing empowers people to report environmental observations such as air quality, biodiversity sightings, water pollution, and climate impacts (Bonney et al., 2009; Haklay, 2013). It enables the collection of data from vast geographic regions, often where professional monitoring resources are uncommon. It reduces the dependence on expensive equipment and specialized personnel and promotes public involvement in environmental issues, fostering awareness and collective action. For example, city planners use air quality data to design low-emission zones, and biodiversity data informs conservation priorities. Additionally, crowd-sourced data often complements official datasets, filling gaps in traditional monitoring systems. Platforms like *AirVisual* and *OpenAQ* use data submitted by individuals and low-cost sensors to map air pollution levels, providing insights into pollution hotspots (AirVisual, 2023). Citizen Science platforms like *iNaturalist* and *eBird* collect data on species distribution and behavior, aiding in the study of biodiversity and ecosystem changes (GBIF, 2024). Crowd-sourced data is critical during natural disasters, such as floods or wildfires, allowing real-time mapping of affected areas. For instance, the *Humanitarian Open Street Map Team* facilitates data collection for disaster recovery efforts.

2.2 Data Analysis : Techniques, Tools, and Models

Environmental analytics employs a diverse range of analytical techniques and tools to process, analyze, and visualize complex environmental data. These methodologies are essential for uncovering patterns, predicting trends, and supporting sustainable decision-making. These techniques and tools enable researchers to derive actionable insights, optimize resource management, and address pressing environmental challenges. They also facilitate collaboration and reproducibility in scientific research. Regression models are powerful statistical and machine-learning tools used to predict future environmental outcomes based on historical and current data. They establish relationships between dependent variables (outcomes) and independent variables, enabling the modeling of trends, patterns, and impacts in environmental systems. Table 1 summarizes the techniques, tools, and models generally used in environmental data analysis (Box et al., 2015; Gorelick et al., 2017; Zhang, 2017).

Table 1 Summarizes the techniques, tools, and models used in environmental data analysis.

Data analysis	*Applications/Examples*
A. Techniques	
• **Geospatial Analysis**	• Uses Geographic Information Systems (GIS) to analyze spatial data related to land use, vegetation, pollution, and climate. • Monitoring deforestation or mapping pollution sources using tools like ArcGIS or QGIS.
• **Remote Sensing Analysis**	• Processing of satellite or drone imagery for vegetation indices (e.g., NDVI), land-cover classification, and climate data. • Sentinel and Landsat imagery is processed with tools like Google Earth Engine.
• **Predictive Modeling (AI and ML)**	• ML and statistical models are used for forecasting environmental trends like climate change impacts or species migration patterns. • Random Forest and neural networks applied to predict wildfire risks.
• **Time-Series Analysis**	• Examines trends and cycles in environmental data over time, such as temperature or precipitation patterns. • Using R or Python libraries for analyzing climate datasets.

Contd.

Table 1 *Contd.*

Data analysis	*Applications/Examples*
• **Multicriteria Decision Analysis (MCDA)**	• Evaluates multiple environmental variables to prioritize actions or resources, such as conservation efforts or pollution control strategies.
B. Tools	
• **ArcGIS and QGIS**	• Geospatial data visualization and analysis.
• **Google Earth Engine (GEE)**	• A cloud-based platform for processing large-scale satellite imagery.
• **Python and R**	• Extensively used for data analysis, visualization, and machine learning. • Libraries like Pandas, SciPy, and ggplot2 are essential for statistical and graphical analysis.
• **MATLAB**	• Used in hydrology and ecosystem modeling for numerical computing and modeling.
• **Hadoop and Spark**	• Used for big data processing in environmental applications.
• **Environmental Data Platforms**	• Tools like Climate Data Store (CDS) and EarthData provide curated environmental datasets for analysis.
C. Models	
• **Linear Regression**	• Models the relationship between a single dependent variable and one or more independent variables assuming linearity. • Predicting air pollutant concentrations based on traffic volume and meteorological factors.
• **Multiple Linear Regression (MLR)**	• Considers multiple predictors to estimate environmental outcomes. • Estimating river flow rates based on precipitation, temperature, and land use.
• **Polynomial Regression**	• Captures nonlinear relationships by extending linear models to higher degrees. • Modeling the growth rate of invasive species as a function of temperature.
• **Logistic Regression**	• Used for binary outcomes, such as predicting whether a region will experience drought. • Classifying areas as prone or not prone to flooding.
• **Generalized Linear Models (GLMs)**	• Extends linear regression to non-normal error distributions for diverse ecological and environmental datasets. • Predicting the abundance of species in varying habitats.
• **Support Vector Regression (SVR)/ Random Forest Regression**	• Nonlinear regression techniques for complex environmental systems where relationships are not explicitly known. • Predicting soil erosion rates or atmospheric carbon levels.
• **Time-Series Regression Models**	• Combines regression techniques with temporal data to forecast trends like temperature rise or pollution levels over time. • ARIMA and SARIMA models for forecasting seasonal variations in water demand.

3. Applications of Environmental Analytics

The applications of environmental analytics are very vast and significant. Table 2 summarizes the applications of environmental analytics in different areas of the environment such as climate change analytics, pollution and emissions monitoring, resource and energy management, biodiversity, and ecosystem services, environmental impact assessment, urban planning and smart cities, disaster risk management, and policy and regulatory decision-making.

3.1 Climate Change Analytics

Climate change analytics leverages data-driven techniques and advanced computational models to understand, monitor, and address the multifaceted impacts of climate change. By analyzing large-scale datasets, such as temperature records and greenhouse gas emissions, satellite imagery helps in identifying patterns to predict future trends. The ML tools and geospatial analysis enable precise modeling of phenomena such as rising sea levels, changing weather patterns, and shifts in ecosystems. Climate change analytics supports the development of effective mitigation and adaptation strategies in optimizing renewable energy deployment, designing resilient infrastructure, and tracking progress toward international climate agreements like the Paris Agreement 2015. It turns complex climatic data into actionable insights which further empowers policymakers, researchers, and industries to make informed decisions and combat the global climate crisis. The predictive modeling for climate change trends employs advanced computational techniques to forecast future environmental conditions by analyzing historical and current data. These models incorporate variables such as atmospheric greenhouse gas concentrations, temperature records, and oceanic patterns to project outcomes like global warming, sea level rise, and the frequency of extreme weather events. The General Circulation Models (GCMs) simulate the Earth's climate system under various emissions scenarios to predict long-term changes in temperature and precipitation. These models inform decision-makers by identifying high-risk areas, guiding mitigation efforts, and evaluating the potential impact of policy measures like the Paris Agreement. Studies published by the Intergovernmental Panel on Climate Change (IPCC) and tools like NASA's Earth System Models show the critical role of predictive analytics in climate science (IPCC, 2022; NASA, 2023). Carbon footprint analysis evaluates the total greenhouse gas emissions, both direct and indirect, generated by individuals, organizations, products, or processes, expressed in terms of carbon dioxide equivalents (CO_2 e). Through data analytics, emission hotspots can be identified in the agriculture, energy, transportation, and manufacturing sectors. The efforts to lower carbon footprints involve shifting to renewable energy, enhancing energy efficiency, adopting sustainable supply chains, and promoting behavioural changes such as reducing meat consumption or opting for public transportation (Pandey et al., 2011). Advanced tools, such as life cycle assessments (LCA) and carbon accounting frameworks, support precise measurements and effective decision-making. Organisations can use LCA to redesign products to lower emissions, while governments leverage data from carbon analysis to implement policies like carbon pricing. These strategies are significant for achieving SDGs (UNEP (2021). Climate change impact assessments on ecosystems and human health, evaluate how rising temperatures, changing precipitation patterns, and extreme weather events affect natural systems and communities. The ecosystems face threats like habitat loss, species migration, and biodiversity loss, as seen in coral reef bleaching and shifts in forest compositions. Climate change also increases risks of heat-related illnesses, vector-borne diseases like malaria and dengue, and respiratory issues to human health (IPBES (2019). Rising sea levels and intensified storms threaten coastal populations, exacerbating food and water insecurity. Such assessments depend on climate models and ecological risk frameworks to identify vulnerabilities and lead to adaptation strategies. The Intergovernmental Science-Policy Platform on Biodiversity and Ecosystem Services (IPBES) and the World Health Organization (WHO) emphasize integrating environmental and health data for comprehensive mitigation and planning. These interconnected impacts can be understood to protect both biodiversity and human well-being, for achieving global sustainability (IPBES, 2019; WHO, 2021).

3.2 Pollution and Emissions Monitoring

Pollution and emissions monitoring data analytics play a vital role in tracking and mitigating environmental degradation caused by pollutants and greenhouse gas emissions. By leveraging real-time data from air and water quality sensors, satellite imagery, and industrial reports, analytics identifies pollution hotspots, trends, and sources. Advanced ML and geospatial analysis methods allow the precise modeling of emission patterns and the evaluation of mitigation strategies. Tools like the European Space Agency's Sentinel satellites monitor global air quality, while predictive algorithms forecast pollution-related risks (ESA, 2023b). This information helps policymakers in shaping regulations, such as emission limits and cleaner production standards, and supports industries in optimizing operations to meet environmental compliance. The publicly accessible platforms like the World Air Quality Index enhance awareness and engagement by providing actionable insights. The effective analysis contributes to global efforts, such as the reduction targets under the Kyoto Protocol and the Paris Agreement, ensuring environmental sustainability and public health protection (World Air Quality Index Project, 2023). Advanced analytics tools integrate data from IoT-enabled sensors, satellite observations, and industrial process records to create comprehensive emissions inventories. Predictive models and ML algorithms analyze this data to forecast emissions trends and evaluate the impact of mitigation strategies. Analytics platforms automate reporting processes to ensure adherence to the Kyoto Protocol, Paris Agreement, and national carbon tax regulations frameworks for regulatory compliance. The tools like Carbon Disclosure Project (CDP) and the Greenhouse Gas Protocol (GGP) provide standardized frameworks for emissions reporting.

By offering transparency and actionable insights, data analytics not only helps organizations meet environmental regulations but also supports their transition to sustainable practices, aligning with global goals for emissions reduction (GGP, 2024; CDP, 2024). In the data analytics field, the Geographic Information Systems (GIS) is a powerful tool for mapping and analyzing pollution across spatial and temporal scales. GIS integrates geospatial data from satellite imagery, ground-based sensors, and environmental databases to visualize pollution sources, distribution patterns, and thereby affected areas. GIS enables precise identification of pollution hotspots and vulnerable communities by layering data on air, water, and soil quality with population density and industrial activity (Longley et al., 2015). The advanced spatial analysis techniques, such as interpolation and hotspot detection, further support monitoring and planning efforts. GIS-based platforms like ESRI ArcGIS and Google Earth Engine have been used to track air pollution in urban areas and model the dispersion of contaminants in waterways. These insights guide policymakers in designing targeted interventions, such as emission reduction zones and waste management systems, while empowering communities with transparent, accessible information. GIS-based pollution mapping is thus critical for effective environmental management and regulatory enforcement (UNEP, 2022).

3.3 Resource and Energy Management

Data analytics plays a critical role in energy consumption analysis and efficiency improvements by providing actionable insights into energy usage patterns and identifying areas for optimization. Advanced analytics techniques, including ML and real-time monitoring, allow for the collection and analysis of data from smart meters, IoT devices, and energy management systems. These insights help industries and households reduce energy wastage, optimize operations, and transition to renewable energy sources. However, predictive analytics enables forecasting of energy demand, assisting utilities in balancing supply and demand while minimizing costs and environmental impacts (IEA, 2022). Google has used AI-driven data analytics to improve energy efficiency in its data centers, achieving significant reductions in electricity consumption. Moreover, governments and organizations leverage energy analytics to track progress toward sustainability goals, such as those outlined in the SDGs (Google AI, 2018). By analyzing data from wind turbines, solar panels, and energy storage systems, advanced analytics enables accurate forecasting of renewable energy generation, which is often variable and weather-dependent. Predictive models and ML algorithms optimize grid operations by balancing renewable energy supply with demand and minimizing reliance on fossil fuels. AI-driven analytics have been used to improve wind energy production by optimizing turbine placement and performance. Furthermore, real-time analytics support smart grid technologies, enabling decentralized energy systems and efficient distribution. These capabilities help utilities and policymakers achieve energy transition goals while reducing greenhouse gas emissions. Tools like the National Renewable Energy Laboratory's (NREL) System Advisor Model exemplify how data-driven approaches facilitate renewable energy planning and deployment (NREL, 2023).

Data analytics plays a growing role in optimizing water usage and waste management, offering insights that enhance efficiency and promote sustainability in resource management. By processing data from smart water meters, sensors, and treatment facilities, analytics uncovers patterns in water consumption, detects leaks, and improves the efficiency of distribution networks. Advanced deep learning models, such as Convolutional Neural Networks (CNN), Deep Belief Networks (DBN), Faster Region-Based Convolutional Networks (Faster R-CNN), Transfer Learning, Stacked Autoencoders (SAE), and Long Short-Term Memory (LSTM), are extensively applied to address challenges in environmental water management (Sun and Scanlon, 2019). In waste management, data-driven approaches enable tracking of waste generation, recycling rates, and landfill usage, supporting more effective strategies like circular economy initiatives. Predictive models also forecast water demand and waste trends, helping planners allocate resources and design infrastructure.

3.4 Biodiversity and Ecosystem Services

Data analytics plays a crucial role in understanding biodiversity and ecosystem services by enabling the collection, integration, and analysis of vast ecological datasets. Tools like remote sensing, geographic information systems (GIS), and machine learning facilitate the mapping of habitats, monitoring of species populations, and assessment of ecosystem health. It is used to identify biodiversity hotspots and predict the impacts of land-use changes and climate shifts on ecosystems. Guisan et al. (2017) and Mace et al. (2018) demonstrate the use of data analytics to quantify ecosystem services, such as carbon sequestration and pollination, aiding in sustainable land management and conservation planning. These insights support evidence-based policies and global initiatives like the Convention on Biological Diversity (CBD) and the SDGs. Analytics strengthens efforts to mitigate biodiversity loss and maintain the benefits of ecosystems by linking ecological data with human well-being. It is essential for valuing ecosystem services and developing effective conservation strategies by providing quantitative assessments of the benefits ecosystems deliver to humans, such as water purification, carbon sequestration, and pollination. Tools like GIS, remote sensing, and econometric models enable the spatial mapping and economic valuation of ecosystem services, helping policymakers prioritize conservation investments (Bagstad et al., 2013). ML algorithms assess the trade-offs between land development and ecosystem service preservation, optimizing decisions for sustainable development. It also aids in simulating the impacts of land-use changes and climate variations on ecosystem functionality. By integrating ecological and socio-economic data, analytics supports global frameworks such as the Convention on Biological Diversity (CBD) and Payments for Ecosystem Services (PES) schemes, ensuring that ecosystem contributions are accounted for in decision-making processes (UNEP-WCMC, 2022). These insights promote the long-term ecological resilience and sustainable management of natural resources.

3.5 Environmental Impact Assessment

Environmental analytics is pivotal in enhancing the accuracy and efficiency of Environmental Impact Assessments (EIA), a critical tool for evaluating the potential environmental consequences of proposed projects. By leveraging Geographic Information Systems (GIS), remote sensing, and predictive modeling, analysts can assess the spatial and temporal impacts of construction, energy projects, mining, and urban development (Glasson et al., 2012). Predictive models simulate pollutant dispersion, enabling planners to mitigate air and water contamination risks. GIS mapping helps in identifying ecologically sensitive areas, ensuring that development avoids critical habitats. Tools like multi-criteria decision analysis (MCDA) integrate diverse datasets, facilitating a comprehensive evaluation of trade-offs between economic benefits and environmental costs (Geneletti, 2013). These analytical approaches also improve stakeholder engagement by providing visualizations and evidence-based insights. Noble (2000) highlights how analytics improves compliance with regulations and enhances the sustainability of development projects by identifying mitigation strategies early in the planning process.

3.6 Urban Planning and Smart Cities

Sustainable city development through environmental data analysis. Use of environmental analytics in smart city infrastructure (e.g., traffic management, energy systems) (Salvati and Carlucci, 2014). Sustainable city development increasingly relies on environmental data analysis to design urban areas that balance ecological integrity with human well-being (Elmqvist et al., 2013). By leveraging data from sensors, Geographic Information Systems (GIS), and satellite imagery, urban planners can optimize land use, improve energy efficiency, and manage natural resources. For instance, air quality monitoring systems provide real-time pollution data, enabling the design of low-emission zones and better public transport networks. Predictive analytics also aid in assessing the impacts of urban expansion on biodiversity and water resources, ensuring that green spaces and ecosystems are preserved (Elmqvist et al., 2013). Additionally, environmental data analysis supports waste management by identifying patterns in waste generation and optimizing recycling systems. Research published in *Sustainable Cities and Society* journal highlights how integrating big data and IoT technologies into urban planning fosters resilience to climate change, reduces carbon footprints, and enhances liveability (Bibri and Krogstie, 2017). As cities strive to meet the United Nations SDGs, environmental analytics plays a crucial role in making urban growth sustainable and inclusive (United Nations, 2023).

3.7 Disaster Risk Management

Predicting and managing natural disasters such as floods, wildfires, and hurricanes has been significantly improved through advancements in environmental data analytics and predictive modeling (Emanuel, 2017; Jain et al., 2020; UNDRR, 2019). By analyzing real-time data from satellite imagery, weather stations, and IoT sensors, scientists can develop early warning systems to predict disaster onset and assess potential impacts. For instance, machine learning algorithms process historical climate data to forecast flood-prone areas based on rainfall patterns, while remote sensing tools like MODIS (Moderate Resolution Imaging Spectroradiometer) detect and monitor wildfire hotspots. Similarly, hurricane tracking models use atmospheric pressure, sea surface temperature, and wind speed data to predict storm intensity and trajectory. These analytics-driven insights enable authorities to implement timely evacuation plans, allocate resources efficiently, and design long-term mitigation strategies such as improved infrastructure.

3.8 Policy and Regulatory Decision-Making

Environmental analytics plays a transformative role in shaping policies for sustainability and environmental protection by providing data-driven insights and evidence-based recommendations. Advanced tools and techniques, such as big data analysis, Geographic Information Systems (GIS), and machine learning, help policymakers evaluate the environmental impacts of existing practices and forecast the outcomes of proposed interventions (Kumar and Jha, 2021). For instance, carbon footprint analysis informs climate policies by identifying key sectors for emission reductions, while predictive models assess the long-term effects of land-use changes on biodiversity. Additionally, analytics supports the development of sustainable urban planning, renewable energy integration, and waste management policies by simulating various scenarios and identifying optimal strategies. Real-time environmental monitoring also aids regulatory compliance and enforcement by tracking pollutants and ecological changes.

Table 2 Application of environmental analytics.

S.No.	*Area*	*Applications*
1.	Monitoring Environmental Changes	• Real-Time Data: Monitoring of air and water quality, deforestation rates, and carbon emissions through IoT devices and satellite data. • Trend Analysis: Detecting long-term changes in temperature, sea level, and biodiversity to understand climate and ecological shifts. • Early Warning Systems: Predicting natural disasters like floods, hurricanes, and droughts to minimize human and economic losses.

Contd.

Table 2 *Contd.*

S.No.	*Area*	*Applications*
2.	Climate Change Mitigation and Adaptation	• Carbon Footprint Analysis: Identifying high-emission sectors and optimizing processes to reduce greenhouse gas emissions. • Renewable Energy: Optimizing the placement and efficiency of solar panels, wind turbines, and other renewable resources using geospatial analytics. • Policy Support: Providing evidence for crafting international agreements, such as the Paris Agreement, and tracking progress toward emission reduction targets.
3.	Sustainable Resource Management	• Water Resources: Analyzing patterns in water availability, usage, and pollution to support equitable distribution and efficient use. • Forestry: Monitoring deforestation and illegal logging to implement conservation strategies. • Agriculture: Using precision agriculture techniques to optimize irrigation, reduce pesticide use, and enhance crop yields sustainably.
4.	Biodiversity Conservation	• Habitat Mapping: Identifying critical ecosystems and biodiversity hotspots for targeted conservation efforts. • Species Monitoring: Tracking populations of endangered species and assessing the effectiveness of conservation programs. • Predictive Models: Anticipating the impacts of climate and land-use changes on ecosystems.
5.	Pollution Control	• Air Quality Monitoring: Identifying pollution sources and assessing the impact of mitigation measures. • Waste Management: Optimizing recycling and waste disposal systems through data analysis. • Marine Protection: Monitoring ocean pollution, including oil spills and plastic waste, to guide cleanup efforts.
6.	Improving Policy and Decision-Making	• Evidence-Based Policies: Providing robust, data-backed evidence for environmental legislation. • Public Engagement: Enhancing transparency and awareness through accessible data visualization tools. • Risk Assessment: Identifying vulnerabilities in ecosystems and urban areas to plan for resilience.
7.	Advancing Sustainability Goals	• Data-Driven SDGs: Tracking progress toward SDGs such as clean water, affordable energy, and life on land. • Circular Economy: Using analytics to design systems that minimize waste and recycle resources efficiently.

4. Challenges and Opportunities

The data may be distributed across different geographic regions. Since environmental changes are expected according to different spaces and times, finding spatial and temporal insights from environmental data is a key challenge (Blair et al., 2019). Figure 2 summarizes the major challenges and opportunities in environmental analytics.

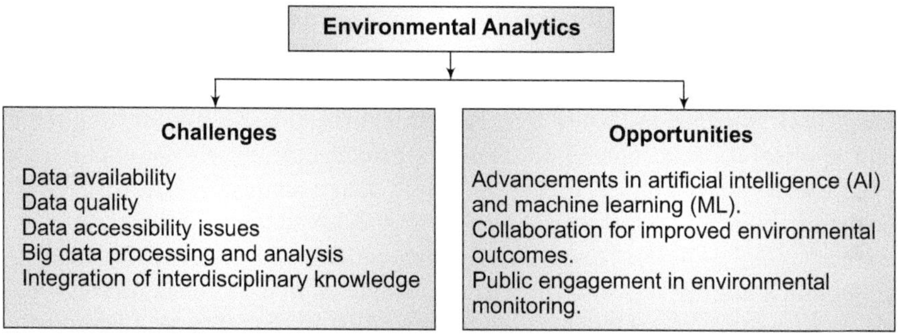

Fig. 2 Challenges and opportunities in environmental analytics.

One of the primary challenges in environmental analytics is the complexity and diversity of data. Environmental data often come from various sources, including satellites, remote sensors, and field observations, leading to issues with interoperability and standardization (Weersink et al., 2018). The heterogeneity of data formats and the lack of standardized protocols can hinder data integration and analysis. Additionally, the sheer volume of data generated, especially from IoT devices and real-time monitoring systems, requires robust infrastructure for storage, processing, and analysis. Another challenge is the uncertainty inherent in environmental modeling. Predictive models in environmental analytics often depend on incomplete or noisy data, which can affect the accuracy of forecasts and policy recommendations (Papadopoulos and Balta, 2022). Coupled with this is the issue of data accessibility, where sensitive ecological or demographic data may be restricted due to privacy concerns or geopolitical reasons. Ethical and societal challenges also come into play. The use of advanced analytics tools must consider equity in decision-making and avoid exacerbating existing inequalities. Moreover, the lack of technical expertise in many parts of the world limits the adoption and application of environmental analytics in regions most affected by environmental crises.

Environmental analytics, the use of advanced data analysis techniques to study and solve environmental issues, is a rapidly evolving field. It encompasses a range of methods, including machine learning, big data analytics, and geospatial technologies, to address global environmental challenges like climate change, biodiversity loss, and resource management. While this field offers numerous opportunities, it also faces significant challenges that must be navigated to maximize its potential. One of the primary challenges in using data science for environmental issues is the quality and availability of data. Environmental data can be highly variable, incomplete, and inconsistent. Factors such as the location, method of data collection, and the technology used can significantly affect data quality. Moreover, environmental data is often collected from remote or inaccessible areas, adding to the complexity of data gathering and maintenance. This variability and inaccessibility can lead to significant gaps in datasets, making it difficult for data scientists to analyze trends accurately and make reliable predictions (Kitchin, 2014). The complexity of environmental issues necessitates collaboration across various disciplines. Data scientists, environmental scientists, and policymakers must work together to ensure that data-driven insights are accurate, relevant, and applicable. Environmental scientists provide the necessary context and expertise about ecological processes, while data scientists bring in advanced analytics skills. Policymakers can use these insights to draft informed, effective environmental policies. However, effective collaboration is often hindered by communication barriers and differing priorities among the disciplines, which can lead to misinterpretations and underutilization of data insights (Turnhout et al., 2020). Using data science in environmental issues also raises several ethical considerations. The privacy of data, especially when gathered from community-monitored devices or apps, is a significant concern. There is also the risk of data being used to support harmful policies or practices that could negatively impact vulnerable ecosystems or communities. Furthermore, there is a moral responsibility to ensure that the interventions based on data analytics do not lead to unintended consequences that could exacerbate environmental problems or displace communities (Boyd and Crawford, 2012). Regression models require high-quality data for accuracy. Overfitting, multicollinearity, and the inclusion of irrelevant predictors can reduce the model's reliability. Advanced techniques like regularization (e.g., Ridge or Lasso regression) address these issues.

Despite these challenges, environmental analytics offers transformative opportunities. The integration of big data analytics and machine learning can significantly enhance the understanding of complex ecological systems. For instance, machine learning models can analyze large datasets to identify patterns and correlations that were previously undetectable, aiding in climate change predictions and biodiversity conservation (Maganathan et al., 2020). IoT and sensor-based technologies present another opportunity by enabling real-time environmental monitoring. These technologies facilitate data collection on air and water quality, soil health, and wildlife movements, contributing to more informed decision-making (Bibri, 2018). Coupled with geospatial analytics, they provide a powerful tool for managing natural resources and monitoring environmental compliance. Additionally, environmental analytics is fostering interdisciplinary collaboration. By bridging data science, engineering, and environmental science, it promotes innovative solutions for sustainable development. For instance, predictive analytics is being used to optimize renewable energy production and reduce waste in manufacturing processes. ML algorithms employ mathematical and statistical methods to identify relationships within a large amount of data without requiring prior assumptions about the data (Jung, 2022). Mapping environmental phenomena with ML can result in high prediction accuracy (Khaledian and Miller, 2020).

5. Future Trends in Environmental Analytics

Environmental analytics is a dynamic field poised to revolutionize how we monitor, understand, and address ecological challenges. As advancements in technology intersect with environmental science, several future trends are emerging, transforming the landscape of sustainability and resource management. The increasing deployment of sensors, satellites, and IoT devices is leading to a data explosion, necessitating robust methods for integration and analysis. Big data analytics, combined with machine learning (ML) algorithms, is expected to play a crucial role in deciphering complex environmental patterns. AI and ML are set to enhance environmental monitoring by identifying patterns that are invisible to traditional analysis methods. For instance, ML models can track biodiversity changes, predict pollution trends, and optimize renewable energy usage. The integration of IoT devices enables real-time environmental monitoring, providing critical data on air quality, water pollution, and deforestation. These technologies are instrumental in creating smart cities that prioritize sustainability. As the cost of IoT devices decreases, their accessibility will expand, allowing even remote regions to benefit from advanced environmental monitoring (Bibri, 2018). Remote sensing technologies, including satellite imagery and drones, are becoming indispensable tools for large-scale environmental monitoring. These technologies are particularly effective in tracking deforestation, glacial melting, and urban sprawl. Future trends also point toward the adoption of green analytical chemistry, which emphasizes the development of environmentally friendly methods for monitoring pollutants. As environmental challenges are global, collaborative platforms for data sharing and analysis are crucial. Open-source platforms and shared databases will enhance international cooperation, enabling scientists worldwide to address issues like climate change and biodiversity loss more effectively.

Conclusion

Environmental analytics plays a significant role in environmental decision-making. It involves the integration of environmental data and a variety of analytical tools for data analysis, which is further significant in decision making to minimize the environmental risks. It facilitates the tracking of progress toward the United Nations Sustainable Development Goals (SDGs) with data-driven insights. It identifies vulnerabilities in ecosystems and resource systems to plan for resilience. The integration of huge, diverse data into environmental sustainability is nothing short of a paradigm shift. It explores new possibilities and approaches to mitigate some of the most persistent environmental challenges. The capacity to gather data from various sources and analyze it in real-time scenarios allows us to develop a comprehensive understanding of environmental variations. The growing importance of environmental data in addressing global environmental challenges and achieving global sustainability goals. It plays an important role in bridging the gap between data and actionable insights. By fostering a better understanding of environmental systems, it empowers global efforts to mitigate environmental degradation, ensure sustainable development, and build a resilient future. The field of environmental analytics possesses immense challenges and opportunities. While issues such as data heterogeneity, modeling uncertainty, and ethical considerations must be addressed, the potential to revolutionize environmental monitoring and management is unquestionable. By leveraging advancements in technology and fostering global collaboration, environmental analytics can play a crucial role in achieving environmental sustainability. The future of environmental analytics is an integration of cutting-edge technologies like IoT, remote sensing, artificial intelligence, machine learning, deep learning, and collaborative efforts. It unlocks exceptional capabilities for addressing and mitigating environmental challenges like global biodiversity conservation, climate change, extreme weather, monitoring of deforestation, glaciers melting in real-time, and natural resource management. Researchers and data scientists can provide actionable insights that drive effective policy-making and sustainable practices by harnessing the power of big data, and modern tools.

References

AirVisual (2024 December). IQAir AirVisual Platform, Retrieved from https://www.iqair.com

Ali, H. and Khattak, W.A. (2022). Big Data for Sustainability: A Holistic Approach to Environmental Conservation and Resource Management. *Journal of Big-Data Analytics and Cloud Computing*, 7(4), 18–30. https://questsquare.org/index.php/JOURNALBACC/article/view/22

Bagstad, K.J., Semmens, D.J., Waage, S. and Winthrop, R. (2013). A comparative assessment of decision-support tools for ecosystem services quantification and valuation, *Ecosystem Services*, 5, 27–39. https://doi.org/10.1016/j.ecoser.2013.07.004

Bibri, S.E. (2018). The IoT for smart sustainable cities of the future: An analytical framework for sensor-based big data applications for environmental sustainability. *Sustainable Cities and Society*, 38, 230–253.

Bibri, S.E. and Krogstie, J. (2017). Smart sustainable cities of the future: An extensive interdisciplinary literature review, *Sustainable Cities and Society*, 31, 183–212.

Blair G.S., Henrys, P., Leeson, A., Watkins, J., Eastoe, E., Jarvis. S. and Young. P.J. (2019). Data Science of the Natural Environment: A Research Roadmap, *Frontiers in Environmental Science*, 7, 1–14.

Bonney, R., Cooper, C.B., Dickinson, J., Kelling, S., Phillips, T., Rosenberg, K.V. and Shirk, J. (2009). Citizen Science: A Developing Tool for Expanding Science Knowledge and Scientific Literacy. *BioScience*, 59(11), 977–984. https://doi.org/10.1525/bio.2009.59.11.9

Box, G.E.P., Jenkins, G.M., Reinsel, G.C. and Ljung, G.M. (2015). *Time Series Analysis: Forecasting and Control.* 5th edition Wiley.

Boyd, D. and Crawford, K. (2012). Critical questions for big data: Provocations for a cultural, technological, and scholarly phenomenon. *Information, Communication & Society*, 15(5), 662–679.

Carbon Disclosure Project (CDP). (2024 December). *CDP Global Environmental Reporting.* Retrieved from https://www.cdp.net

Chalmeta, R. and Barqueros-Muñoz, J.E. (2021). Using big data for sustainability in supply chain management, *Sustainability*, *13*(13), 7004. https://doi.org/10.3390/su13137004

Ciobanu, R., Ţuclea, C.E., Holostencu, L.F. and Vrânceanu, D.M. (2022). Decision-making factors in the purchase of ecologic products. *Sustainability*, 14, 9558. https://doi.org/10.3390/su14159558.

Drusch, M., DelBello, U., Carlier, S., Colin, O., Fernandez, V., Gascon, F., Hoersch B., Isola, C., Laberinti, P., Martimort, P., Meygret, A., Spoto, F., Sy, O., Marchese, F. and Bargellini,

P. (2012). Sentinel-2: ESA's optical high-resolution mission for GMES operational services. *Remote Sensing of Environment*, 120, 25–36.

Elmqvist, T., Fragkias, M., Goodness, J., Güneralp, B., Marcotullio, P.J., McDonald, R.I., Parnell, S., Schewenius, M., Sendstad, M., Seto, K.C. and Wilkinson, C. (2013). *Urbanization, Biodiversity and Ecosystem Services: Challenges and Opportunities.* Springer.

Emanuel, K. (2017). Will global warming make hurricane forecasting more difficult? *Bulletin of the American Meteorological Society*, 98(3), 495–501.

Environment Statistics (2024 December). Environment Statistics Section United Nations Statistics Division, DC2-1418 New York, http://unstats.un.org/unsd/environment/newsletters.htm

European Space Agency (ESA) (2023a). *Copernicus Open Access Hub.* Retrieved from https://scihub.copernicus.eu

European Space Agency (ESA) (2023b). *Sentinel-5P Monitoring Air Quality.* Retrieved from https://www.esa.int

Geneletti, D. (2013). Multicriteria analysis for sustainability assessment: Concepts and case studies, *Environmental Impact Assessment Review*, 38, 23–34.

Glasson, J. and Therivel, R. (2012). *Introduction to Environmental Impact Assessment.* 4th edition, Routledge. https://doi.org/10.4324/9781315881218

Global Biodiversity Information Facility (GBIF) (2024). *Citizen Science Contributions to Biodiversity Monitoring.* Retrieved from https://www.gbif.org

Google AI (2018). Blog: *Machine Learning for Energy Efficiency.* Retrieved from https://ai.googleblog.com

Gorelick, N., Hancher, M., Dixon, M., Ilyushchenko, S., Thau, D. and Moore, R. (2017). Google Earth Engine: Planetary-scale geospatial analysis for everyone. *Remote Sensing of Environment*, 202, 18-27. https://doi.org/10.1016/j.rse.2017.06.031

Greenhouse Gas Protocol (GGP) (2024 December). *A Corporate Accounting and Reporting Standard.* Retrieved from https://ghgprotocol.org

Guisan A., Lausanne U.D., Thuiller, W. and Zimmermann, N.E. (2017). Habitat Suitability and Distribution Models with Applications in R. Cambridge University Press, https://doi.org/10.1017/9781139028271

Haklay, M. (2013). Citizen Science and Volunteered Geographic Information –overview and typology of participation. In Sui, D. Z., Elwood, S. and Goodchild, M. F. (eds.) Crowdsourcing Geographic Knowledge: Volunteered Geographic Information (VGI) in Theory and Practice. Springer, 105–122. DOI: 10.1007/978-94-007-4587-2_7

International Energy Agency (IEA). (2022). *Renewables 2022: Analysis and Forecast to 2027.* Retrieved from https://www.iea.org

IPBES (2019). *Global Assessment Report on Biodiversity and Ecosystem Services.* Retrieved from https://www.ipbes.net

IPCC (2022). *Climate Change 2022: Impacts, Adaptation, and Vulnerability.* Sixth Assessment Report, Retrieved from https://www.ipcc.ch

Jain, P., Coogan, S.C.P., Subramanian, S.G., Crowley, M., Taylor, S. and Flannigan, M.D. (2020). A review of machine learning applications in wildfire science and management. *International Journal of Wildland Fire*, 28(4), 301–313.

Jung, A., (2022). Components of ML. *In: Machine Learning: The Basics. Springer Nature Singapore*, Singapore, 19–56. https://doi.org/10.1007/978-981-16-8193-6_2.

Khaledian, Y. and Miller, B.A. (2020). Selecting appropriate machine learning methods for digital soil mapping. *Applied Mathematical Modelling*, 81, 401–418. https://doi.org/10.1016/j.apm.2019.12.016.

Kitchin, R. (2014). The Data Revolution: Big Data, Open Data, Data Infrastructures and Their Consequences. SAGE Publications.

Kumar, A. and Jha, S. (2021). Big data for environmental protection: Applications and case studies. *Sustainable Cities and Society*, 65, 102654.

Longley, P.A., Goodchild, M.F., Maguire, D.J. and Rhind, D.W. (2015). *Geographic Information Systems and Science.* 4th Edition, Wiley.

Mace, G.M., Barrett, M., Burgess, N.D., Cornell, S.E., Freeman, R., Grooten, M. and Purvis, A. (2018). Aiming higher to bend the curve of biodiversity loss. *Nature Sustainability*, 1(9), 448–451.

Maganathan, T., Senthilkumar, S. and Balakrishnan, V. (2020). Machine learning and data analytics for environmental science: A review, prospects and challenges. *IOP Conference Series: Materials Science and Engineering*, 955, 012107, DOI 10.1088/1757-899X/955/1/012107.

NASA. (2023). *Earth System Models.* Retrieved from https://www.nasa.gov.

National Renewable Energy Laboratory (NREL) (2023). *System Advisor Model (SAM).* Retrieved from https://sam.nrel.gov

Nativi, S., Mazzetti, P., Santoro, M., Papeschi, F., Craglia, M. and Ochiai, O. (2015). Big Data challenges in building the global earth observation system of systems. *Environmental Modelling & Software*, 68, 1–26. https://doi.org/10.1016/j.envsoft.2015.01.017

NOAA (2023). *National Centers for Environmental Information.* Retrieved from https://www.ncei.noaa.gov

Noble, B.F. (2000). Strengthening EIA through adaptive management: A systems perspective. *Environmental Impact Assessment Review*, 20(1), 97–111.

Pandey, D., Agrawal, M. and Pandey, J.S. (2011). *Carbon footprint: Current methods of estimation. Environmental Monitoring and Assessment*, 178(1–4), 135–160.

Papadopoulos, T. and Balta, M.E. (2022). Climate Change and big data analytics: Challenges and opportunities. *International Journal of Information Management*, 63(4), 102448. https://doi.org/10.1016/j.ijinfomgt.2021.102448

Salvati, L. and Carlucci, M. (2014). The economic and environmental performances of urban districts: Evidence from a metropolitan area. *Sustainable Cities and Society*, 10, 145-152.

Sun, A.Y. and Scanlon B.R. (2019). How can Big Data and machine learning benefit environment and water management: a survey of methods, applications, and future directions. *Environmental Research Letters*, 14(7), 073001. DOI 10.1088/1748-9326/ab1b7d

Turnhout, E., Dewulf, A. and Hulme, M. (2020). What does policy-relevant global environmental knowledge do? The cases of climate and biodiversity. *Current Opinion in Environmental Sustainability*, 42, 15–21.

UNDRR (2019). *Global Assessment Report on Disaster Risk Reduction.* United Nations Office for Disaster Risk Reduction.

UNEP (2021). *The Emissions Gap Report 2021.* Retrieved from https://www.unep.org

UNEP (2022). *Mapping Pollution with GIS Technology.* Retrieved from https://www.unep.org

UNEP-WCMC (2022). *Biodiversity and Ecosystem Services Data Tools.* Retrieved from https://www.unep-wcmc.org

United Nations (2023). Sustainable Development Goals Report. UNSDG Knowledge Platform. https://unstats.un.org/sdgs/report/2023/

Weersink, A., Fraser, E., Pannell, D., Duncan, E. and Rotz, S. (2018). Opportunities and challenges for big data in agricultural and environmental analysis. *Annual Review of Resource Economics*, 10, 19–37. https://doi.org/10.1146/annurev-resource-100516-053654

WHO (2021). *Climate Change and Health.* Retrieved from https://www.who.int

World Air Quality Index Project (2023). *Global Air Quality Data.* Retrieved from https://waqi.info

World Environment Situation Room (WESR) (2024 December), Harnessing Data to Impact on People, Countries and our Planet, https://wesr.unep.org/

Wulder, M.A., White, J.C., Goward, S.N., Masek, J.G., Irons, J.R., Herold, M., Cohen, W.B., Loveland, T.R. and Woodcock, C.E. (2008). Landsat continuity: Issues and opportunities for land cover monitoring. *Remote Sensing of Environment*, 112(3), 955–969. https://doi.org/10.1016/j.rse.2007.07.004

Zhang, Z. (2017). Environmental Data Analysis: Methods and Applications, DeGruyter Germany. https://doi.org/10.1515/9783110424904

6 Application of Artificial Intelligence-based Technologies for Evaluating a Healthcare Provider's Maturity in the Healthcare Worker Onboarding Process

Naga Ravi Shankar Mulagaleti[1], Arpana Kumari[2*] and Anuja Shukla[3]

The healthcare industry is increasingly adopting artificial intelligence. AI has successfully contributed to various functions in the healthcare industry because of its quick and consistent outcomes. The purpose of the present paper is to examine the usage of AI in the onboarding process for new employees. The application of AI-driven solutions, such as machine learning, natural language processing, and predictive analytics, to help evaluate and enhance the onboarding procedure for medical professionals is covered in this paper. It will examine how AI is being used to automate hiring processes and evaluate skills, training efficacy, and compliance tracking while ensuring staff retention. Healthcare businesses may use these tools to learn more about the maturity of onboarding, create possibilities for development, and help healthcare workers have a positive, customized experience. The results of the study exhibit that AI and machine learning are successful tools to enhance the effectiveness of the onboarding process of employees in the healthcare industry.

1. Introduction

The costs of employee turnover in healthcare have an impact on an organization's operating budget and patient care. UMASS Amherst's (2022) study on onboarding young workers states that even before the COVID-19 pandemic propelled the bargaining power of new labor market entrants, employers paid a high price for failed hires. In addition to the hiring and onboarding costs, there is lost productivity from empty positions and considerable time and frustration invested in recruiting and interviewing new workers. Also, labor force participation has been steadily declining since 1999, Pellegrini, Lawrence C. (2017), the turnover has only been exacerbated by COVID-19. COVID-19 had an overwhelming impact on the mental health of healthcare workers and brought the focus toward the current working conditions in healthcare systems, Donley, J. (2021b); Sharma et al., 2023). Leaders

[1] 0009-0000-1166-007X
[2] 0000-0001-8752-674X
[3] 0000-0002-3001-5962
* Corresponding author: arpanasngh15@gmail.com

DOI: 10.1201/9781003514282-6

in Healthcare organizations have understood the need to use technology to simplify employee experiences, in the areas of new employee onboarding, retention, and patient care.

Talent management includes an onboarding procedure that is vital for successful job performance, job satisfaction, and low employee turnover. Kumar and Pandey (2017) mentioned the significance of onboarding programs in measuring organizational fit and getting new hires fast and productive (Shufutinsky and Cox, 2019). Finding the right talent is important for a company, thus businesses and their HR departments spend money on hiring, recruiting, and relocating staff, but for some reason, they don't put as much effort into ensuring that these workers are properly onboarded. As a result, bright new workers with a lot of potential occasionally leave their new companies within months (Krasman, 2015).

On average, the expense of replacing an employee can range from 20 to 250 percent of that employee's annual salary (Byrnes and Cascio, 1984; Petrucelli, 2017; Shufutinsky and Cox, 2019) in addition to the lost institutional knowledge and cultural factors that cannot be easily calculated. Thus, it makes sense for employers to want to retain effective staff. Although studies are limited, the literature does show that there are direct and indirect links between onboarding and employee retention. Effective multi-stage employee onboarding helps assist employees to integrate into the organization and join the organization's culture (Kumari and Saharan, 2021), and when orientation and socialization processes are well planned, retention may be improved, Chapman (2009). Onboarding and socialization is a pivotal stage of the employer-to-employee relationship (Nobel, C. 2013) and leader-member exchange, and it is an excellent opportunity to foster new employee self-expression and innovation. Understanding the employee onboarding journey, bottlenecks in onboarding, and launching effective onboarding programs thus becomes critical. Healthcare employee onboarding is complex and requires factoring in multiple areas – demographics, specialty/expertise, organizational context, etc.

1.1 Objectives of the Study

1. To explore the usage of modern technologies such as Artificial Intelligence, specifically Machine Learning, to help organizations predict issues with their existing onboarding program(s), in the health care industry.
2. To provide suggestions for changes to the onboarding programs and improve these programs using data-driven insights.
3. To analyze historical data and identify patterns on how AI can predict and address potential challenges in onboarding for efficient and effective programs.

2. Review of Literature

Artificial Intelligence Methods used in Healthcare Employee Onboarding, based on the results of previous research, it was determined appropriate to use the Artificial

Intelligence technique of Machine Learning to improve the efficiency and efficacy of onboarding procedures in healthcare provider organizations Wiljer and Hakim, (2019). This research will outline the technical solution that encompasses data collection, preprocessing, predictive machine learning model development, causal analysis using causal inference techniques, training, and evaluating such models, generating suggestions using techniques such as cluster analysis, deployment of such models, and user interface design to display the insights and key metrics from the models. If textual data (e.g., feedback comments) is available, leverage Natural Language Processing (NLP) techniques such as advanced pretrained transformer models such as BERT (Bidirectional Encoder Representations from Transformers) or GPT (Generative Pre-trained Transformer) (A. Gillioz, et al., 2020) models can be customized based on finetuning tasks and that can be integrated into our classification/regression models.

3. Possible Solutions with the Applied ML Approach

3.1 Real-time Predictions

Real-time predictions integrate ML models into the onboarding workflow to provide real-time predictions of onboarding risks and personalized recommendations for each provider. Why is it important? Physicians, nurses, and medical professionals change organizations, facilities within an organization, roles, and responsibilities within a facility, and are re-hired into the organization, and accordingly, the onboarding workflows are different in the organization. These workflows include credentialing/re-credentialing, verification of existing or new certificates, re-looking at the access controls within the physical facilities such as badging/re-badging, health screen scheduling, tailored mandatory or optional professional training/learning paths depending on the role/state/county, and understanding of the various Standard Operating Procedures by the professional as part of their onboarding journey. Given the complexity of the onboarding process as described above, these workflows require real-time monitoring by fetching data from multiple technology systems, which include HR Information Systems, Credentialing systems, Identity Management systems, Electronic Health Records, Badging systems, etc. The advantage of this method of monitoring and predictive capabilities is to identify hotspots or process bottlenecks and accordingly address them through technology systems or process interventions including automation. This requires investments into Healthcare Informatics & Analytics 2.0 (Pramanik I., et al, 2020) in the HR organization.

3.2 Adaptive Onboarding

Adaptive Onboarding implements a dynamic onboarding process that adapts to individual provider needs and predicted risks, ensuring a more efficient and effective onboarding experience. This requires organizations to intimately understand the specific needs at an individual level, the needs of the job using the job description, and personalize the approach to support the incoming talent. This approach requires investment into the right technologies such as developing mobile applications, customized notifications based on onboarding progress and interests of the individual, online scheduling software to track both pre-boarding activities and post-onboarding activities such as 30-60-90-day plans, prioritized onboarding activities based on role, Generative AI Prompt Engineering, (Korzynski et al., 2023) based chat bots to retrieve document summaries, etc. for easier consumption of administrative information and tasks. While this approach provides significant caregiver experience advantages and a good onboarding experience for the individual, it requires significant technology awareness to build these technology systems, and the potential need to implement ethical AI guard rails in the organization in dealing with Generative AI technologies.

3.3 Data-driven Decision Support

Data-driven Decision Support provides data-driven insights and decision support tools (Mishra et al., 2024; Shukla et al., 2024a; Mishra et al., 2023) to onboarding managers using prescriptive analytics, allowing them to make informed decisions based on predictive modeling and recommendation generation. Due to a shortage in healthcare staffing, we have seen that onboarding new quality healthcare professionals has become expensive. To reduce these costs, prescriptive analytics analyzes health care data to recommend optimal changes to the onboarding experience (Nagpal et al., 2020; Rana et al., 2023). Some of the areas include optimized personalized training plans based on the professionals' skills/expertise, experience, and certification levels, optimized scheduling considering the availability and life situations of the professional, recommended health screening and credentialing sequences, analyze the risk of retention through simplified satisfaction surveys periodically during the onboarding process and prescribe the next best actions such as recommending mentorship programs, employee special interest group networks, recommend resource allocations to unblock the onboarding process such as licenses or accesses, suggest milestone-based learning progressions to ensure competency levels are reached for critical tasks in-time, suggest cost-effective strategies such as group training to ensure competency level or quality needs, recommend personalized benefit(s) to employees (Latorre, P., et al, 2024; Rakshit and Sharma, 2020), etc. in turn improving employee engagement and retention.

3.4 Continuous Monitoring and Updating

Monitoring Metrics

For a healthcare provider system that is multi-site or has both acute and ambulatory facilities, there are hundreds of complex scenarios that develop when the Machine Learning computing models are implemented. These include organizational policy changes, regulatory changes, population health situations such as epidemics or pandemics, healthcare services mix, etc. These changes impact the type of talent that is onboarded, the onboarding processes and task flows, and could potentially impact the model performance over time. To ensure the Machine learning models are working as designed, they must be monitored regularly. The f ollowing are some of the activities that can be placed to achieve the same.

Monitor model performance in real-time

Performance metrics for a machine learning model include accuracy, precision, recall, AUC-ROC, data drift in input data, concept drift, etc. Data drift can be detected by comparing statistical distributions of features in current data with the historical dataset and identifying any significant changes. This is possible if a major process intervention happened in the employee onboarding process, or any major technology modernization happened upstream. For instance, during COVID-19, many additional mandatory statutory steps related to medical staff health were introduced in the employee onboarding process, such as additional vaccination tracking, online platform technology training for remote/virtual care, training for COVID-19 sanitation compliance requirements, etc. These factors must be additionally added to the features of the predictive models to improve their accuracy.

Set up automated monitoring systems to track model predictions and system behavior

On performance metrics such as accuracy, precision, recall for classification, and Root Mean Square Error (RMSE) for regression, Machine Learning Operations (ML Ops) software pipelines can be built to frequently monitor the model performance comparing it with the baseline metrics (accuracy, precision, recall, F1-score) and detect any deviations from the same. This ML Ops software pipeline detects model deviations using the metric calculation component that calculates these relevant metrics on live data streams in production environments and performs data quality checks on input data for aspects such as missing values, schematic changes in the data, and outliers in values if any. An anomaly detection component can be built using statistical deviation analysis to identify outliers in metrics. Alternatively, Machine learning models for anomaly detection can also be built for large, unusual pattern detection.

Track onboarding success rates and identify deviations.

Some of the success rate metrics are laid out in the Key Metrics section below.

Establish alert mechanisms for significant deviations in onboarding success rates.

Ensure thresholds for alerts are set to detect when metrics deviate significantly from the expected values. These alerts could be delivered into system logs with classifications of type information or warning or error, severity high, medium, or low, and high severity errors can be relayed in real time as emails or through incident management systems to the operations personnel and through the reporting user interface to the end users of these insights.

4. Model Updating

The model performance monitoring drives actions into the model engineering process for fine-tuning the models based on the changed circumstances or assumptions. Following are some of the activities that can be put in place to achieve the same.

- **Implement periodic model retraining.** Issues found through the model performance monitoring or additional requirements gathered from HR business partners regarding changes to the employee onboarding process provide triggers to the process of retraining the model either with the updated dataset or feature engineering.
- **Schedule regular updates to retrain the model with new data.** Automation is key in monitoring the metrics for the model(s). As discussed earlier, ML Ops pipelines can be built to retrain the model(s) with new data.
- **Incorporate feedback mechanisms for continuous improvement.**
- **Collect user feedback on model suggestions and use it to enhance the model.**

4.1 User Interface

Dashboard Design

Key Metrics: Large provider systems have hundreds or thousands of caregivers joining them every month. They are multi-site and have various types of care that include acute, ambulatory, palliative, hospice, community, home, etc. Acute care deals with hospitals and in-patient services. Ambulatory care deals with clinics and out patient as well as testing services, Hospice and Palliative care deals with life-threatening illnesses and their long-term management. Community care is usually a holistic approach to healthcare that supports people's health in their communities by helping people with disabilities, mental illness, and nutrition, along with treatment of illnesses that include endemic, epidemics, and pandemics. Home care is for patients with special needs who cannot reach other centers of care outside their home and need sustained care including housekeeping, personal care, cooking, and money management, along with healthcare including oxygen supply, medication, etc.

Given the complexity of healthcare as laid out above, various roles support these various types of care. These include general physicians, specialists such as cardiologists or neurologists, family medicine doctors, allergists and immunologists, nutritionists, psychologists, etc. Similarly, nurses are of various types including Licensed Practical Nurse (LPN), Registered Nurse (RN), Advanced Practice Registered Nurse (APRN), Clinical Nurse Specialist (CNS), Travel nurses, and specialist nurses such as Oncology Nurses or Certified Registered Nurse Anesthetist (CRNA) or Psychiatric-mental health nurse practitioner (PHNP), etc. LPN nurses perform basic medical tasks such as checking vital signs or changing bandages. RNs provide hands-on patient care support to doctors in hospitals, medical offices, or nursing homes to provide treatment and administer nursing interventions. Travel nurses work on short-term contracts and are employed through staffing agencies. APRNs are RNs with advanced graduate-level education and are usually nurse practitioners, certified midwives, clinical nurse specialists, CRNAs, PHNPs, etc.

Many other supporting roles are onboarded for health administration, clinical support, and technology needs with a Healthcare Provider organization. Apart from the doctors and nurses as laid out above, there are shared services personnel such as HR or Recruitment, Finance or Compliance or Procurement, medical coders, medical assistants, dieticians, Lab technicians, medical record clerks, etc.

US Healthcare is highly regulated with various laws and regulations governing it. The major laws include Healthcare Insurance Portability and Accountability Act (HIPAA), The Health Information Technology for Economic and Clinical Health Act (HITECH), Emergency Medical Treatment and Labor Act (EMTALA), Affordable Care Act (ACA), Anti-Kickback Statute (AKBS) and Stark Law, The Patient Safety and Quality Improvement Act (PSQIA), False Claims Act, Medicare Access and CHIP Reauthorization Act, Medicare, Medicaid, etc. These laws mandate various compliance controls and assessments that need to be in place and reported to various government agencies such as the Centers for Disease Control and Prevention (CDC), Food and Drug Administration (FDA), Health Resources and Services Administration (HRSA), National Institute of Health (NIH), U.S. Public Health Service Commissioned Corps for public health, etc. Many of these controls and associated measurements start right from the time an employee is hired and onboarded to a Healthcare Provider organization.

There are also different types of employment within the organization. These include Full-time employees (traditional employment type with fixed work hours and fixed salary), Contract employees (contracted for a specific period), Leased employees (contracted with a staffing agency), Independent contractors (one who is not on a company's payroll), Contingent employee (a temporary job arrangement dependent on specific needs or periods of demand).

Given the diversity of medical practitioners' roles, the various compliance needs of US healthcare right from the hiring time, their type of employment with the Provider organization, and the measurement of their onboarding tasks, success becomes significantly important. When Machine Learning is employed to improve onboarding success, the measurement of pre- and post-conditions becomes even more significant. Below are a few key metrics that the organization can measure. The following is not an exhaustive list but is provided to give an idea of the significance of this measurement.

- **Display onboarding success rates** using measures such as Preboarding completion rate (completion of preboarding tasks), Preboarding engagement rate (percentage of new hires who actively interact with preboarding materials sent to them that includes emails, online forms, videos, etc.), Time to complete preboarding (Average time taken by new hires to finish preboarding process that is from offer acceptance to Day 1 at work), Preboarding feedback score (feedback from new hires on their experience with the preboarding process), Number of required tasks/documents submitted on time (tracking necessary compliance tasks or paperwork), Acceptance rate of employment offers (percentage of candidates who accepted the job offer after going through preboarding), Time to Productivity (how long it takes new hires to become productive and start contributing to the organization), Retention rate (calculate the Number of employees at the end of the measurement period divided by the number at the start), Turn-over rates (Measure the turnover rates for different groups or departments or roles of new hires, such as those who started in a particular year or location or facility or department or under a core leader), New-hire satisfaction (Survey new-hires at various intervals during their first 90 days or an year and learn about their pain points), Engagement rates for new employees, Job satisfaction of new employees, Training completion rate, Recruiting ROI (Cost to Company at recruitment and associated retention rate or employee productivity), employee morale, etc.

- **Design visually appealing dashboards showcasing key performance indicators.** It is essential to devise a preboarding and onboarding dashboard with various filters including state, county, facility, department, Job role, employment type, hire status, onboarding status, duration of employment, etc. This helps CHROs (Chief HR Officers), Recruitment specialists, Core leaders, and Chief Executives in the various facilities to track the preboarding and onboarding of candidates or employees in both strategic and tactical views. These dashboards are pivotal in identifying bottlenecks in onboarding and remediating them. It is also essential to track system readiness for onboarding that shows the status of provisioning the variously identified birthright accesses required for a particular employee role or type, such as in EMR, ERP, HRM, Credentialing, Badging, Physical security, and other systems. Machine Learning models can help identify hot spots or variations in targeted experience and help such leaders as mentioned above with decision support to configure/re-configure their onboarding programs based on insights developed through this process. Assessment of the onboarding programs in organizations that operate at scale is complex and can only be done effectively using machine learning algorithms that sift through various data elements that are essential to track the overall onboarding processes and is especially helpful for provider organizational leadership to institute strategic business and technology programs such as automating or simplifying credentialing processes, re-wiring employee outreach and engagement programs, optimize the channels and content of communication with candidates and new hires, etc.
- **Showcase model predictions and recommendations.** This paper detailed extensively the type of machine learning models that can be built to track the onboarding processes of the organization. This includes near real-time

predictive models showcasing the health of the various onboarding states and predicting any critical hotspots that need to be remediated such as access delays or health screening delays, etc. NLP and Generative AI-based models to process the vast amount of unstructured and textual knowledge (Shukla et al., 2024b) base that exists within the organization, including professional training documentation, policies, standard operating procedures, HR and operational process documentation, user manuals, etc. These models help identify redundancy in such documentation, help summarize and standardize the format of such documentation for easier human consumption, analyze sentiments from candidate or employee survey responses, and accordingly suggest changes to the associated documentation. They also help identify grammatical and other errors in new and existing documentation, in turn, help correct them in time.

- **Integrate charts and visualizations illustrating model suggestions.** Standard data-based reports can be annotated with an insight summary with suggestions or recommendations coming from the Machine Learning models.

4.2 User Interaction

- **Feedback Mechanism.** Inferences drawn or suggestions from Machine Learning models have broad implications on the decisions, including the Next Best Action (NBA), taken by the organization's Leadership. While historical data is used to provide such recommendations, their correctness is limited by the accuracy of the models and sometimes need additional parameters or features built into them to improve their accuracy. To enable this process, a feedback mechanism needs to be built where the consumer of these insights can annotate their feedback on the specific insights as well as more generically.
- **Implement user-friendly interfaces for users to submit feedback on model suggestions.** Mechanisms such as surveys or chatbots can be built into the user-facing interfaces to provide feedback on the suggestions/recommendations/insights provided by the Machine Learning models.
- **Incorporate user-driven adjustments into the system.** The user-collected feedback can be relayed back to the business analysts or data scientists building the models so they can triage the feedback, optionally speak to the feedback initiator, identify patterns of feedback, and harden the engineering process of these models.
- **Use user feedback to adapt and improve the model over time.** The process to improve the models based on the feedback may involve multiple steps that include identifying false positives or false negatives in the feedback, sending further surveys or interviewing users to get more precise information on the issues, and performing deeper analysis on the models that include identifying additional data attributes that may be required, further cleaning the existing or new data sets, if new data sets are not necessary then better feature engineering by creating new features from existing data, and data

augmentation by generating additional data variations to increase training data diversity.

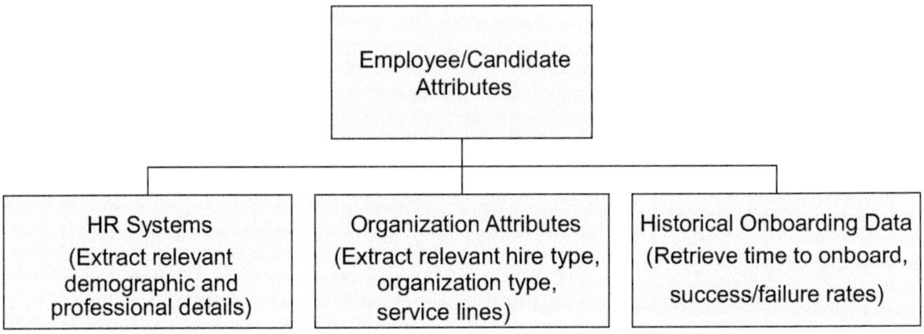

Fig. 1 Flow chart for Data Collection and Preprocessing.
Source: Author's own

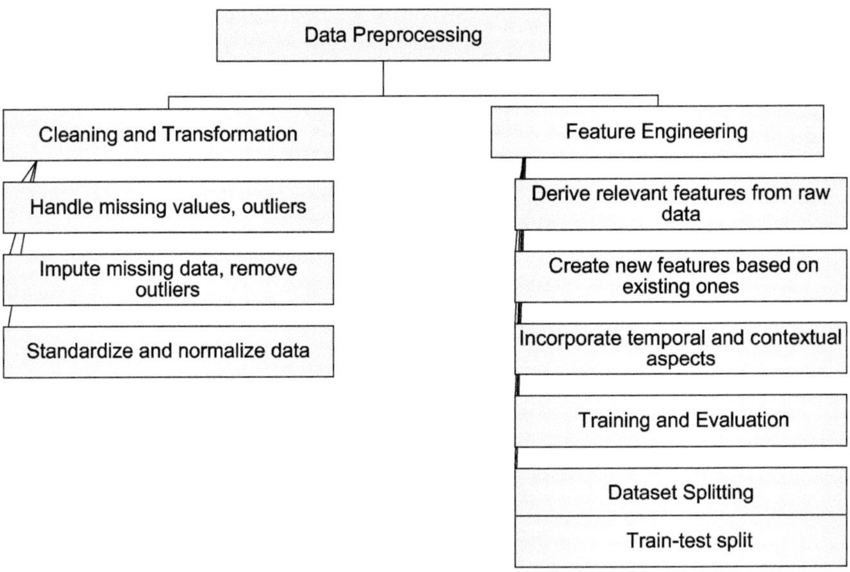

Fig. 2 Flow chart for Model selection.
Source: Author's own

Overview

Through the literature review, we can conclude that Artificial Intelligence techniques can be used to identify Healthcare provider onboarding issues. Leveraging the insights drawn from this analysis can significantly reduce the time and effort needed by a Healthcare Provider organization to respond to and resolve these issues and arrest further caregiver attrition due to onboarding experience issues.

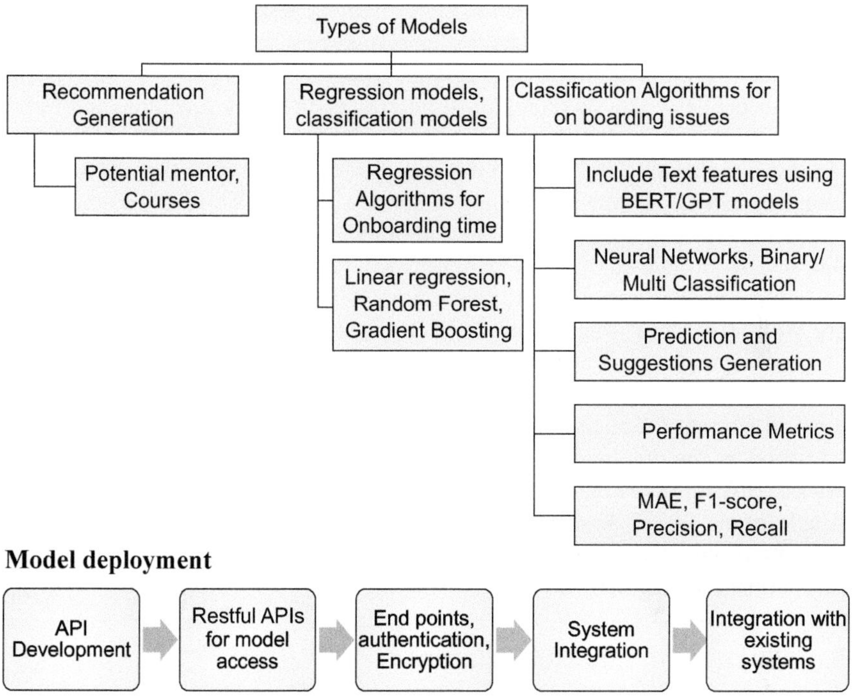

Fig. 3 Flow chart for model development.
Source: Author's own

In this section, we present what the literature has been telling us about applying technology to these problems.

Digital Health and Health Informatics

Health Informatics, along with Digital health, has been a rapidly growing discipline in the sustainability of health service provision in the last 2 decades. The COVID-19 pandemic has seen an unprecedented swift adoption of digital health technologies, Brommeyer et al. (2022). Artificial Intelligence in Healthcare will play a critical role in turning around the Industry towards better efficiencies and improving patient and provider employee experiences (Sharma et al. 2024). Safavi and Kalis (2019) estimate that "AI applications could create up to $150 billion in annual savings for U.S. healthcare by 2026." The different branches or subfields of AI in Healthcare are comprised of (1) RBES (rule-based expert systems), (2) RPA (robotic process automation), (3) NLP (natural language processing), (4) ML (machine learning), and (5) DL (deep learning) Wiljer and Hakim, (2019). Healthcare Providers can leverage the Machine Learning (ML) subfield to better understand the challenges in their onboarding programs and improve the experience for new employees.

Healthcare Employee Onboarding Objectives

Tomaskovic-Devey (2022) interviewed many workforce specialists and identified typical mistakes employers make, sometimes repeatedly. They found that employers must focus on their (Al-Alawi et al., 2021) training practices, increase the speed at which new hires become productive team members, and reduce the high dollar and emotional costs of turnover from failed hires.

The primary aim of onboarding new hires involves forming long-term, high-performance relationships between them and their respective organizations (Kumar and Pandey, 2017). Some workers may decide to leave the company too soon because of the way onboarding is done, which has a significant impact on both employee performance and the durability of those connections in the long run. A well-formulated, multi-step process will help employees to be accepted as part of the organizational culture of the firm. When orientation and socialization are well-structured, then retention is likely to improve (Chapman, 2009). Onboarding and socialization form an essential stage that sets up the employer-employee relationship (Nobel, C. 2013) and leader-member exchange, creating an ideal context for eliciting innovative expressions from new hires. This early transition in the employee lifecycle, particularly during onboarding, has been associated with immediate high performance and improved employee retention (Cable et al., 2013). Knowing this importance is crucial as the baby boomer generation approaches retirement (Chapman, 2009) and millennials emerge as the predominant demographic in the workforce.

5. Discussion and Conclusion

5.1 Discussion

There is limited research available on using Artificial Intelligence in healthcare employee onboarding. With the increase in the use of Artificial Intelligence in various areas of Healthcare, this chapter attempts to provide a framework for healthcare provider organizations to use Machine Learning models that can suggest Healthcare Human Resources and hospital leadership to better construct their onboarding programs. This research provides a foundational framework for efforts to increase understanding of the effectiveness of onboarding programs in healthcare.

Effective multi-stage employee onboarding helps employees integrate into the organization and its culture. When orientation and socialization processes are well planned, retention may improve (Chapman, 2009). The model helps organizations identify budgeting priorities based on the suggested areas of improvement in the onboarding process.

Healthcare provider employee onboarding analysis requires understanding the various employees/candidates, provider attributes, and their influence on the effectiveness of employee onboarding programs. For instance, *Candidate Attributes* that are relevant for recruitment and Onboarding include demographic information such as Race, Ethnicity, Gender Identity, Sexual Identity, Age Category (Millennials, Gen X, etc.), Candidate health profile, Disability status, Socioeconomic standing, Location, Language; professional information such as Experience, Qualification, Specialty, Certificate of Insurance, etc.

Provider organization Attributes that are considered for recruitment and Onboarding include Hire Type (Full time, Contingent, Traveler, etc.), Organization Type (Hospital, Clinic, Laboratory, Pharmacy, Home care, Virtual healthcare), Organization sub-type (Not for Profit, For Profit), Organization category (Faith-based, Secular), Service Lines (departments), regulatory constraints, Average Current Onboarding Time (4–6 m, 2–4 m, 1–2 m), etc.

Conclusion

It has been established through previous research that effective employee onboarding is a critical area for employee satisfaction, workforce retention, and productivity. It has also been established that there is a need for healthcare providers to better understand the effectiveness of their onboarding programs, particularly in light of the extensive workforce crisis following COVID-19, and to improve the effectiveness of such programs using machine learning-based Artificial Intelligence insights.

References

Al-Alawi, M., Naureen, E.I., AlAlawi and Naser Al-Hadad, A.A. "The Role of Artificial Intelligence in Recruitment Process Decision-Making," *2021 International Conference on Decision Aid Sciences and Application (DASA)*, Sakheer, Bahrain, 2021, pp. 197–203, https://doi.org/10.1109/DASA53625.2021.9682320

Brommeyer, M., Whittaker, M., Mackay, M., Ng, F. and Liang, Z. (2022). Building health service management workforce capacity in the era of health informatics and digital health – A scoping review. *International Journal of Medical Informatics, 169*, 104909. https://doi.org/10.1016/j.ijmedinf.2022.104909

Byrnes, J.F. and Cascio, W.F. (1984). Costing Human Resources: The financial impact of behavior in organizations. *Academy of Management Review*, *9*(2), 370. https://doi.org/10.2307/258455

Cable, D.M., Gino, F. and Staats, B.R. (2013). Reinventing employee onboarding. *MIT Sloan Management Review*, 54(3), 23. Reinventing employee onboarding.

Chapman, C. (2009), "*Retention begins before day one: orientation and socialization in libraries*", New Library World, vol. 110 No. 3/4, pp. 122–135. https://doi.org/10.1108/03074800910941329

Donley, J. (2021b). The Impact of Work Environment on Job Satisfaction. *Nurse Leader, 19*(6), 585–589. https://doi.org/10.1016/j.mnl.2021.08.009. The experience people have at work impacts their personal well-being as well as job satisfaction.

Gillioz, J., Casas, E., Mugellini and O.A. Khaled, "Overview of the Transformer-based Models for NLP Tasks," *2020 15th Conference on Computer Science and Information Systems (FedCSIS)*, Sofia, Bulgaria, 2020, pp. 179–183, doi: 10.15439/2020F20. https://ieeexplore.ieee.org/stamp/stamp.jsp?tp=&arnumber=9222960&isnumber=9222848

Korzynski, P., Mazurek, G., Krzypkowska, P. and Kurasinski, A. (2023). Artificial intelligence prompt engineering as a new digital competence: Analysis of generative AI technologies such as ChatGPT. *Entrepreneurial Business and Economics Review*, 11(3), 25–37.

Krasman, M. (2015). Three Must-Have Onboarding Elements for New and Relocated employees. *Employment Relations Today*, *42*(2), 9–14. https://doi.org/10.1002/ert.21493

Kumar, N.V. and Pandey, N.S. (2017). New employee onboarding process in an organization. *International Journal of Engineering Development and Research*, *5*(1), 198–206. https://www.ijedr.org/download.php?file=IJEDR1701032.pdf

Kumari, A. and Saharan, T. (2021). Organisational culture as a stimulant to knowledge management practices: an empirical analysis on Indian real estate companies. *International Journal of Knowledge and Learning, 14*(4), 360–384.

Latorre, P., López-Ospina, H., Maldonado, S., Guevara, C.A. and Pérez, J. (2024). Designing employee benefits to optimize turnover: A prescriptive analytics approach. *Computers & Industrial Engineering*, [online] 197, p.110582. https://doi.org/10.1016/j.cie.2024.110582

Mishra, A. and Shukla, A. (2023). Gyan fresh: digital transformation of dairy business with resilience and technology innovation. *FIIB Business Review*, 12(1), 20–30.

Mishra, A., Malik, N. and Shukla, A. (2024). Decoding individual motivations and responses to misinformation: insights from thematic analysis. *Journal of Research in Interactive Marketing*.

Nagpal, S., Athavale, V.A., Saini, A.K. and Sharma, R. (2020, November). Indian health care system is ready to fight against COVID-19 a machine learning tool for forecast the number of beds. In *2020 Sixth International Conference on Parallel, Distributed and Grid Computing (PDGC)* (pp. 61–65). IEEE.

Nobel, C. (2013). First minutes are critical in new-employee orientation. *Harvard Business School Working Knowledge. https://hbswk.hbs.edu/item/first-minutes-are-critical-in-new-employee-orientation*

Nunkoo, R., Shukla, A. and Mishra, A. (2024). The role of temporal distance, conspicuous consumption, and multisensory interfaces on consumer judgments in metaverse. *Journal of Retailing and Consumer Services*, 81, 104037.

Onboarding Young Workers in a Post-Pandemic World | Center for Employment Equity | UMass Amherst. (n.d.). https://www.umass.edu/employmentequity/onboarding-young-workers-post-pandemic-world

Pellegrini and Lawrence C., "Labor Force Participation, Disability, and Implications for Healthcare Utilization and Employment" (2017). Doctoral Dissertations. 897. https://doi.org/10.7275/9466050.0https://scholarworks.umass.edu/dissertations_2/897

Petrucelli, T. (2017). Winning the "cat-and-mouse game" of retaining millennial talent. *Strategic Hr Review, 16*(1), 42–44. https://doi.org/10.1108/shr-08-2016-0078

Pramanik, I., Lau, R.Y.K., Azad, A.K., Hossain, S., Hossain, K. and Karmaker, B.K. (2020). Healthcare Informatics and Analytics in Big Data. *Expert Systems with Applications*, 152(1), p.113388. https://doi.org/10.1016/j.eswa.2020.113388

Rakshit, P. and Sharma, R. (2020). Building Healthcare 4.0 with Smart Workforce. In *ICRMAT* (pp. 91–94).

Sharma, R., Harish, V., Saini, A.K. and Rana, G. (2023). "Smart Healthcare: Harnessing IoT Technologies for Efficient Medical Services," *2023 Seventh International Conference on Image Information Processing (ICIIP)*, Solan, India, pp. 254–260, doi: 10.1109/ICIIP61524.2023.10537775

Sharma, R., Harish, V. and Rana, G. (2024). Navigating Risk in the Age of Artificial Intelligence: Assessing And Identifying Risks With AI Strategies. *Educational Administration: Theory and Practice*, *30*(4), 2130–2140.

Safavi, K., Kalis, B. How AI can Change the Future of Health Care.Harv. Bus. Rev.2019. p. 1.

Shufutinsky, A. and Cox, R. (2019). Losing Talent on Day One: Onboarding Millennial Employees in Health Care Organizations. *ResearchGate*. https://www.researchgate.net/profile/Anton-Shufutinsky-3/publication/333745337

Shukla, A., Jain, E. and Kumari, A. (2024a). Enhancing Organizational Performance in the Green Metaverse: An Analysis of Artificial Intelligence and Knowledge Management. In *Green Metaverse for Greener Economies* (pp. 92–108). CRC Press.

Shukla, A., Mishra, A., Rana, N.P. and Banerjee, S. (2024b). The future of metaverse adoption: A behavioral reasoning perspective with a text-mining approach. *Journal of Consumer Behaviour*.

Tomaskovic-Devey, D. (2022, May 12). The key to retaining young workers? better onboarding. *Harvard Business Review*. https://hbr.org/2022/05/the-key-to-retaining-young-workers-better-onboarding

Wiljer, D. and Hakim, Z. (2019). Developing an Artificial Intelligence–Enabled Health Care Practice: Rewiring Health Care Professions for Better Care. *Journal of Medical Imaging and Radiation Sciences*, *50*(4), S8–S14. https://doi.org/10.1016/j.jmir.2019.09.010

7 Utilizing Analytic Hierarchy Process (AHP) to Prioritize Factors Impacting Financial Inclusion for Economic Advancement

Shuchita Singh[1*] and Nidhi Srivastava[2]

Numerous research endeavors have delved into the nexus between financial inclusion and economic growth, underscoring the pivotal role of fintech firms in enhancing inclusion. Since 2010, following the acknowledgment of the importance of financial inclusion (FI) by G20 leadership, there has been a discernible upswing in worldwide attention from regulators and policymakers. According to the World Bank (2021) update of this database, derived from extensive surveys encompassing around 128,000 adults across 123 economies during the COVID-19 pandemic, it offers updated metrics covering formal and informal financial services, along with digital transactions. However, the sluggish pace of user adoption begs the question: are the contributing factors adequately elucidated and prioritized? To address this, our study identifies, prioritizes, and proposes an index for practical application in both corporate and academic spheres. Drawing upon a literature review and insights gleaned from discussions with banking and finance professionals, we identify nine pertinent factors, which are further analyzed using the Analytical Hierarchical Process (AHP). Through AHP analysis we discerned the relative importance of hierarchical factors using the Consistency Ratio (CR). Our findings serve as a benchmark for institutions, facilitating self-evaluation and strategic alignment with fintech for enhanced financial inclusion.

1. Introduction

Financial inclusion is represented by access and usage of financial services by individuals and businesses at affordable costs. According to the World Bank, "Financial inclusion means that individuals and businesses have access to useful and affordable financial products and services that meet their needs – transactions, payments, savings, credit and insurance – delivered responsibly and sustainably." Ensuring higher finances has become a priority as a fundamental pillar of economic growth. At the G20 leadership summit, in 2010, member nations accepted

[1] 0000-0003-0691-026X
[2] 0000-0002-3350-0728
* Corresponding author: shuchita.singh@glbimr.org

DOI: 10.1201/9781003514282-7

the importance of financial inclusion (FI), and then onwards there has been a remarkable rise at a global level. However, it is noteworthy that the popularity of digital finance greatly depends upon the cyber security structure in an economy. It may be a possibility that the financial development in an economy is sufficient, but sufficient financial inclusion is not there. Regulators and policymakers have shown interest in research inputs given by many researchers in the field of Financial Inclusion (FI). According to studies, digital finance has been a catalyst to improve the accessibility of financial services to foster economic growth. Although the fundamental goal of FI is to enable all segments of society to have easy access to financial services and products, it is still in many developing economies. Financial Inclusion empowers individuals to seize development opportunities (Ozili, 2020a). Despite great global efforts to enhance FI, significant disparities persist among developing regions (Demirguc-Kunt, Klapper, Singer, and Van Oudheusden, 2015; Lyons and Kass-Hanna, 2021). There is a wealth of literature on the topics affecting financial inclusion, which might shed light on the reasons for inconsistencies. Digital finance, income level, socio economic environment, inflation, and internet usage are identified as major factors as suggested by (Mohammad O. Al-Smadi, 2022). In their 2021 study, Gálvez-Sánchez et al. conducted a bibliometric analysis to identify research trends, revealing a growing interest in promoting financial inclusion, especially through digital financial technologies like Fintech. The 2021 iteration of the Global Findex Database, drawn from comprehensive surveys involving approximately 128,000 adults across 123 economies amid the COVID-19 pandemic, highlights the role of digital finance for financial inclusion. This chapter covers a wide spectrum of countries and regions, including Latin America and the Caribbean, India, Bangladesh, Spain, the United Arab Emirates (UAE), the Middle East and North Africa (MENA) region, emerging economies, the United Kingdom (UK), South Africa, Ghana, China, Europe, and Sub-Saharan Africa. This geographical diversity highlights the global significance of financial technology (fintech) and digital financial inclusion, demonstrating that these subjects are pertinent and meaningful to both developed and developing nations worldwide. Researchers delve into regional intricacies, regulatory frameworks, socio-economic landscapes, and cultural influences, contributing to a thorough comprehension of fintech's impact on financial inclusion in varied contexts.

Digital finance is widely accepted by fintech companies and the role of fintech companies can't be ignored towards financial inclusion. Fintech, short for financial technology, is the integration of technology into financial services. Fintech companies are innovative intermediaries using technology to transform business models, operations, and service delivery in the financial sector. The inception of fintech dates back to the early 1990s, coinciding with the rise of the internet. The internet has been instrumental in driving the growth of fintech, offering opportunities to improve transparency, accessibility, and affordability of financial services. Fintech poses challenges to traditional financial institutions like banks and insurers, driving them to adapt to changing landscapes. One key factor contributing to fintech's expansion is its regulatory environment, which allows flexibility for experimentation within regulatory frameworks, fostering the creation of novel financial products and services. Seven out of the seventeen Sustainable Development Goals are thought to be significantly facilitated by financial inclusion. Among account users, there is a strong demand for a wide variety of financial services. Many would like to obtain credit, insurance, and savings products. Digital financial services delivered via mobile phones offer a particularly strong potential for expanded access to and use of accounts, as well as other financial services. Some of the most exciting innovations in digital financial services are coming from Fintech, leveraging technology to provide low-cost, scalable, and innovative financial services both directly and indirectly. Much of Fintech is still uncharted territory, and commencing our understanding of its impact on financial inclusion is incredibly important for guiding how it can be nurtured to best help the poor and the unbanked.

Academic interest in fintech has surged across various disciplines, particularly in understanding its role in promoting financial inclusion. Recent research highlights the transformative potential of digital currencies in enhancing financial accessibility, as well as the relationship between financial inclusion and economic growth. Scholarly investigations have also focused on related areas such as microfinance and the impact of information and communication technology (ICT) on financial inclusion efforts. Technological studies have examined the influence of digital innovation and fintech on banking operations. However, there remains a gap in the comprehensive literature review regarding fintech's impact on financial inclusion. Along with the study of the role of fintech for financial inclusion, the synthesis of existing literature underscores the significance of socio-economic, technological, and policy-related factors in shaping financial inclusion. It contributes to advancing financial inclusion theory by examining various dimensions of access to financial services and the role of fintech in addressing barriers to inclusion.

1.1 Structure of Chapter

The current chapter seeks to expand our understanding of how various factors like digital finance, cyber security, Internet usage, Income level, Inflation, Socio economic Environment, Service trust, and Mindset affect financial inclusion and finally the economic growth of an economy. Therefore, this chapter seeks to probe the factors affecting financial inclusion.

This chapter is structured in various sections. Section I sets the introduction, as stated above.

Section I is followed by Section II, Review of Literature, and further followed by Section III, which covers Research Objectives.

Section IV covers Research Methodology and Section V covers Data Collection and Analysis.

Finally, the conclusion, limitation, novelty, and scope of the study are mentioned in section VI.

The Analytic Hierarchy Process (AHP) has been used as the primary analytical technique for arranging the sequence of various factors and structuring the research further. AHP is a widely used research tool, developed by (Saaty, 1980). The pairwise comparison matrix of independent factors has been used to construct the questionnaire, to be served to experts in areas of finance and banking.

2. Review of Literature

Before embarking on an investigation path to construct the current chapter, previous research is evaluated. In this case, various studies cover a wide range of factors, reflecting the multifaceted nature of financial technology (fintech), and digital financial inclusion research was considered alongside other studies that have primarily highlighted technological, economic, and behavioral reasons for promoting financial inclusion in any economy. These constructs include economic indicators such as income per capita and income inequality. Also, educational factors, characteristics of the financial sector, the cost of financial services, the availability and accessibility of financial services, along with internet and cellular network coverage, are critical dimensions. Legal frameworks, regulatory environments, and trade-offs between financial inclusion and profitability also feature prominently. In some studies, formal account ownership, savings, and borrowing, are specifically attributed to government-to-person (G2P) and business-to-person (B2P). According to Demirguc-Kunt et al. (2015), individuals' engagement with the financial system enhances their capacity to initiate, manage, and navigate financial access to them. As a result, international organizations such as the World Bank have urged global policymakers to develop strategies that improve financial inclusion, thereby promoting sustainable economic growth and reducing poverty (Rana et al., 2024). In support of this, Ozili (2020b) has also proposed the theory of public good of financial inclusion, which argues that financial services should be universally accessible for all sectors of society. He also suggested that higher financial inclusion will promote the engagement of the public in economic progress, also enabling them to harness the benefits of development, Ozili (2020a). Ghosh (2016) investigated the effects of mobile penetration on economic growth in major Indian states during 2001–2012.

Research shows that an absence of a developed financial system is correlated with both income inequality and reduced economic expansion (Anand et al., 2012; Beck T., Demirguc-Kunt A., and Peria M.S.M., 2007; Sarma, M., Pais J., 2011). Scholars such as Galor et al. (1993) and Honohan (2004) have proposed that progress in financial inclusion shows that all stakeholders within the economic framework can readily access formal financial services, including bank deposits, credit, investment, and insurance. Behavioral aspects, such as intentions, social influence, trust, usability, and user behavior in fintech for financial inclusion are central to several investigations. Furthermore, variables related to infrastructure access, including bank and digital infrastructure, are considered by Khera et al., (2021). Additionally, the availability and accessibility of financial services, along with internet and cellular network coverage, are critical dimensions (Shen et al., 2021). The cost of financial services, legal and regulatory framework was explored in detail by Wang et al. (2021) and Ozilii et al. (2023). Alesina et al. (2016) have studied that India has the world's highest rate of account inactivity according to the 2017 survey. Furthermore, only around 39 percent of the survey participants reported engaging in domestic remittance transactions via financial institutions, Odhiambo, and Nicholas (2019). In 2020, Mhlanga's study explains the influence of artificial intelligence (AI) on digital financial inclusion and supports broader AI adoption to support the engagement of marginalized communities. In some studies, formal account ownership, savings, and borrowing, specifically attributed to government-to-person (G2P) and business-to-person (B2P) Niankara and Islam, (2023) electronic transfers, are scrutinized. In 2022, Demir et al. examined the complex relationship between FinTech, financial inclusion, and income inequality across multiple countries, highlighting the crucial role of fiscal policies alongside financial inclusion initiatives. Yang and Zhang (2020) scrutinized the relationships between various measures of financial inclusion and their drivers. Senyo and Osabutey's (2020) research focused on the innovation of mobile money within FinTech, identifying key factors influencing its adoption and providing insights relevant to enhancing financial inclusion. Additionally, the availability and accessibility of financial services, along with internet and cellular network coverage, are critical dimensions, Gershenson et al. (2021). Hasan et al. (2022) and Gabor et al. (2017) investigated the effect of financial knowledge on rural consumers' access to financial technology services, offering practical insights into rural finance and the role of FinTech in economic development. Koomson, I., Villano, R., and Hadley, D. (2020) explored the concept of poverty and per capita income to study the socio economic aspect of financial inclusion. Demirgüç-Kunt et al. (2020) emphasized the World Bank's initiatives to advance financial inclusion through the Global Findex database, a comprehensive tool that monitors adults' access to financial services and their use of financial technology. Additionally, Arner et al. (2020) highlighted FinTech's crucial role in enhancing financial inclusion and promoting sustainable, balanced development. They identified four essential pillars to leverage this potential, including digital identity and electronic payment systems. The availability and accessibility of financial services, along with internet and cellular network coverage, are critical dimensions, Gallego-Losada et al., (2023). Van et al. (2021) studied the role of economic factors towards financial inclusion in emerging markets. Educational factors, characteristics of the financial sector, and the cost of financial services are interrelated, as studied by Garg et al. (2023). Additionally, variables encompass socio-demographic factors, such as age, gender, location, literacy levels, income, and knowledge of financial services (Lagna and Ravishankar, 2022) and Hasan, M. et al. (2022), Okello Candiya Bongomin et al. (2018)). This diverse array of variables allows researchers to comprehensively explore the intricate dynamics among each one of them. Tram et al. (2023) used to construct new indices to study the various factors impacting financial inclusion. Together, these studies provide a thorough understanding of the complex interplay between financial inclusion, technology, and development, offering valuable insights for policymakers, researchers, and financial industry practitioners.

2.1 Research Objectives

Research objectives clarify the purpose of a study, contribute to the overall coherence and rigor of the study, and contribute to its overall coherence. Defining a research objective is an important step in developing a research plan. To give the right direction and structure to the chapter, after chapter. The research objectives are defined as follows:

RO1. To determine the hierarchy among various factors.
RO2. Classify the factors based on their driving and dependence characteristics, and identify the most impactful ones.
RO3. To suggest that business organizations set up a conducive environment for higher financial and inclusion-driven economic growth.

3. Research Methodology

The research methodologies utilized in previous studies demonstrate a broad range of analytical approaches. Numerous studies employ quantitative methods like regression analysis to examine the connections between various aspects of financial inclusion and their determinants. Qualitative methods are also embraced, with in-depth interviews serving as a primary data collection tool in certain projects. Bibliometric analyses are employed to dissect the intellectual structure of research streams, while cross-sectional panel data analyses and principal component analysis (PCA) are utilized to develop new indices. Confirmatory factor analysis (CFA) and structural equation modeling (SEM) are applied to evaluate relationships between variables. Spatial data and techniques, as well as a generalized method of moments (GMM), are employed for specific investigations. Discussions and desktop research facilitate broader explorations, while statistical techniques such as quantile regression and logistic regression uncover nuanced insights in this multifaceted field of research. These varied methodologies mirror the intricate and interdisciplinary nature of studies on financial technology and digital financial inclusion.

Primarily, in the present chapter, before conducting the AHP analysis, Bibliometric analyses are employed to dissect the intellectual structure of research streams and identification of variables, Maia et al. (2019). Liu et al. (2022) suggested techniques to find out the interrelationship among identified variables. Ben Khelifa et al. (2022) studied how to construct new indices using the variables identified in bibliometric analysis. Bibliometric analyses are employed to dissect the intellectual structure of research streams, Gálvez-Sánchez et al. (2021). After conducting bibliometric analysis, the systematic review is done using the databases of Scopus and Web of Science (WOS) databases. The selection of the Scopus and WOS database for this study was done to extend the scope of previous research with similar objectives. The search was refined further for Article or Review document types and restricted to the English language. The time frame was from 1980 to 2023. From the extensive repositories of the Scopus and Web of Sciences databases, a total of 4360 research papers. Initially, a total of ninety-seven documents (97) were acquired through screening, signifying their relevance to our research topic. However, these identified studies were further narrowed to 45 as these were inclined to the objective of identifying variables that ensure economic rigor by promoting financial inclusion. TCCM (Theory, Context, Characteristics, and Method) was also referred to for a deep understanding of the theories, contextualized with countries and services, highlighting the key variables and methods used by leading researchers across the globe in Fintech-driven financial inclusion.

As the research objectives of the study revolve around prioritization and identification of the most impactful factor toward sustainable economic growth, the Analytic Hierarchy Process is considered an appropriate statistical tool for the study. AHP is one of the multiple criteria decision-making methods that were originally developed by Prof. Thomas L. Saaty (1977). It is a mathematical model that generates a weighting value. This method is designed to reflect how people think. An AHP questionnaire is developed by the researchers to seek the opinion of experts in the field of finance and banking. The AHP Procedure is applied in the following steps:

Step 1: Identification of important competencies was done based on a review of literature and discussions held with experts.

Step 2: Data was collected from experts in a qualitative questionnaire's pairwise comparison matrix of variables.

Step 3: The pair-wise comparison of various factors is generated into a square matrix.

Step 4: The Normalized Principal Eigen Vectors (weights) of factors/competencies were computed.

Step 5: The Consistency Index (CI) was calculated to describe the consistency of respondents.

Step 6: Relative Index and Cumulative Relative Index are developed to assign the percentages to factors under consideration.

Step 7: Further, a Pareto chart is prepared to design an index to propagate accelerated economic growth.

Step 8: To identify the driving and dependence characteristics of various variables, MIC MAC analysis was also run.

3.1 Data Collection and Analysis

To achieve the objective of the study, the primary data was collected from 10 respondents, i.e. Finance and Banking professionals from various corporates (public and private) located in north and west India. Particularly the respondents

Table 1 Pairwise Comparison Matrix of Factors Affecting Financial Inclusion.

	Financial Inclusion	*Digital Finance*	*Income Level*	*Inflation*	*Internet Usage*	*Socio Economic Environment*	*Service Trust*	*Cybersecurity*	*Mind set*
Financial Inclusion	**1**	2.49	2.50	2.00	2.88	2.00	2.00	2.65	5.00
Digital Finance	0.40	**1**	2.41	1.64	2.00	2.00	2.00	4.37	5.00
Income Level	0.40	0.42	**1**	3.08	1.88	2.00	2.00	3.00	3.17
Inflation	0.50	0.61	0.32	**1**	2.45	2.00	2.00	2.18	3.25
Internet Usage	0.35	0.50	0.53	0.41	**1**	2.00	2.00	4.65	4.36
Socio-Economic Environment	0.50	0.50	0.50	0.50	0.50	**1**	2.00	3.96	4.66
Service Trust	0.50	0.50	0.50	0.50	0.50	0.50	**1**	2.41	4.16
Cybersecurity	0.38	0.23	0.33	0.46	0.22	0.25	0.42	**1**	4.54
Mindset	0.20	0.20	0.32	0.31	0.23	0.21	0.24	0.22	**1**
Total	**4.23**	**6.45**	**8.41**	**9.89**	**11.65**	**11.97**	**13.66**	**24.43**	**35.14**

Source: Author's Compilation

Table 2 Normalization Matrix of Factors Affecting Financial Inclusion.

	Financial Inclusion	*Digital Finance*	*Income Level*	*Inflation*	*Internet Usage*	*Socio Economic Environment*	*Service Trust*	*Cybersecurity*	*Mind set*	*Total*	*Normalized Principal Eigenvector (average)*
Financial Inclusion	0.24	0.39	0.3	0.2	0.25	0.17	0.15	0.11	0.14	1.95	0.22
Digital Finance	0.09	0.16	0.29	0.17	0.17	0.17	0.15	0.18	0.14	1.52	0.17
Income Level	0.09	0.06	0.12	0.31	0.16	0.17	0.15	0.12	0.09	1.27	0.14
Inflation	0.12	0.09	0.04	0.1	0.21	0.17	0.15	0.09	0.09	1.06	0.12
Internet Usage	0.08	0.08	0.06	0.04	0.09	0.17	0.15	0.19	0.12	0.98	0.11
Socio Economic Environment	0.12	0.08	0.06	0.05	0.04	0.08	0.15	0.16	0.13	0.87	0.10
Service Trust	0.12	0.08	0.06	0.05	0.04	0.04	0.07	0.1	0.12	0.68	0.08
Cybersecurity	0.09	0.04	0.04	0.05	0.02	0.02	0.03	0.04	0.13	0.46	0.05
Mindset	0.05	0.03	0.04	0.03	0.02	0.02	0.02	0.01	0.03	0.25	0.03
								CI	**0.1**		
								RI	**1.01**		
								CR	**0.099**		

Source: Author's Compilation

belong to the middle to senior levels of management, as they hold an important position in policy making and implementing in their respective organizations. A survey method with the use of a questionnaire (pairwise comparison matrix) was used for the data collection. Table 1 shows the average value of responses given by all 10 respondents. At the end of the table, the total of each column was calculated and the average value was divided by the respective column sum to calculate the normalized value, shown in Table 2.

Further, the relative significance of each factor was calculated by taking the row average and representing the same in percentage value. This is represented in Table 3 as below:

Table 3 Relative Significance Table.

Financial Inclusion	21.49%
Digital Finance	16.76%
Income Level	14.19%
Inflation	11.76%
Internet Usage	10.86%
Socio-Economic Environment	9.70%
Cybersecurity	7.56%
Service Trust	5.01%
Mindset	2.66%

Source: Author's Compilation

As clear from the analysis so far, "Financial inclusion" is the most important factor to ensure economic growth, followed by "Digital Finance", "Income Level", "Inflation", "Internet Usage", "Socio Economic Environment", "Cybersecurity", "Service Trust" and "Mindset". Now, to categorize the importance of sequenced factors, the cumulative relative significance table is calculated, as mentioned in Table 4. As classified in the cumulative relative importance table, the factors are divided into "High", "Moderate" and "Low" importance. Clear from the table, "Financial Inclusion", "Digital Finance" and "Income Level" are high-priority factors for economic growth. Further, as mentioned in the table, "Inflation", "Internet usage" and "Socio Economic Environment" are moderately intense variables contributing to economic growth. Lastly, "cyber security", "service trust" and "mindset" are listed as the least priority variables for economic growth. This is to reiterate here that the outcomes are based upon the views of experts, who are interviewed for this chapter.

Now, when we have the highlights "Financial Inclusion", "Digital Finance" and "Income Level" are the most important factors to ensure economic growth. It is noteworthy that each factor belongs to the financial, economic, and technical spheres of economic growth. Further, "Inflation", "Internet Usage" and "Socio

Table 4 Cumulative Relative Importance Table

Factor	*Relative Importance*	*Cumulative Relative Importance*	*Impact*
Financial Inclusion	21.49%	21.49%	**High**
Digital Finance	16.76%	38.25%	**High**
Income Level	14.19%	52.44%	**High**
Inflation	11.76%	64.19%	**Moderate**
Internet Usage	10.86%	75.06%	**Moderate**
Socio Economic Environment	9.70%	84.76%	**Moderate**
Cybersecurity	7.56%	92.33%	Low
Service Trust	5.01%	97.34%	Low
Mindset	2.66%	100.00%	Low

Source: Author's Compilation

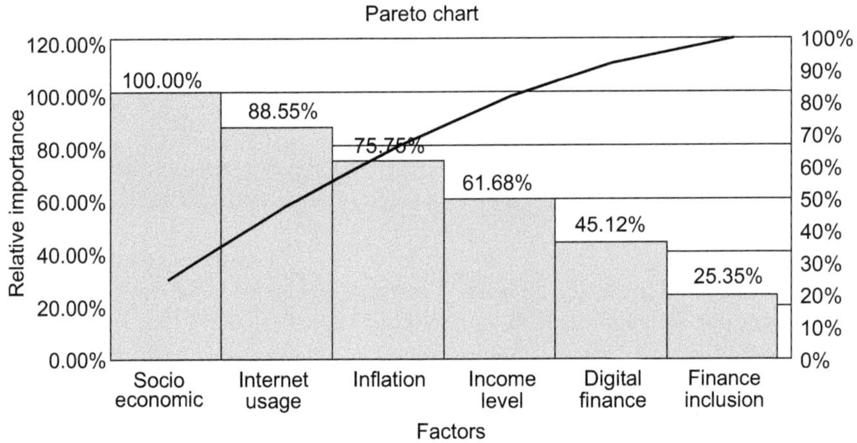

Fig. 1 Copulative Relative Importance.
Source: Author's Compilation

Economic Environment" are moderately significant factors towards economic growth, we have moved to create a Pareto chart for further prioritization of the various factors. Figure 1 has considered the top six variables that impact economic growth to 80% and dropped the variables that impact accelerated economic growth by just up to 20%.

Further following the 80:20 rule of Pareto analysis, factors having a high impact on economic growth were identified. The priority-wise sorted factors are presented in Table 5. The factors with high priority are arranged in a sequence of high priority to lesser priority, even within the category of "high" and "moderate". "Financial Inclusion", "Digital Finance", Income Level", "Internet Usage" and

"Socioe conomic Environment" are shown with the percentage of impact towards economic growth.

Table 5 High-Priority Factors towards Economic Growth.

Financial Inclusion	25.35%
Digital Finance	45.12%
Income Level	61.86%
Inflation	75.73%
Internet Usage	88.55%
Socio-Economic Environment	100.00%

Source: Author's Compilation

Figure 2 shown further, gives a graphical representation of the top six variables for better understanding of the readers.

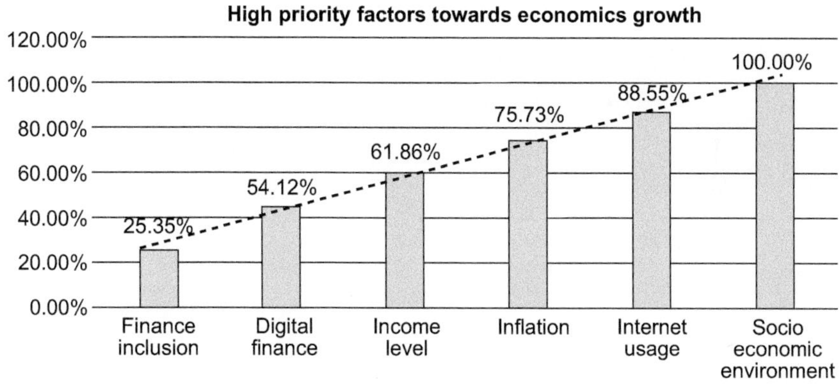

Fig. 2 High Priority factor towards Economic Growth.
Source: Author's Compilation

Finally, an index is proposed based on the high-priority factors contributing towards the use of fintech for financial inclusion and higher economic growth. It is shown in the table that Financial Inclusion has a cumulative impact of 25.3% on economic growth, Digital Finance has a cumulative impact of 45.12%, Income levels have a cumulative impact of 61.86%, Inflation has a cumulative impact of 88.55% and Socioeconomic Environment has a cumulative impact of 100%.

25.35% *Financial Inclusion + 45.12% Digital Finance + 61.86% Income Level + 75.73% *Inflation + 88.55% *Internet Usage + 100% * Socio Economic Environment

As these top 6 factors are responsible for up to 80% of economic growth, the nature (deriving or dependence) is identified using MIC MAC analysis.

The results are shown in Fig. 3, below:

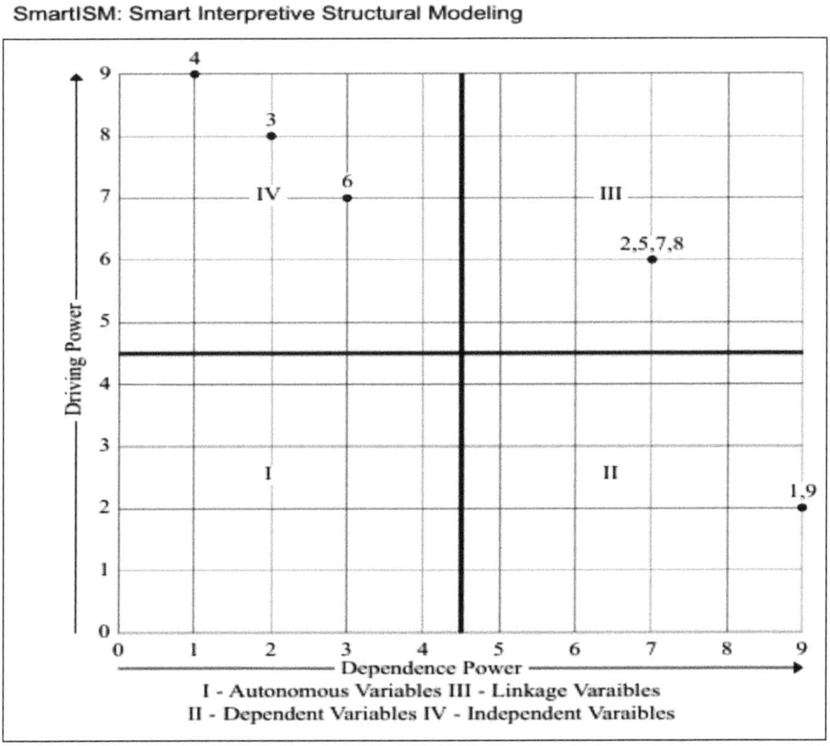

Fig. 3 Smart Interpretive Structure Modeling.
Source: SMART ISM software output

As mentioned in Fig. 3, four quadrants represent four categories, Autonomous Factors, Dependent Factors, Linkage Factors, and Independent Factors.

Here, (1) is represented for "Financial Inclusion", which is a dependent variable on "Digital Finance" represented as (2) and "Internet Usage" represented as (5) has "moderate driving" along with "moderate dependence" nature. "Income Level" represented as (3), "Inflation" represented as (4), and "Socio-Economic Environment" represented as (6) factors are "High Driving" towards economic growth, by driving financial inclusion.

Conclusion

The conclusion of every study highlights major findings and insights, which makes it a critical portion. During the conclusion, the researcher summarizes the findings, links them to the research objectives, and emphasizes the study's implications or contributions to the larger field. Well-written conclusions emphasize the importance of the research in expanding knowledge or understanding of the topic, leaving a lasting impression on the reader. This chapter answers the research objectives by identifying the nine factors prima face, which contribute towards higher economic growth. These factors are financial inclusion, digital finance, income level, inflation, internet usage, socio economic environment, cybersecurity, service trust, and mind set, as pointed out through the AHP Questionnaire. Further, the results reveal that "financial inclusion", "digital finance" and "income level" are the most important factors which affect economic growth through financial inclusion. These factors are followed by other factors like "inflation", "internet usage" and "socio economic environment". All nine factors are then used to run MIC MAC analysis to identify the deriving and dependence power of each. The outcome of this analysis highlights "Financial Inclusion" to be dependent on "Income Level", "Inflation" and "Socio Economic Environment" which are identified as high driving factors. These results are in line with the cumulative relative strength index, and this way results of AHP are verified using MIC MAC analysis.

In conclusion, through the analysis of a broad spectrum of constructs and variables, researchers have explored economic indicators, educational factors, regulatory environments, infrastructure access, behavioral aspects, and socio-demographic factors. This thorough analysis allows for a nuanced understanding of fintech adoption and its implications for financial inclusion across different regions and contexts. The global significance of this relationship is evident, as they are of relevance to both developed and developing nations worldwide. Moving forward, continued interdisciplinary research efforts will be essential in addressing the complex challenges and opportunities presented by fintech in advancing financial inclusion and economic empowerment on a global scale.

4. Limitation and Scope of the Study

This study concentrates on analyzing perspectives on financial inclusion in North and West India. To broaden its scope and address its limitations, future research could include other regions of India and extend internationally. Furthermore, it's crucial to consider a wider array of factors affecting financial inclusion and economic growth. Educational institutions should adjust their training programs based on these findings.

This research can provide valuable insights for enhancing government financial initiatives and developing applications by corporations in the BFSI sector. It can also serve as a reference for further studies in this field. The findings can guide companies and financial firms in formulating policies for financial inclusion. Additionally, these firms and banks can encourage their existing customers to increasingly use FinTech platforms for various financial transactions.

References

Alesina, Alberto, Stelios Michalopoulos and Elias Papaioannou, (2016), "Ethnic Inequality," *Journal of Political Economy*, vol. 124, No. 2, pp. 428–488. 10.1086/685300.

Al-Smadi, M.O. (2023). Examining the relationship between digital finance and financial inclusion: Evidence from MENA countries. *Borsa Istanbul Review*. 10.1016/j.bir.2022.11.016.

Arner, D.W., Buckley, R.P., Zetzsche, D.A. and Veidt, R. (2020). Sustainability, FinTech and Financial Inclusion. *European Business Organization Law Review*, *21*(1), 7–35. 10.1007/s40804-020-00183-y

Anand S., Kodan K. and Kuldip S.C. (2012) "A Theoretical and Quantitative Analysis of Financial Inclusion and Economic Growth." *Management and Labour Studies* 2:103–133. 10.1177/0258042X13498009.

Beck T., Demirguc-Kunt A. and Peria M.S.M. (2007) "Reaching out: Access to and Use of banking Services Across Countries." *Journal of Financial Economics* 85:234–266. 10.1016/j.jfineco.2006.07.002

Cardoso, M.S., Carvalho de Benedicto, G., Willer do Prado, J., Robb, D.A., Neto de Almeida Bispo, O. and José de Brito, M. (2019). "Mapping the Literature on Credit Unions: A Bibliometric Investigation Grounded in Scopus and Web of Science." *Scientometrics* 120 (3): 929–960. 10.1007/s11192-019-03165-1

Demirguc-Kunt, A., Klapper, L.F., Singer, D. and Van Oudheusden, P. (2015). The global findex database 2014: Measuring financial inclusion around the world. *In World Bank Policy Research Working Paper* (7255). 10.1596/1813-9450-7255

Demirgüç-Kunt, A., Klapper, L., Singer, D., Ansar, S. and Hess, J. (2020). The Global Findex Database 2017: Measuring Financial Inclusion and Opportunities to Expand Access to and Use of Financial Services. *World Bank Economic Review*, *34*, S2–S8. 10.1093/wber/lhz013

Demir, A., Pesqué-Cela, V., Altunbas, Y. and Murinde, V. (2022). Fintech, financial inclusion and income inequality: a quantile regression approach. *European Journal of Finance*, *28*(1), 86–107. 10.1080/1351847X.2020.1772335

Gabor, D. and Brooks, S. 2017. "The Digital Revolution in Financial Inclusion: International Development in the Fintech Era." *New Political Economy* 22 (4): 423–36. 10.1080/13563467.2017.1259298.

Gallego-Losada, Jesús, M., Montero-Navarro, A., García-Abajo, E. and Gallego-Losada, R. (2023). "Digital Financial Inclusion. Visualizing the Academic Literature." *Research in International Business and Finance 64* (January). 10.1016/j.ribaf.2022.101862

Gálvez-Sánchez, Jesús, F., Lara-Rubio, J., José Verdú-Jóver, A. and Meseguer-Sánchez., V. (2021). "Research Advances on Financial Inclusion: A Bibliometric Analysis." *Sustainability* (Switzerland) 13 (6). 10.3390/su13063156

Garg, G., Shamshad, M., Gauhar, Tabash, I.R., Hamouri, B. and Daniel, M. (2023). "A Bibliometric Analysis of Fintech Trends: An Empirical Investigation." *International Journal of Financial Studies* 11 (2): 79. 10.3390/ijfs11020079

Gershenson, D., Herrera, L., Lambert, F., Ramos, G., Rousset, M., Torres, J., Werner, A. et al. (2021). "Fintech and Financial Inclusion in Latin America and the Caribbean. The Authors would like to thank Pelin Berkmen, for initiating the work on this topic, Patricia Alonso-Gamo for Her Guidance and Support."

Ghosh, S. (2016). Does mobile telephony spur growth? Evidence from Indian states. *Telecommunications Policy*, 40(10–11), 1020–1031.

Galor O. and Zeira J. (1993) "Income Distribution and Macroeconomics." *The Review of Economic Studies* 60: 35– 52. 10.2307/2297811

Hasan, M., Noor, T., Gao, J., Usman, M. and Abedin, M.Z. (2022). Rural Consumers' Financial Literacy and Access to FinTech Services. *Journal of the Knowledge Economy*. 10.1007/s13132-022-00936-9

Honohan P. (2004) "Financial Development, Growth and Poverty: How close are the links?" *World Bank Policy Research Working Paper* 3203:1–31

Khelifa, S.B., Hmaied, Ouda, O.L., Ayadi, R. and Makni, R. (2022). "Financial Inclusion: A New Multi-Dimensional Index and Determinants – Evidence from the Union for the Mediterranean Countries." *Journal of Economic and Administrative Sciences*, December. 10.1108/jeas-12-2021-0266.

Khera, P., Ng, S., Ogawa, S., Sahay, R., Adrian, T., Čihák, M., Von Allmen, U.E., et al. (2021). "Measuring Digital Financial Inclusion in Emerging Market and Developing Economies: A New Index IMF Working Paper Monetary and Capital Markets Department Measuring Digital Financial Inclusion in Emerging Market and Developing Economies: A New Index"

Koomson, I., Villano, R. and Hadley, D. (2020). Effect of Financial Inclusion on Poverty and Vulnerability to Poverty: Evidence Using a Multi-Dimensional Measure of Financial Inclusion. *SSRN Electronic Journal*. 10.2139/ssrn.3518908

Lagna, A. and Ravishankar, M.N. (2022). "Making the World a Better Place with Fintech Research." *Information Systems Journal* 32 (1): 61–102. 10.1111/isj.12333

Liu, D., Zhang, Y., Hafeez, M. and Ullah, S. (2022). Financial inclusion and its influence on economic-environmental performance: demand and supply perspectives. *Environmental Science and Pollution Research*, 29(38), 58212–58221. 10.1007/s11356-022-18856-1

Lyons, A.C. and Kass-Hanna, J. (2021). Financial inclusion, financial literacy and economically vulnerable populations in the Middle East and North Africa. Emerging Markets Finance and Trade, 57(9), 2699–2738. 10.1080/1540496X.2019.1598370

Mhlanga, D. (2020). Industry 4.0 in finance: The impact of artificial intelligence (AI) on digital financial inclusion. *International Journal of Financial Studies*, 8(3), 1–14. 10.3390/ijfs8030045

Niankara, I. and Islam, A.R.M. (2023). "The Impact of B2P Electronic Payroll and G2P Digital Welfare on Formal Financial Inclusion in the Global Open Economy." *Journal of Open Innovation: Technology, Market, and Complexity* 9 (2). 10.1016/j.joitmc.2023.100034

Odhiambo and Nicholas. (2019). How Enhancing Gender Inclusion Affects Inequality: Thresholds of Complementary Policies for Sustainable Development.

Okello C.B., Ntayi, G., Munene, J.M., J.C. and Malinga, C.A. (2018). Mobile Money and Financial Inclusion in Sub-Saharan Africa: the Moderating Role of Social Networks. *Journal of African Business*, 19(3), 361–384. 10.1080/15228916.2017.1416214

Ozili, P.K. (2020a). Financial inclusion research around the world: A review. Paper presented at the Forum for social economics. 10.1080/07360932.2020.1715238

Ozili, P.K. (2020b). Theories of financial inclusion. *In Uncertainty and challenges in contemporary economic behaviour*. Emerald Publishing Limited. 10.1108/978-1-80043-095-220201008

Ozili, Peterson K. (2023). "Munich Personal RePEc Archive Financial Inclusion Research around the World: A Review Ozili, Peterson K Financial Inclusion Research around the World: A Review." Accessed August 31, 2023. https://mpra.ub.uni-muenchen.de/101809/

Rana, G., Sharma, R. and Sharma, B. (2024). Impact of Financial Inclusion for Achieving Sustainable Development for Removing Poverty in all Forms: A Review Study from India. *E-Financial Strategies for Advancing Sustainable Development: Fostering Financial Inclusion and Alleviating Poverty*, 339–350.

Saaty, T.L. (1980). The analytic hierarchy process. McGraw-Hill: New York. 10.1016/0270-0255(87)90473-8

Sarma, M., and Pais J. (2011) "Financial Inclusion and Development." *Journal of Development* 23: 613–628

Senyo, P.K. and Osabutey, E.L.C. (2020). Unearthing antecedents to financial inclusion through FinTech innovations. *Technovation, 98*. 10.1016/j.technovation.2020.102155

Shen, Yan, Wenxiu Hu and Hueng. C.J. (2021). "Digital Financial Inclusion and Economic Growth: A Cross-Country Study." *In Procedia Computer Science*, 187:218–223. Elsevier B.V. 10.1016/j.procs.2021.04.054

World Bank. (2021). The Global Findex Database 2021: Financial inclusion, digital payments, and resilience in the age of COVID-19. Washington, DC: World Bank. https://doi.org/10.1596/978-1-4648-1843-2

The World Bank Global Findex Database (2017)

Tram, Thi Xuan Huong, Tien Dinh Lai and Thi Truc Huong Nguyen. (2023). "Constructing a Composite Financial Inclusion Index for Developing Economies." *Quarterly Review of Economics and Finance 87* (February): 257–265. 10.1016/j.qref.2021.01.003

Van, Loan Thi Hong, Anh The Vo, Nhan Thien Nguyen and Duc Hong Vo. (2021). "Financial Inclusion and Economic *GROWTH*: An International Evidence." *Emerging Markets Finance and Trade* 57 (1): 239–63. 10.1080/1540496X.2019.1697672

Wang, Y., Xiuping, S. and Zhang, Q. (2021). Can fintech improve the efficiency of commercial banks?—An analysis based on big data. *Research in International Business and Finance*, 55. 10.1016/j.ribaf.2020.101338

Yang, L. and Zhang, Y. (2020). Digital financial inclusion and sustainable growth of small and micro enterprises-evidence based on China's new third board market listed companies. *Sustainability* (Switzerland), 12(9). 10.3390/su12093733

8 Digital Transformation in Family-Owned Enterprises: Challenges and Opportunities for Business Analytics Adoption

Arun Bhatia[1*], Aditi Sharma[2] and Geeta Rana[3]

Digital Transformation offers tremendous opportunities and poses significant challenges for family-owned enterprises (FOEs) worldwide. The global trends, elements, and tactics for integrating the technological advancements such as Artificial Intelligence analytics, and the Internet of Things, successfully in FOEs have been examined in this chapter. Environmental forces, organizational outlook toward change, owners' perceptions towards innovation, and technological preparedness are some of the factors that affect the family's adoption of new technologies. Financial constraints, issues related to governance, and aversion to risk are common problems for FOEs. Nonetheless, business analytics offers chances to enhance consumer engagement levels, operational effectiveness, and strategic decision-making in these firms. Nonetheless, the young blood in these firms is emphasizing the creation of an organizational culture that is prepared for digitalization, incorporating innovation as well as balancing family values, and customizing technological solutions to meet the specific requirements of FOEs. Unlocking the full potential of digital transformation and business analytics in family-owned businesses requires an understanding of and ability to navigate the complexity of governance structures, resource limits, and family dynamics while promoting an innovative and change-oriented culture.

1. Introduction

The digital revolution has put family-owned firms, which have long been a pillar of the global economy, at a crossroads. As technology transforms entire industries and redefines consumer behavio r, these businesses must embrace digital transformation to be competitive (Tirdasari et al., 2022; Sharma et al., 2019). As a result of digital transformation and advancements in information and communication technologies, enormous amounts of data being produced worldwide. Businesses now aim to extract value from these propagated data pools, both organized and unstructured

[1] 0000-0003-3464-6128
0000-0001-5774-528X
0000-0002-3733-5176
* Corresponding author: arunbhatiacuhp@gmail.com

DOI: 10.1201/9781003514282-8

(Bayraktar et al., 2023). Rearranging procedures can alter a company's business logic or value-generation process. The emergence of big data and new digital technologies such as AI, blockchain, IoT, and robots is projected to have a substantial influence on organizations (Verhoef et al., 2021; Sharma et al., 2021). The "winner takes all" era of the digital economy is currently changing the nature of market competition. The family CEO can quickly establish a consensus within the organization to achieve high pragmatism and develop suitable digital strategies once the market opportunities and competitive advantages brought about by DT are identified (He et al., 2024). These advantages are driven by a strong sense of organizational identity, psychological belonging, and family mission (Ano and Bent, 2022). Organizations are constantly looking for ways to boost their performance and get a competitive edge in the market by developing business models and evaluating the data they have gathered. Consequently, every industrial sector uses business analytics (BA) tools and approaches to manage the enormous amount of data that is flooding into organizations (Bayraktar et al., 2023).

Digital transformation presents both challenges and opportunities for family-owned enterprises, particularly in the adoption of business analytics. Implementing digital technologies presents unique challenges for family-owned, low-technology small and medium-sized enterprises (SMEs), owing to their distinctive organizational attributes and "families" characterized by the concentration of decision-making authority among senior family members, some of whom may exhibit reluctance to adopt technological technologies.

1.1 Background and Significance of Digital Transformation (DT) and Business Analytics (BA)

Digital Transformation (DT) involves significant technological shifts wherein companies are confronted with the task of reengineering their operations to keep pace with emerging technologies. This adaptation aims to enhance the overall efficiency of business processes, enhance customer satisfaction, and develop novel value propositions. Family firms are globally dominant but tend to invest less in innovation compared to non-family firms. This reluctance is a result of conservative methods, limited resources, and difficulties associated with generational shifts. It's interesting to note that some family businesses are the most innovative in the world, and even though they invest less in R&D, some of these businesses produce more original products and exhibit higher levels of creativity than non-family businesses (Heider et al., 2022b).

BA has been shown to have a lot of promise for resource and capacity planning and can be used to gain a competitive advantage. Utilizing digital technologies, including business analytics, is crucial to assist the business model transition of modern digital SMEs since they have few alternatives for innovation and business model transformation aside from leveraging technology (Dereli et al., 2020). Zamani et al. (2022) examined the role of Business Analytics in supporting a Greek start-up, TrackApp, to adapt and innovate its business model in response to the

COVID-19 crisis, allowing the company to sense, seize, and transform its approach under pressure. The authors proposed that BA can empower SMEs to sense the need for business model transformation, mobilize resources for adaptation, and ultimately transform their business models in response to exogenous shocks. Silaban (2022) highlighted the importance of organizational culture, executive buy-in, talent capability, financial resources, data availability and quality, data management, and stakeholder collaboration as critical factors influencing the successful adoption of BA.

A key component of digital transformation for family-owned businesses, business analytics (BA) has enormous potential for leveraging opportunities and making better decisions that lead to better firm performance (Sharma et al., 2017).

1.2 The Unique Characteristics of Family Owned Enterprises

Up to 90% of firms are family-owned or controlled, making family-owned businesses common around the world (Tirdasari et al., 2022). Because family relationships and company operations are intertwined, these businesses have special qualities that present both opportunities and difficulties. Competitive advantages, quicker decision-making, and less bureaucracy are characteristics of family-owned firm governance (Alderson, 2012). A distinctive multi-business portfolio structure known as "grupo," which consists of multiple companies owned by a family or group of families, is frequently adopted by family businesses in Latin America (Botero and Vélez, 2018). Family-owned companies are renowned for their tenacity and long-term focus, especially in challenging economic times. The conservative approaches, scarce resources, and challenges brought on by generational changes are the causes of this hesitancy. It's fascinating to note that some family firms are the most creative in the world; despite spending less on research and development, some of these companies create more unique goods and are more creative than non-family enterprises (Harris snd Ozdemir, 2020).

2. Global Patterns of Technology in FOE

Olubusola (2024) highlighted the necessity of context-specific techniques by comparing the digital revolution in company growth in the US and Africa. While Africa has demonstrated resiliency and inventiveness in overcoming challenges, the US has upgraded its technology infrastructure and economic processes. The incorporation of cutting-edge technology, ecosystem-based business models, and societal and environmental sustainability are some of the trends in digital transformation. While the US has embraced cutting-edge technology like cloud computing, data analytics, and artificial intelligence, African enterprises must contend with issues including regulatory complexity, a lack of digital skills, and restricted access to digital infrastructure. Prior knowledge, the degree of IT usage, business location, and community size/type were significant determinants in the choice to adopt in research on small family businesses in the US (Niehm et al., 2010).

In an emerging economy like India, a study by Pujari et al. (2024) highlighted the importance of integrating next-generation leadership with tech-savvy successors, government initiatives like Digital India and Make in India, and external expertise to enhance professionalism and improve technology adoption. The author underscored the need for proactive strategies and collaboration to navigate the digital landscape, emphasizing successful digital transformation initiatives and context-specific approaches. In a study of Chinese A-share listed companies, He et al. (2024) found that family CEOs play a decisive role in promoting digital transformation.

3. Challenges in Adopting Business Analytics in FOES

Examining the factors that influence technological inertia in family-run SMEs in India and the obstacles they confront while adopting new technologies, Pujari et al. (2024) highlighted fear of change, conservative leadership, a lack of expertise, and simplicity of use as primary barriers to technology adoption. The authors emphasized the future potential and beneficial influence of technology on family-owned SMEs, namely AI, ChatGPT, and social media, and argued that integrating these technologies can increase customer interaction, decision-making, and corporate efficiency. Gudfinnsson et al. (2019) highlighted several causes of the lack of motivation for small and medium-sized manufacturing enterprises (SMMEs) to adopt business intelligence and analytics (BI&A). Some of the causes include ignorance of the potential of BI&A, lack of understanding of how to leverage analytics for business value, lack of managerial support, and skepticism, especially about IT investments. Silaban (2022) analyzed the challenges in implementing business analytics using the Technology-Organization-Environment (TOE) architecture. It identifies three main challenges: environmental, technological, and organizational. Organizational hurdles include establishing a data-driven culture, securing leadership support, promoting good communication, and finding data-savvy personnel. Technological challenges include data availability, quality, and resource preparedness. Environmental challenges involve privacy concerns and moral implications due to data gathering and use. It was suggested that given the complexity and interrelated nature of these challenges, firms need to address these for the successful implementation of business analytics.

3.1 Resistance to Change and Risk Aversion

Digital adoption and family businesses have a negative relationship. The reason for this is that family businesses are risk averse. As a result, they are less inclined to make R&D and innovation investments. Furthermore, it usually has a traditional deed. As a result, high-risk tactics like investing in technology may be put on hold (Fernández and Nieto, 2006).

Family firms are generally more risk-averse than non-family firms, which impacts their investment decisions, capital structure, and business growth. Factors such as family involvement, environmental competition, and family members' goals and behaviors influence the risk aversion of family firms (Hiebl, 2012).

3.2 Complexity in Family Dynamics and Governance Structures

Family-owned businesses exhibit complex dynamics and governance structures that significantly impact their performance and longevity. These firms, which constitute up to 90% of businesses worldwide, including a third of S&P 500 and Fortune 500 companies, face unique challenges and opportunities (Alderson, 2012). The interplay between family dynamics and governance mechanisms is crucial for sustained performance in family firms. Effective governance structures, well-managed family dynamics, and the adoption of best practices correlate with improved financial performance, reduced agency conflicts, and enhanced access to resources (Trebicka and Tartaraj, 2023). However, the familial nature of these businesses can complicate decision-making processes and hinder strategic planning, particularly in areas such as succession planning and conflict resolution. Family business governance models prioritize short-term, guaranteed advantages over long-term profits from disruptive technology, favoring incremental (exploitative) breakthroughs over experimental ventures (Ceipek et al., 2021).

One distinctive feature of family businesses in Latin America is the "grupo" structure, a multi-business portfolio holding company owned by a family or group of families (Botero and Vélez, 2018). This structure adds another layer of complexity to governance issues. The development of governance structures and mechanisms plays a vital role in mediating conflicts between family, property, and management, as evidenced by a case study demonstrating the implementation of governance mechanisms addressing these three dimensions (Bertucci et al., 2009). Interestingly, while family ownership, leadership, and governance structure significantly influence firm performance metrics like ROA and ROE, the impact of family ownership is less consistent (Kabiraj and Siddik, 2016; Rana et al., 2019).

4. Opportunities Presented by Business Analytics in FOES

4.1 Enhanced Strategic Decision-Making

Vlasic (2023) does provide evidence and importance of enhanced strategies for decision making which differ for both family and nonfamily firms. Barros-Contreras, Morales-Serazzi, Torres-Toukoumidis, and Palma-Ruiz (2022) outline the importance of business analytics for enhanced strategic decision-making but also believe that family-owned enterprises also need to inculcate the feeling of cooperation among the family members along with improving on characteristics of information sharing and usage of family members' expertise. Kathuria et.al (2023) using panel regression estimations on the publicly listed family firms does affirm the findings that Indian family-owned firms depict conservative strategic decision-making to invest in the IT resources for the future, but if adapted to modern change, can make the organizations competitive in comparison to other similar firms. Decision-making for the better future of family owned enterprises is the main concern but looking into other dimensions also needs to be taken into account (Sawang and Kivits, 2024) highlight the different shades of decision-making for strategic intervention for business analytics usage where family owned enterprises need to work on ethical concerns including by highlighting data privacy along with transparency to ensure fair practices.

4.2 Improving Operational Efficiency

The answer to an important question arises from addressing an important issue that the operational efficiencies of family-owned firms improve using business analytics or they are previously well efficient as far as operations are concerned, that's why they could sustain and create an extended position in the business dynamic environment. The usage of the power of business analytics and technology intervention has been admitted as a correct strategy to improve upon operational efficiencies in family owned firms but many family firms are not proactive in integrating the technologies in the business thereby largely hindering the business's operational efficiency, business processes and last but not the least customer engagement. Anticipation to benefit from the technology but reluctance to adapt to change is the major cause of not improving operational efficiencies of the family-owned businesses.

4.3 Customer Engagement and Market Adaptability

Modern businesses must remain competitive with better customer engagement, by improving the processes, working on strategies, training the people, and making operations efficient to create mass marketing, segment-based, and loyalty-based programmes for customers. Engaging the customers with the business with customer loyalty has been at the center point of the family-owned businesses. Van Scho (2024) states that customer engagement in family firms is a direct outcome of share of community bonds, personalized services, importance to product quality, etc. Actively engaging with the concept of customer engagement for market adaptability (Pereira et al., 2024) in the emerging market of India and family firms brings forth a greater view point that usage of social media platforms with customer feedback and reviews, managing the reputation of the family firm and managing the trustful relationship is the top priority for the firms. However, Indian family firms have to be highly dependent on the third party services for content creation and digital marketing. Customer engagement in the contemporary world is largely depicted with the use of modern business analytics with tools like Chatbots. Family firms that highly engage in digital integration in their firm businesses and want to leverage the same are using chatbots for customer engagement and market adaptability (Rizomyliotis et al., 2022). Natasha, Angelia, and Susilo (2023) carried out the research in the tourism sector, bringing that customer engagement with the social media strategy and specifically with the usage of Instagram has been highly beneficial for the family-owned firms in the tourism sector.

5. Strategies for Successful Business Analytics Adoption in FOES

Adoption, absorption, and action are the 3A's that highlight the successful strategies for successful business analytics in family-owned enterprises. Strategic intervention for successful business analytics is an important area of discourse in family and non-family owned enterprises making them more agile, competitive, and cooperative in modern times. Drawing inferences from the power of ICT in data integration and data mining along with statistical power are highly influential in helping business analytics move from non-family firms to family firms.

5.1 Building a Digital-Ready Organizational Culture

While accepted that the use of digital technology can rejuvenate the success of business enterprises in the long run, overcoming the gap in organizational adaptation to change with a digitally ready organizational culture is a topic of study. Baalmans, Broekhuizen, and Fabian, (n.d) admit the change in the organizational digital ready culture is due to two very specific reasons where at one end of the continuum is the change in the customer centric behavior towards digital technologies and their usage and on the other end investment in the digital technologies help in four areas for the firms like improvement in processes, well defined structures, adapting to new systems and lastly activities associated with it. Borsano et al. (2024) study of Thai firms in the developing nation contributes to the study area that the lack of an innovative ecosystem is one of the prime reasons for non-adoption and redesigning of the internal organizational structure. Grover, Tseng, and Pu (2022) also admit that organizations are competing and technology is changing, and to remain competitive, the organizations must admit the importance of change in the organizational culture with more adoption of technological interventions and form a digital culture. Digital-ready organizational culture is embedded in the transformational change in the organization, with key areas like an introduction to digitalization, along with digital objects, helped by digital resources (Grover, Tseng, and Pu, 2022).

5.2 Integrating Innovation through Tradition

Integrating Innovation through Tradition is mostly talked about as ITT largely focuses on getting the benefits from integrating tradition into the modern innovation process in the firms (Heider et al., 2022b; De Massis et al., 2016) rather than discarding the same. The major part of the integration lies in getting value from the products or services manufactured/offered and remaining tied to the roots of tradition in modern international society. Most of the family-owned business firms affirm to stay competitive while showcasing unique, destination-based products with attraction to gain an advantage from the increased demand due to its value at the international level for traditional knowledge-based products. Due to more competition from a large number of firms with similar products (Della Corte, Zamparelli, and Micera, 2013), admit that to stay in business for several years of internationalization with the philosophy of cooperation and networking has been the right path to showcase the tradition to the international audience. On the contrary, the same (Sheikh et al., 2019) provides evidence that tension does erupt while creating a changing society, both for traditions and innovation, where fear (ITT) of losing tradition from innovation is at the heart of the Integrating Innovation through Tradition discourse. Erdogan, Rondi, and De Massis (2020) in the research article, highlighting the context discourse, came up with the construct of temporal symbiosis, admitting two major key highlights of adoption and retrospection for the family firms to be innovative in modern society and keep intact with the values and belief system.

5.3 Tailoring Technology Solutions to FOEs

Admitting to the current times, it becomes imperative for the firms to remain competitive, along with ideas for nonclosure of the family-owned firms, which have a legacy from the roots of the family and the region. Tailoring technology for family owned enterprises has been spotted in the three key areas where family firms' technology acceptance has to be understood firstly process digitalization, secondly digitalizing the product or service, and lastly, business model digitalization with the constraints from both the technological and human resources with resource constraints (Soluk, and Kammerlander, 2021). Pikkemaat and Zehrer (2016) submit that in the modern day time, the first and foremost need for small business firms is to drive attention toward the collection of customer data, manage the data, and study the data to identify and prioritize the customer needs and wants. Rizomyliotis et al., (2022) highlights the use of technology in family firms and recognizes the importance of tech-driven businesses where the usage of chatbots has increased in the marketing strategies of these kinds of firms and it has also been understood that getting tech-driven solutions to the customers add value for the family firm and the business together. Batt, et al., (2020) in the research article "The Digitalization of Family Firms: A Research Agenda" tries to show the future of tailoring digital technologies for family firms where it is addressed that technology intervention with the family firms is destination based and the knowledge-based society therefore more understanding to the topic can lead to more tailor-made ICT products for the family firms. Alberti and Pizzurno (2013) believe that technological interaction has been a key area for the family firms, but still, the family firms affirm with an idea of knowledge sharing among the members of the family rather than knowledge availability along with working on new product development.

6. Future Directions for Digital Transformation in FOES

Future Directions for Digital Transformation in Family Owned Enterprises attract the vision, mission, and objectives of the organization, individual, and the government to make the digital ecosystem run and operate, making a mark at the international level. Zapata-Cantu, Sanguino, Barroso, and Nicola-Gavrilă (2023) demonstrate the two most important aspects to be crucial in the Digital Transformation for family owned enterprises that are to remain competitive in the environment and secondly to work on sustainable advantage along with knowledge management. Digital transformation is interdependent and reciprocal to the number of variables working simultaneously, where leadership ability, along with usage of internet marketing, technology utilization, and artificial intelligence, play a critical role in the future use of digital transformation. But largely as outlined by Bhatia (2020), most of the technological intervention by enterprises will work on the 3C formula, i.e. Content, Community, and Commerce. Within the discussion in the discourse, the future of digital transformation is also an outcome of other factors like the dialogue and cohesion among the members to adopt the digital transformation while attaching the feeling of pride to the products produced (Ranaldo, Dicuonzo, and Diéguez-Soto, 2024).

6.1 Emerging Technologies and Their Potential Impact

Most people's lives have been impacted by technological advancements, and because of their widespread use, neither individuals nor businesses can ignore them. To ensure that interactions with the company and individuals are highly personalized and driven by needs and demands in the direction of an organizational growth model, businesses in the modern day have digitally driven all of the touch points. Organizations have started using technology from small operations to high-tech-driven technology where Artificial intelligence, big data, and digital elasticity have made a big mark in business operations. Tech-driven highly efficient robots have changed the face of the hospitality industry, where robots have showcased all the possible behaviors like a human being, Bowen, J., and Morosan, (2018). Similar outcomes have been shown in other areas where digital technologies have also been seen in the development of new social media channels, virtual reality, the internet of things, etc. Cloud-based software has touched organizations with better usage of data storage capabilities of the firm offering the service, rather than identifying a service provider to store the data and provide backup. Moving further digital technologies provide not only the option of product choice by the customer rather with also work on its key features of Key Performance Indicators i.e. KPI and Just in Time i.e. JIT. Blockchain technology benefitting the financial model of data storage and protection does provide great opportunities for the future (Brandon, 2016).

6.2 Support Mechanisms for FOEs in Digital Transformation

In the current climate, support mechanisms are becoming increasingly vital for all family-owned businesses that are hesitant to become digitally controlled and operated, as well as corporations that are in the early stages of technological intervention. In a larger sense, digital transformation in family-owned businesses is a collaborative effort to focus on the technological side of business as well as knowledge management. In a developing country like India, an initiative like GEM (Government e-marketplace) is the perfect way to allow a large number of product makers to put their products on the internet and reach every corner of the country. Another method in the same direction is the establishment of ONDC (Open Network for Digital Commerce). It also offers opportunities for people to post their things and sell to a broad audience. Mahila e-Haat is a direct connection between small family-owned companies and customers of traditional products from all states of the country via a system for displaying, interacting, and selling goods/services. Other important specified products as a government endeavor in India include the "Technology Center Framework," "Digital India Mission," collaboration with non-banking financial companies, value addition while partnering with technology suppliers, and so on.

Conclusion

Digital transformation has become increasingly crucial for family businesses to maintain their competitiveness and ensure long-term success in today's rapidly evolving business landscape. Family businesses, which form a significant part of global economic activity and job creation, face unique challenges in adapting to digital technologies. The importance of digital transformation for family businesses lies in its potential to increase efficiency, improve customer experiences, and create innovative value propositions. Studies have shown that family businesses can benefit from a holistic approach to digitalization, encompassing process digitalization, product digitalization, and digital transformation of business models and organizational identities. While product digitalization appears to be independent of family influence, process digitalization benefits from family ability, and digital transformation is closely tied to the family's willingness to pursue change (Heider et al., 2022a; Soluk and Kammerlander, 2021). Interestingly, family businesses possess unique strengths that can aid in their digital transformation journey. Their agile decision-making processes, long-term vision guided by consolidated family values, and strong commitment to workers and employees can be leveraged to implement digital strategies. However, it is essential to note that family businesses may face challenges related to the stereotype of being unprofessional or centralized, which can hinder their digital transformation efforts (Harish et al., 2023). To overcome these challenges and maximize the benefits of digital transformation, family businesses should focus on developing dynamic capabilities, implementing appropriate technologies, and fostering a culture of innovation and change (Kaban and Tukiran, 2023).

Adopting digital transformation and business analytics presents special potential and problems for family-owned businesses (FOEs). Although advanced technology adoption is most prevalent in wealthy economies, businesses in poor nations are making progress despite limited resources. Environmental forces, organizational preparedness, and technological compatibility are some of the factors that affect the adoption of new technologies. Financial obstacles, complicated family relationships, and aversion to change are common problems for FOEs. Nonetheless, business analytics offers chances to improve consumer engagement, operational effectiveness, and strategic decision-making. Creating an organizational culture that is prepared for digitalization, incorporating innovation through tradition, and customizing technological solutions to meet the specific requirements of FOEs are all components of successful adoption strategies. Future directions emphasise the significance of support systems for FOEs in their digital transformation journey as well as the possible influence of developing technology. For FOEs to use digital technologies for long-term success and growth in a global economy that is becoming more and more competitive, a thorough strategy that takes into account both internal and external elements is essential. Adoption of business analytics and digital transformation present special opportunities and challenges for family-owned businesses. While poor regions make progress despite resource limits, developed economies are at the forefront of adopting sophisticated technologies. Environmental forces, organizational preparedness, and technological compatibility are some of the factors that affect the adoption of new technologies. Financial obstacles, complicated family dynamics, and aversion to change are problems for family-owned businesses. Better client interaction, more operational efficiency, and greater strategic decision-making are all made possible by business analytics. Creating a culture prepared for digitalization, incorporating innovation through tradition, and customizing technological solutions are all components of successful adoption strategies.

References

Alberti, F.G. and Pizzurno, E. (2013). Technology, innovation and performance in family firms. *International Journal of Entrepreneurship and Innovation Management*, *17*(1– 3), 142–161.

Alderson, K.J. (2012). *Effective Governance in the Family Owned Business* (pp. 399–414). https://doi.org/10.1007/978-3-642-31579-4_17

Ano, B. and Bent, R. (2022), "Human determinants influencing the digital transformation strategy of multigenerational family businesses: a multiple-case study of five French growth-oriented family firms", *Journal of Family Business Management*, vol. 12 No. 4, pp. 876–891.

Baalmans, B.S., Broekhuizen, T.L.J. and Fabian, N.E. 9N.D. *Digital Transformation*.

Barros-Contreras, I., Morales-Serazzi, M., Torres-Toukoumidis, A. and Palma-Ruiz, J.M. (2022). Knowledge integration and organisational performance of data analytics in the family business. *Journal of Open Innovation: Technology, Market, and Complexity*, *8*(3), 135.

Batt, C.E., Cleary, P., Hiebl, M.R., Quinn, M. and Rikhardsson, P.M. (2020). The digitalization of family firms: A research agenda. *A Research Agenda for Family Business*, 247–260.

Bayraktar, Erkan and Tatoglu, Ekrem and Aydıner, Arafat and Delen, Dursun. (2023). Business Analytics Adoption and Technological Intensity: An Efficiency Analysis. *Information Systems Frontiers*. 26. 1–18. 10.1007/s10796-023-10424-3

Bertucci, J.L.D.O., Campos, E.Á.D.S., Pereira, R.D. and Pimentel, T.D. (2009). Governance Mechanisms and Processes of Succession: a study on the influence of the elements of corporate governance in the direction of the succession process in a family business. *Review of Business Management, 11*(31), 152–167. https://doi.org/10.7819/rbgn.v11i31.524

Bhatia, A. (2022). Determinants of ICT adoption for digital inclusiveness in hill state of Himachal Pradesh-tourists perspective. *Atna Journal of Tourism Studies*, 17(1), 1–31. https://doi.org/10.12727/ajts.27.1

Borsano, P., Marozzo, V., Bonaglia, M., Di Minin, A. and Crupi, A. (2024). Digital transformation and digital dynamic capabilities improvement in low-medium technology sector: evidence from Thai family firms. *Asian Business & Management*, 1–30.https://doi.org/10.1057/s41291-024-00281-6

Botero, I.C., and Vélez, D.G. (2018). *Ownership Structure and Governance in Latin American Family Firms* (pp. 19–24). routledge. https://doi.org/10.4324/9781315098630-4

Bowen, J. and Morosan, C. (2018). Beware hospitality industry: the robots are coming. *Worldwide Hospitality and Tourism Themes, 10*(6), 726–733.

Brandon, D. (2016). The blockchain: The future of business information systems. *International Journal of the Academic Business World, 10*(2), 33–40.

Ceipek, R., Hautz, J., DE Massis, A., Matzler, K. and Ardito, L. (2021), "Digital transformation through exploratory and exploitative internet of things innovations: the impact of family management and technological diversification", *Journal of Product Innovation Management*, vol. 38 No. 1, pp. 142–165.

De Massis, A., Frattini, F., Kotlar, J., Petruzzelli, A.M. and Wright, M. (2016). Innovation through tradition: Lessons from innovative family businesses and directions for future research. *Academy of Management Perspectives, 30*(1), 93–116.

Della Corte, V., Zamparelli, G. and Micera, R. (2013). Innovation in tradition-based firms: dynamic knowledge for international competitiveness. *European Journal of Innovation Management, 16*(4), 405–439.

Dereli, S.E., Korthaus, A., Scheepers, H. and Ahmed, A. (2020). The Role of Business Analytics for Competitive Advantage of Small and Medium Enterprises. Australasian Conference on Information Systems 2020. Wellington. https:// aisel. aisnet. org/ cgi/ viewc ontent.cgi?article=1078&context=acis2020

Erdogan, I., Rondi, E. and De Massis, A. (2020). Managing the tradition and innovation paradox in family firms: A family imprinting perspective. *Entrepreneurship theory and practice, 44*(1), 20–54.

Fernández, Z. and Nieto, M.J. (2006) 'Impact of ownership on the international involvement of SMEs', *Journal of International business studies*, 37(3), pp. 340–351.

Grover, V., Tseng, S.L. and Pu, W. (2022). A theoretical perspective on organizational culture and digitalization. *Information & Management, 59*(4), 103639.

Gudfinnsson, K., Rose, J. and Aggestam, L. (2019). Tackling Lack of Motivation in Aspirational Analytics Companies. *International Journal of Business Intelligence Research, 10*(1), 1–18. https://doi.org/10.4018/ijbir.2019010101

Harish, V., Sharma, R. and Rana, G. (2023). Unveiling the Digitalization Dilemma: A Study of Business Analytics Adoption in Small and Mid-Sized Manufacturing Firms Post-COVID-19. In *International Conference on Business Data Analytics* (pp. 340–359). Cham: Springer Nature Switzerland.

Harris, P. and Ozdemir, O. (2020). Turkish delight a public affairs study on family business: The influence of owners in the entrepreneurship orientation of family-owned businesses. *Journal of Public Affairs, 20*(1). https://doi.org/10.1002/pa.2082

He, Q., Lassala, C. and Currás-Móstoles, R (2024). Can family CEOs promote enterprises' digital transformation? An analysis based on ability-willingness paradox. *Int Entrep Manag J* **20**, 3181–3205. https://doi.org/10.1007/s11365-023-00941-7

Heider, A., Clauss, T., Gerken, M., Rüsen, T. and Hülsbeck, M. (2022a). Blood is Thicker Than Water: The Role of Family Willingness and Family Ability. *Academy of Management Proceedings, 2022*(1). https://doi.org/10.5465/ambpp.2022.10612

Heider, A., Hülsbeck, M. and von Schlenk-Barnsdorf, L. (2022b). The role of family firm specific resources in innovation: an integrative literature review and framework. *Management Review Quarterly*. 72. 10.1007/s11301-021-00256-3

Hiebl, M.R.W. (2012). Risk aversion in family firms: what do we really know? *The Journal of Risk Finance, 14*(1), 49–70. https://doi.org/10.1108/15265941311288103

Kabiraj, S. and Siddik, Md. N.A. (2016). Family-Owned Firms between Agency Conflicts and Stewardship: Corporate Governance Factors Driving Firm Performance. *Journal of Business and Management Research, 1*(2), 33–47. https://doi.org/10.3126/jbmr.v1i2.15662

Kathuria, A., Karhade, P.P., Ning, X. and Konsynski, B.R. (2023). Blood and water: information technology investment and control in family-owned businesses. *Journal of Management Information Systems, 40*(1), 208–238.

Maria Kaban, L. and Tukiran, M. (2023). Dynamic Capability for Digital Transformation in Family-Owned Company: A Systematic Review. *International Journal of Research and Review, 10*(5), 465–473. https://doi.org/10.52403/ijrr.20230553

Natasha, J.B., Angelia, C.R., and Susilo, D. (2023). Social Media Marketing Strategy in Increasing Customer Engagement of Taman Safari Ndonesia Bogor (Case Study on Instagram@ Taman_Safari). *Indonesian Journal of Business Analytics, 3*(2), 351–408.

Niehm, L.S., Fitzgerald, M.A., Shelley, M.C. and Tyner, K. (2010). Technology Adoption in Small Family-Owned Businesses: Accessibility, Perceived Advantage, and Information Technology Literacy. *Journal of Family and Economic Issues, 31*(4), 498–515. https://doi.org/10.1007/s10834-010-9197-0

Olubusola, O., Ajayi-Nifise, A., Mhlongo, N., Falaiye, T. and Daraojimba, E. (2024). Digital transformation in business development: A comparative review of USA and Africa. *World Journal of Advanced Research and Reviews, 21*(2), 1958–1968. https://doi.org/10.30574/wjarr.2024.21.2.0443

Pereira, V., Laker, B., Bamel, U., Sharma, G.D. and Paul, H. (2024). Customer engagement strategies within family businesses in emerging economies: A multi-method study. *Journal of Business Research, 174*, 114508.

Pikkemaat, B. and Zehrer, A. (2016). Innovation and service experiences in small tourism family firms. *International Journal of Culture, Tourism and Hospitality Research, 10*(4), 343–360.

Pujari, P., Arora, M., Kumar, A. and Pandey, A. (2024). Understanding factors influencing technical inertia in family-run SMEs : A study on technology adoption challenges. *Journal of Statistics and Management Systems, 27*(1), 155–168. https://doi.org/10.47974/jsms-1208

Rana, G., Sharma, R. and Goel, A.K. (2019). Unraveling the power of talent analytics: Implications for enhancing business performance. *Business governance and society: Analyzing shifts, conflicts, and challenges*, 29–41.

Ranaldo, S., Dicuonzo, G. and Diéguez-Soto, J. (2024). Multiple-generation involvement in family businesses: an in-depth exploration of the effects on digital transformation process. *International Journal of Digital Culture and Electronic Tourism, 4*(3), 189–212.

Rizomyliotis, I., Kastanakis, M.N., Giovanis, A., Konstantoulaki, K. and Kostopoulos, I. (2022). "How may I help you today?" The use of AI chatbots in small family businesses and the moderating role of customer affective commitment. *Journal of Business Research, 153*, 329–340.

Sawang, S. and Kivits, R.A. (2024). Revolutionizing family businesses with artificial intelligence: a perspective article. *Journal of Family Business Management, 14*(4), 802–807.

Sharma, R., Singh, S.P. and Rana, G. (2019). Employer branding analytics and retention strategies for sustainable growth of organizations. *Understanding the role of business analytics: Some applications*, 189–205.

Sharma, R., Saini, A.K. and Rana, G. (2021). Big data analytics and businesses in industry 4.0. *Design Engineering*, 238–252.

Sharma, R. and Rana, S. (2017). The Use of Management Control Systems in the Pharmaceutical Industry. International Journal of Engineering Technology, Management and Applied Sciences, Vol. 5, Issue 6, pp. 23–29.

Sheikh, F.A., Ferreiro, M.D.F., De Sousa, C., Reidolf, M. and Bhaduri, S. (2019). Tradition and innovation: Between dynamics and tensions. *African Journal of Science, Technology, Innovation and Development*, *11*(5), 533–542.

Silaban, D.P. (2022). Challenges in Business Analytics Implementation: A Comprehensive Review Using Toe Framework. *Milestone: Journal of Strategic Management*, *2*(2), 131. https://doi.org/10.19166/ms.v2i2.5953

Soluk, J. and Kammerlander, N. (2021). Digital transformation in family-owned Mittelstand firms: A dynamic capabilities perspective. *European Journal of Information Systems*, *30*(6), 676–711.

Tirdasari, N.L., Arief, N.N. and Dhewanto, W. (2022). Digital Adoption in the Family Business: A State-of-the-art Review. *European Conference on Innovation and Entrepreneurship*, *17*(1). https://doi.org/10.34190/ecie.17.1.478

Trebicka, B. and Tartaraj, A. (2023). Governance mechanisms, family dynamics, and sustained performance: Insights from family firms. *Corporate Governance and Organizational Behavior Review*, *7*(4, special issue), 334–341. https://doi.org/10.22495/cgobrv7i4sip11

Van Scho, M.J. (2024). *Investigating consumers' brand loyalty in small family-owned businesses in the Stilfontein area, North West Province* (Doctoral dissertation, North-West University (South Africa)).

Verhoef, Peter, Broekhuizen, Thijs & Bart, Yakov & Bhattacharya, Abhi & Dong, John & Fabian, Nicolai and Haenlein, Michael. (2021). Digital transformation: A multidisciplinary reflection and research agenda. *Journal of Business Research*. 122. 10.1016/j.jbusres.2019.09.022

Vlasic, G. (2023). Comparative analysis of the role of strategic orientation, strategic performance metric focus and strategic audacity in driving firm performance: family businesses vs nonfamily businesses. *Journal of Family Business Management*, *13*(1), 7–25.

Zamani, E.D., Griva, A. and Conboy, K. (2022). Using business analytics for SME business model transformation under pandemic time pressure. *Information Systems Frontiers*, *24*(4), 1145–1166.

Zapata-Cantu, L., Sanguino, R., Barroso, A. and Nicola-Gavrilă, L. (2023). Family business adapting a new digital-based economy: Opportunities and challenges for future research. *Journal of the Knowledge Economy*, *14*(1), 408–425.

9 Role of Business Intelligence and Big Data in Higher Education

Nidhi Shridhar Natrajan[1*] and Rutuparan Anil Pandey[2]

Technology has become an inherent part of every industry, including the education sector. The strategy is formulated with the technology infrastructure and technological advancements in mind. Education, specifically higher education, has transformed from traditional face-to-face delivery to the blended approach heavily relying on technology for combining offline and online pedagogical approaches. From an institutional perspective, this calls for robust policymaking and decision-making. Business Intelligence has emerged as a new approach with a blend of data storage and data analysis tools to support better decision-making. So, no more are decisions based on the knowledge and experience of the board members, but it is all data-driven decision making. Apart from this, business intelligence tools also facilitate the optimization of the education process, the development of relevant curricula, and higher student engagement (Parashar et al., 2023). With the usage of online educational portals, online classes, cloud-based tools like Google Meet and Zoom to conduct classes, use of multiple device connectivity for students and corresponding application of IOT devices, blending social media tools for feedback and discussion forums, a huge amount of data is generated. This combined with the basic data of students and marks enables the application of big data analytics to generate better insights for enhancing quality education. The current chapter is based on a mixed-method approach to analysing data. The results of the study will imply three important stakeholders: administration, teachers, and learners.

1. Introduction

Data analytics and business intelligence (BI) are becoming essential tools in higher education, providing insightful information that can change how institutions function, instructors instruct, and students learn. These cutting-edge technologies enable schools and universities to use enormous volumes of institutional data to guide data-driven decisions that increase overall institutional effectiveness, expedite processes, and boost academic results. In higher education, business intelligence (BI) and data analytics play a role in a wide range of applications that support the entire institution's academic, administrative, and operational demands,

[1] 0000-0001-8629-4799

[2] 0009-0004-7785-2030

* Corresponding author: nidhi.natrajan@scmsnoida.ac.in

DOI: 10.1201/9781003514282-9

going beyond simply monitoring student performance or institutional metrics. The BI tools facilitate the better allocation of resources, both technological and others. The smart pedagogies for the new way of delivering education and better student engagement are also supported through it (Al-khateeb, 2024; Mukul and Büyüközkan, 2023; Harish et al., 2023a; Mansor, Abdullah, and Rahman, 2020). The efficient utilization of these tools determines the success of an effectively data-driven decision-making process, optimization of operations, and fulfillment of the needs of all the stakeholders in the higher education context. BI is an efficient choice of technology for fact-based strategic planning. This is possible through the data analysis of student enrolment, student performance, and overall institutional outcome. This set of data helps in making decisions based on trends and facts. It is also possible to predict the future trends related to student dropout, transfer, attrition, and popular choices of courses among the learners. Based on these reports, future policies can be efficiently formulated. The performance of the students can be studied to identify slow learners and accordingly, remedial plans can be made to support them. Quality education is ensured through strong reporting and documentation, regular academic audits, and adherence to the accreditation standards. BI tools make this entire process seamless and reduce administrative workload.

From the perspective of the teachers, the use of BI tools enables the modification of the pedagogical approach, increasing the student outcome (Lorenzen, 2021). The course material can be updated and improved using reports based on the performance, engagement, and feedback of the learners. The understanding of learner groups and individual performances also aids in designing special sessions for improving the performance of students who are lagging behind o thers. These tools also enable effective communication among the academic advisor, teacher, and student to improve the outcome in various academic aspects. Combining BI tools and a learning management system supports the teacher to properly focus on better teaching methods while utilizing the analytical reports (Harish et al., 2023b).

Data-driven BI tools are also beneficial for learners in many ways. In the context of outcome-based education, the learner gets the flexibility to set the pace of learning and asses the gradual step-up in performance. These tools also provide an environment of student-centric learning. The learner can not only monitor attendance, submit assignments, and monitor grades but also get real-time feedback. Setting long-term goals makes it possible for them to stay engaged and focused for better performance. Through predictive analytics, the learner can understand the real cause of low performance and plan for overall enhanced performance. They can also get advice on extra tutoring, counselling, and other academic interventions in case of consistently low performance. BI tools assist the learner in choosing the right courses, aligning with their larger career goals.

BI tools help higher education institutions to transform the academic process from manual to automated. And there is a shift from personal experience or gut feeling-based decision-making to the data-driven decision-making culture. BI and its allied technologies provide a clear picture to the institutes regarding their strengths, opportunities, and the areas in which they require to improve. There will be a few challenges in embedding BI and data analytics in the working of higher education in terms of technical skills available, resources & budgets, appropriate training, and data security. However, the benefits outweigh these challenges and provide a long-term enhancement in performance for all the stakeholders. The current chapter studies the role of Business Intelligence and the benefit of big data analysis in higher education, focusing on three important stakeholders: administration, teachers, and learners (Al-khateeb, 2024; Patel, 2024).

1.1 Chapter Organization

The focus is on the effectiveness of BI tools in higher education. It begins by introducing the concept of integrating technology, specifically BI, into operations and decision-making processes of higher education institutions. Following the introduction, the chapter provides a detailed overview of the work conducted by previous researchers, primarily referencing papers from the last ten years. The literature review yields three significant perspectives: administrator, teacher, and learner. These perspectives are captured through data. The data collection and analysis occur in two stages: qualitative and quantitative. The mixed-method approach enhances the insights reflected in the data analysis. Finally, the conclusion connects the proposed objectives to the findings, highlighting the importance of BI in the higher education system and the factors that contribute to effective integration.

1.2 Objectives of the Study

The current chapter highlights the importance of BI tools in the higher education sector with the following objectives:

- To explore the utility of BI tools in the higher education sector.
- To understand the administrator's, teachers', and learners' perspectives on the efficient application of BI tools in higher education.

2. Literature Review

Business Intelligence refers to the technology, algorithms, methods, and analytical tools to store and analyze data effectively. The analyzed results and the reports are further used for data-driven decision making. Simplistically put, the wise decision making based on data is BI. Data mining algorithms, data analytics, reporting, and visualization are the most important aspects of BI-based applications. It provides meaningful insights, uncovering the hidden pattern and trends in the data. The strength lies in analyzing a huge amount of varied data coming from multiple sources. With time BI has become the important aspect of strategic planning in the organization and proved its utility across many sectors (Mobo, 2021).

The education sector has witnessed the evolution and adoption of BI and related applications in the last decade. A vast amount of data is created and recorded by the use of computer-based teaching and learning tools as well as the integration of social media, the internet, and the internet of things. This provides a seamless integration of BI in the education system. To assess and act upon the huge volumes of data that educational institutions today have access to—from financial reports and institutional goals to student performance indicators and attendance records—effective business intelligence (BI) systems are required. The administration and the teachers are greatly supported through BI for data-driven decision-making. These decisions are available in short timelines and thus the problem is solved and future planning is foolproof (Mukul and Büyüközkan, 2023).

In the education system, a wide variety of data is available; students' enrolment data, student performance data, data regarding use of online portals, student engagement in terms of online content, etc. All of this data is analyzed to help administration and teachers improve student performance, which in turn improves higher education success. There are a certain set of students who may require financial support, BI enables the smooth and speedy process for loan or funding for such learners. There is a lot of potential that the education sector can leverage by integrating BI in its working system. There can be a lot of challenges in doing so, however, the proper planning and technology strategy can make this a seamless process.

2.1 Understanding BI

Business Intelligence has been defined and described by many researchers. According to (Yoon, Ghosh, and Jeong, 2014) BI as the "combination of devices that include such as data analysis and query that produces robust reports presentations which enable a strong precision in decision-making". In a recent research by Al-khateeb (2024), the description is that "BI covers things such as technologies, applications, including data analysis which facilitate decision". The importance of BI and big data analytics in higher education has been described as the ultimate approach to deliver quality education (Sorour and Atkins, 2024).

2.1.1 Component of BI

The BI applications are an assembly of various components. All these applications have three basic aspects, the first is to store the voluminous data that comes from a variety of sources. The collection of mining tools for analyzing data using different algorithms is the second crucial component. And the last one is related to reporting and visualization to support data-driven decision making. The components under these three characteristics are explained in detail in Fig. 1. The BI applications harness the strength of big data analytics along with the suitable data models to generate reports and dashboards which give a bird's eye view and real insights for better decision making. Figure 1 elaborates on the above-discussed component of BI (Miranda, et al., 2021).

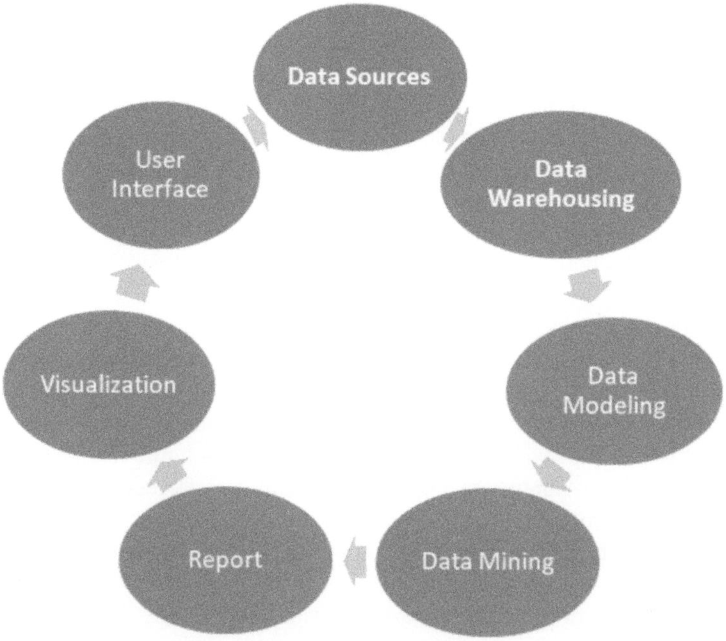

Fig. 1 Component of BI applications.
Source: Author Creation

2.1.2 Sources of Data in Education System

One of the biggest strengths and challenges of the BI-based applications is the variety of data sources and different categories of data. To manage this complex data, the data warehouse is created by an extensive process of extraction, cleaning, and transformation. This enables successful application of mining tools to generate hidden trends and insights. Concerning the education sector, following are the important sources of data:

- Learners' data—The student database is updated as soon as the student enrolls. Additionally, it keeps track of information about performance, behavior, health, and academics. Exams and assessments serve as a vast data warehouse that aids in determining how well pupils are progressing and meeting learning objectives.
- Curriculum related data—There are special learning management systems to keep records of the curriculum offered to every batch. The detailed course material and other details related to projects and internship are also maintained. Most higher education institutes also facilitate international exposure and hence data related to this is also recorded.
- Institutional data—There is a huge amount of data stored related to the infrastructure, the student, the faculty, staff, and related operational data.

- Teacher data—In the operational front, the subjects taught and the corresponding feedback also provide meaningful insights.
- Behavioral data—It is very important to keep a record of the disciplinary actions taken on the students for maintaining the sanctity of the educational environment. In addition, a record of the student's conduct with the instructor and peer group in personal interactions as well as through collaborative technologies is kept.
- Learning environment and student engagement—The learning environment data includes both physical and technology-based facilities. This will include information about classrooms, labs, libraries, and classroom equipment. Additionally, the list of e-content databases and other e-resources is kept up to date for operational, financial, and decision-making reasons. The way the learner navigates through the content of the courses, assessments, and other e-resources reflects their level of engagement. Analyzing this type of data benefits both the learner and the teacher in planning future learning paths.
- Alumni data—Alumni are important stakeholders for education institutes. The carrier progression of the pass out students in terms of further higher education, job, or self-employment is very important for the higher education institutes.
- Others—There are many sets of external data that are useful for the educational institutes. There are many entrance tests both at the national, state, and institute levels, which are obtained and used by these institutes. A lot of analysis is done on the student demographic data for the admission purpose. Aside from this, the institutions prioritize providing high-quality instruction through highly qualified teachers, so their hiring process involves a significant amount of data management. The institutes can not run without a required budget, hence financial data is also maintained accurately by the institutes.

All the above data constitutes the need for operational and decision-making needs and makes it possible to effectively use BI (Nguyen, Gardner and Sheridan, 2020).

2.1.3 Challenges of Incorporating BI tools

BI tools are powerful means of taking good decisions based on facts, however, their application and usage are associated with many challenges. BI requires storage of a large amount of historical data and connectivity with the real-time data. To store this voluminous data, the warehouse is used. The very first challenge is integration of data coming from multiple sources and ensuring its quality. There are issues of fragmented and missing data to be taken care of before these applications are used successfully. Inconsistent, inaccurate, and un convertible data also is a problem area during the creation of the data warehouse. Another challenge is in terms of a lack of skilled workforce to implement this application, specifically in the context of the education sector. Even when the education institute outsources the work of implementing the BI application, there has to be an expert within the organization to have an apt fitment with the education process requirement. Unlike big corporate houses, higher education institutes may face budget constraints to deploy a BI application and invest in the related technology infrastructure. While the administration plan for this, the ongoing cost should also be considered. Data privacy and security are another concern. The sensitive student data, performance data, and other financial data are susceptible to security breaches. By and large, most of the higher education institutions have common academic processes, however, there may also be specific needs for which the BI applications must be customizable. The high cost of such an application is a barrier to customized BI products. The training and change management too are complexities that the institutes face during upgrading the existing system to the BI-based applications. Although BI technologies have a lot of potential to promote data-driven decision-making and boost productivity in higher education, overcoming these obstacles calls for a carefully considered approach (Sorour and Atkins, 2024).

2.2 Stakeholders Perspective

The importance of BI in the education sector has been recognized in the last decade. As per the research in this area, BI supports in delivering quality education and management in the higher education institutes (Pérez-Pérez, 2018; Patel, 2024). Many articles emphasize the student's perspective, however, the current chapter explores a three-dimensional perspective (Sorour and Atkins, 2024; Singh and Miah, 2020). The three important stakeholder perspectives are discussed in detail, namely the administrator's, the teacher's, and the learner's perspective.

2.2.1 Administrator's Perspective

The integration of technology and specifically BI requires a strong strategic planning at the management level. Well planned deployment of BI tools facilitates the effective utilization of this integration by the end users.

The first and foremost step is to chalk down the requirement in terms of functionality and the corresponding choice of BI tool as per the budget planned. In the next stage it is vital to plan and facilitate for data integrity. Along with it, data quality and accuracy also need to be ensured. The data-driven decision making is successfully utilized only when the data is clean, complete, and accurate. Timely and pertinent decision-making depends on the data utilized in BI systems being current and reflecting actual institutional processes. A proper integration of data across all the departments and removing all possibilities of data silos must be done for creating a strong data warehouse. In doing so, strong collaboration and engagement of all stakeholders is vital. To keep clear communication among all the departments, a visionary and transformational leadership is required at the higher education institutions. Since the users and specifically students would need the facility to access resources, etc. anytime-anywhere, the interface of the application should be user-friendly and easy to access. There should be proper training for all the departments and stakeholders to utilize the BI-based educational system

(Moraes, et al., 2023). Whenever a new technology is adopted in any organization, the management must have a strategy to manage change and develop a technology-friendly and technology-driven culture. Training being one of the ways to do so. Deciding the key performance indicators is an important task at the administrative level. The key performance indicators (KPI's) and the educational metrics help in decision making through effective dashboards and analytical reports. These reports reflect administrative performance along with teaching-learning performances. When dealing with online portals and online analytical systems, data security and privacy are of utmost importance. Compliance with the regulations, access authorization, data encryption and protection are important aspects involved in this. While planning for technology integration, efficient resource planning and management, and budgeting are vital. The institution's budget must be in line with the implementation of BI. It is important to keep track of both the initial and continual expenditure on the technology infrastructure. It's critical to regularly assess the return on investment of BI technologies (Salaki and Ratnam, 2018). The administration of the higher education can make the successful use of BI tools by focusing on quality data, collaboration among stakeholders, managing KPI's, ensuring data security and privacy, and planning well, keeping the latest technology in mind.

2.2.2 *Teacher's Perspective*

Teachers play an important role in the successful adoption of technology in the higher education system. All the strategies and policies formulated by the management and administration are communicated to the learners through teachers. Teachers can be important stakeholders for providing feedback in the requirement gathering stage to build BI-based applications. The data collected regarding the learning management system helps the teachers to update the curriculum, improve pedagogies, harness the benefits of technology, and analyze the student performance to take necessary remedial actions. With the implementation of new technology in the higher education sector, appropriate training is required for the teachers to use the BI-based application to its fullest. The BI application should have an effective user interface and should provide a great user experience. Based on the grades, the teachers should be able to identify early during the session regarding the low performing students (Harden, 2007). The analytical tools help them to take corrective measures and provide personalized attention and academic support to the students. The study's greatest beta value further supports the conclusion that students are the ultimate beneficiaries of the overall educational system. Apart from this, the education sector has evolved from traditional classroom teaching to a blended learning model and moving to futuristic smart education. In this case, student information can be found on social media-based teaching-learning platforms, learning management systems, and other popular sources for discussion boards and assignments (Hung and Yuen, 2010). Integrating all such platforms and sources is vital for effective analysis, providing important insights to the teachers.

Such tools not only help the teachers to support their students but also enhance their professional ability.

2.2.3 Learners Perspective

Technology has significantly changed the way students choose their learning path, get engaged, and utilize the learning resources. Use of BI applications has further added to these benefits. The use of BI and data analytics provided academic support through personalized learning. Learners can easily identify at the early-stage academic challenges if any based on the grades, attendance, and engagement data. BI tools can provide recommendations for both advanced and low learners. These tools suggest courses and paths that align with the larger academic goals. Learners have access to the real-time data and can track their progress. The transparency of these applications enables the students to monitor their academic success. The BI tools very efficiently highlight the progress of the learners in terms of the peer performance and outcome expectation of the course itself. Online material consumption, assignment completion, and discussion forum involvement are other indicators of student engagement (Scholtz, Calitz, and Haupt, 2018). The BI tools give proactive support to the learner by predicting the overall future performance based on the history and current performance of the student. They also get suggestions regarding any interventions required. Since a lot of data is stored and analyzed for data-driven student performance monitoring, the students must be educated regarding data security concerns. There is a strong collaboration with the faculty and the peer group. This advanced course site now offers peer feedback in addition to teacher remarks. Thus, the peer learning process is also the highlight of BI applications for education. With these kinds of performance tracking and feedback systems, the ownership of the students increases. This leads to the student-centric teaching-learning process. According to students, real-time data, transparency, and customisation are essential for the effective use of BI tools in higher education since they enable students to take control of their education. Students can attain better results, enhance their educational experience, and make better decisions regarding their academic and professional pathways when BI tools offer proactive support, actionable insights into academic performance, and improved communication with advisors and faculty (Moscoso-Zea, Castro, Paredes-Gualtor and Luján-Mora, 2019).

As seen in Figure 2, the literature review was completed in the following steps. This procedure produced the list of variables found in Table 1.

Based on the literature, the following variables have been identified for the effective application of BI in the higher education sector.

3. Method and Data Analysis

The current chapter focuses on understanding the stakeholder perspective on the effective application of BI in the higher education sector. The literature provided the variables to be further tested through the data collected. The data collection

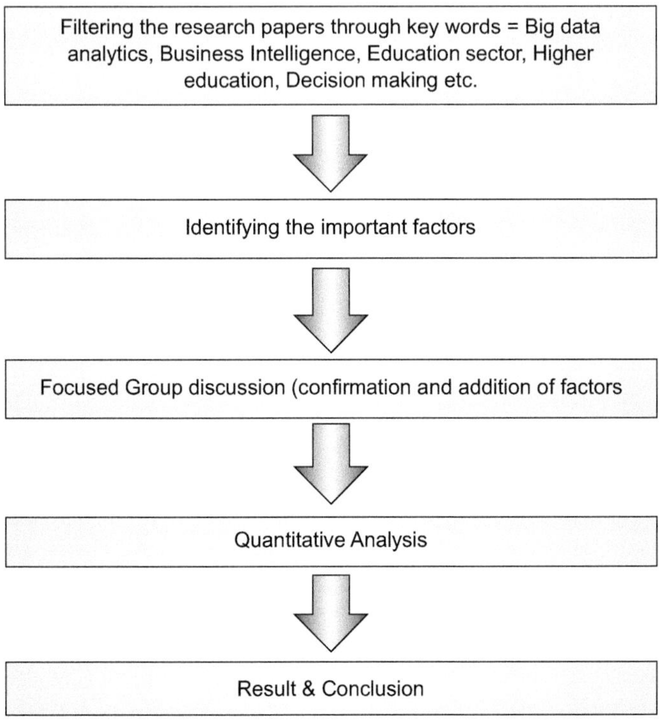

Fig. 2 The flow of literature review (*Source:* Author creation)

and methodology of the current study were divided into two stages: qualitative and quantitative.

3.1 Qualitative Stage

The qualitative study was done through a focused group discussion with the experts. The experts were the members of the administration and senior teachers in the five higher education institutions. A total of ten respondents participated. From the administration perspective, a new variable in terms of financial requirement for the BI infrastructure was highlighted. Thus, the fifth variable was added to the study as a result of the qualitative study.

3.2 Quantitative Stage

The effective application of BI in the higher education sector was studied from the perspective of administration, teachers, and students. After the qualitative study, the variables were finalized and a survey instrument was prepared to record these perceptions measured on a Likert scale. The total of 220 data were collected out of which 205 were further usable for analysis. PLS-Sem was applied using Warp PLS to analyse the data.

Table 1 List of variables

Factors	*Variables*	*Citation*
Administrators Perspective	Data quality & integration	Daniel (2019), Moraes, et al., (2023)
	Stakeholder collaboration & training	Al-khateeb (2024), Mukul and Büyüközkan (2023)
	Data privacy & security	Fischer, et al. (2020), Salaki and Ratnam (2018)
	Change management & decision making	Doleck, et al. (2020), Patel (2024)
Teachers Perspective	User friendly interface	Englund, et al. (2017), Nguyen, Gardner, and Sheridan (2020)
	Integration with learning platform	Díaz-Parra, et al. (2022), Shah (2022).
	Improved assessments and feedback	Alkhalil, et al. (2021), Singh and Miah (2020)
Students Perspective	Personalized learning and academic support	Baig, et al. (2020), Gaardboe and Jonasen (2018)
	Engagement and improved learning environment	Bond, et al. (2020), Keser and Semerci (2019)
	Peer support and learning	Bonfield, et al. (2020), Sorour and Atkins (2024)
	Setting learning path and learning goal	Foster and Francis (2020), Ifenthaler and Gibson (2020)

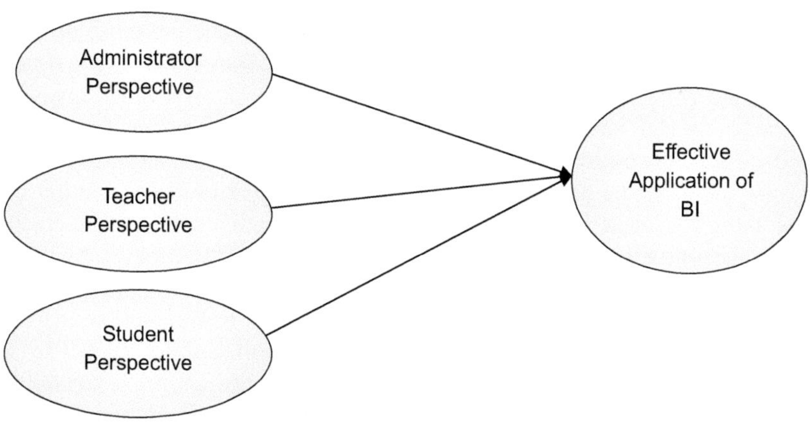

Fig. 3 Research Model.
Source: Author creation

The research model of the study is elaborated in Fig. 3. The results of the SEM analysis are provided in Table 2.

Table 2 Dependent Variable: Effective Application of BI.

Predictor	*Construct validity*	*Beta value*	*t-value*	*p-value*	*R square value*
Administrators Perspective	0.85	0.30	3.40	0.011	0.63
Teachers Perspective	0.76	0.28	2.10	0.030	
Students Perspective	0.82	0.44	4.62	0.001	

From the result given in the result, it is apparent that all three perspectives are a significant contribution in explaining the effectiveness of the Application of BI in the higher education. The study's greatest beta value further supports the conclusion that students are the ultimate beneficiaries of the overall educational system.

4. Discussion and Conclusion

The integration of technology in the education sector, specifically the higher education, has transformed it from the traditional face-to-face teaching to the blended model in the Education 4.0 era. With the integration of new technology, the way the education process is conducted and the teaching and learning is carried out requires new strategies in terms of pedagogy and student engagement activities. All these factors constitute a great decision-making challenge. As a result, BI plays a crucial part in the higher education institution's strategic planning and policymaking. The post-COVID era is witnessing more and more use of online platforms for the delivery of the sessions, which varies from Zoom, Google Meet, WebEx, etc. Not only do these platforms facilitate synchronous sessions, but digital and internet portals have also seen an evolution in learning resources. The e-books, the databases, the e-resources, online submission and evaluation of assignments, and the social media-based discussion forums have led to the generation of a huge amount of data. The teachers can also create synchronous yet interactive videos using prompts and quizzes. Which further adds to the variety and amount of data. With this enormous data available, the applicability of BI tools becomes seamless. There are many challenges in implementing BI-based solutions in higher education. However, all of the advantages of BI tools can be utilized by the institutes if these issues are resolved, including the use of appropriate technology, training and change management inside the institute, data privacy, and security.

Further to this, if the higher education institutes must manage these challenges from a specific stakeholder's perspective. It is imperative to take a holistic approach and blend the need and challenges from the administrative, teacher, and student perspective. To support teachers and students, educational administrators must create explicit policies that encourage the efficient use of technology and BI tools. They must also make sure that training and infrastructure are available. To improve their ability to understand data for better instructional techniques, teachers must embrace technology as a tool to enhance learning and adjust to new pedagogical approaches. The student needs to shift their understanding of the outcome-oriented education. The BI tools support them by giving them insights regarding their performance and engagement with the teaching content. The personalized curriculum and guidance further lead them to the student-centric learning approach. It can be concluded that the use of BI tools in higher education supports all aspects of teaching, learning, and all the stakeholders efficiently.

References

Al-khateeb, B.A.A. (2024). Business Intelligence (BI): A Critical Strategy for University Success and Sustainability. *International Journal of Asian Business and Information Management* (IJABIM), 15(1), 1–15.

Al-Khateeb, B.A.A. (2024). Business Intelligence (BI). International Journal of Asian Business and Information Management, 15(1), 1–15. https://doi.org/10.4018/ijabim.340387

Alkhalil, A., Abdallah, M.A.E., Alogali, A. and Aljaloud, A. (2021). Applying big data analytics in higher education: A systematic mapping study. *International Journal of Information and Communication Technology Education (IJICTE)*, 17(3), 29–51.

Baig, M.I., Shuib, L. and Yadegaridehkordi, E. (2020). Big data in education: a state of the art, limitations, and future research directions. *International Journal of Educational Technology in Higher Education*, 17, 1–23.

Bond, M., Buntins, K., Bedenlier, S., Zawacki-Richter, O. and Kerres, M. (2020). Mapping research in student engagement and educational technology in higher education: A systematic evidence map. *International Journal of Educational Technology in Higher Education*, 17, 1–30.

Bonfield, C.A., Salter, M., Longmuir, A., Benson, M. and Adachi, C. (2020). Transformation or evolution?: Education 4.0, teaching and learning in the digital age. *Higher Education Pedagogies*, 5(1), 223–246.

Daniel, B.K. (2019). Big Data and data science: A critical review of issues for educational research. *British Journal of Educational Technology*, 50(1), 101–113.

Díaz-Parra, O., Fuentes-Penna, A., Barrera-Cámara, R.A., Trejo-Macotela, F.R., Ramos-Fernández, J.C., Ruiz-Vanoye, J.A.,... and Rodríguez-Flores, J. (2022). Smart education and future trends. *International Journal of Combinatorial Optimization Problems and Informatics,* 13(1), 65.

Doleck, T., Lemay, D.J., Basnet, R.B. and Bazelais, P. (2020). Predictive analytics in education: a comparison of deep learning frameworks. *Education and Information Technologies*, 25, 1951–1963.

Englund, C., Olofsson, A.D. and Price, L. (2017). Teaching with technology in higher education: understanding conceptual change and development in practice. *Higher Education Research & Development*, 36(1), 73–87.

Fischer, C., Pardos, Z.A., Baker, R.S., Williams, J.J., Smyth, P., Yu, R.... and Warschauer, M. (2020). Mining big data in education: Affordances and challenges. *Review of Research in Education*, 44(1), 130–160.

Foster, C. and Francis, P. (2020). A systematic review on the deployment and effectiveness of data analytics in higher education to improve student outcomes. *Assessment & Evaluation in Higher Education*, 45(6), 822–841.

Gaardboe, R. and Jonasen, T.S. (2018). Business intelligence success factors: a literature review. *Journal of Information Technology Management*, 29(1), 1–15.

Harden, R.M. (2007). Outcome-based education: the future is today. *Medical Teacher*, 29(7), 625–629.

Harish, V., Sharma, R., Rana, G. and Nayyar, A. (2023a). Artificial Intelligence in Sustainable Education: Benefits, Applications, Framework, and Potential Barriers. In *The Role of Sustainability and Artificial Intelligence in Education Improvement* (pp. 219-236). Chapman and Hall/CRC.

Harish, V., Sharma, R., Rana, G., Nayyar, A. and Parashar, B. (2023b). Expanding the Scope of Learning Analytics in Blended Learning Environments. In *Sustainable Blended Learning in STEM Education for Students with Additional Needs* (pp. 317–340). Singapore: Springer Nature Singapore.

Hung, H.T. and Yuen, S.C.Y. (2010). Educational use of social networking technology in higher education. *Teaching in Higher Education*, 15(6), 703–714.

Ifenthaler, D. and Gibson, D. (Eds.). (2020). Adoption of data analytics in higher education learning and teaching. Cham, Switzerland: Springer.

Keser, H. and Semerci, A. (2019). Technology trends, Education 4.0 and beyond. *Contemporary Educational Researches Journal*, 9(3), 39–49.

Lorenzen, M. (2021). Using outcome-based education in the planning and teaching of new information technologies. *Information Technology Planning*, 141–152.

Mansor, N.A., Abdullah, N. and Rahman, H.A. (2020). Towards electronic learning features in education 4.0 environment: literature study. *Indonesian Journal of Electrical Engineering and Computer Science*, 19(1), 442–450.

Miranda, J., Navarrete, C., Noguez, J., Molina-Espinosa, J.M., Ramírez-Montoya, M.S., Navarro-Tuch, S.A.... and Molina, A. (2021). The core components of education 4.0 in higher education: Three case studies in engineering education. *Computers & Electrical Engineering*, 93, 107278.

Mobo, F. (2021). The role of emerging trends in education. *International Journal of Multidisciplinary: Applied Business and Education Research*, 2(10), 909–913.

Moscoso-Zea, O., Castro, J., Paredes-Gualtor, J. and Luján-Mora, S. (2019). A hybrid infrastructure of enterprise architecture and business intelligence & analytics for knowledge management in education. *IEEE Access*, 7, 38778–38788.

Moraes, E.B., Kipper, L.M., Hackenhaar Kellermann, A.C., Austria, L., Leivas, P., Moraes, J.A.R. and Witczak, M. (2023). Integration of Industry 4.0 technologies with Education 4.0: advantages for improvements in learning. *Interactive Technology and Smart Education*, 20(2), 271–287.

Mukul, E. and Büyüközkan, G. (2023). Digital transformation in education: A systematic review of education 4.0. *Technological forecasting and social change*, 194, 122664.

Nguyen, A., Gardner, L. and Sheridan, D. (2020). Data analytics in higher education: An integrated view. *Journal of Information Systems Education*, 31(1), 61.

Parashar, B., Sharma, R., Parashar, V., Rana, G., Nayyar, A. and Harish, V. (2023). An Overview of the Accessibility and Need of AI Animation Tools for Specially Abled Students. *AI-Assisted Special Education for Students With Exceptional Needs*, 1–22.

Patel, K. (2024). An analysis of quality assurance for business intelligence process in education sector. *IJNRD-Int. J.* Nov. Res. Dev, 9(9), a884–a896.

Pérez-Pérez, Y.M., Rosado-Gómez, A.A. and Puentes-Velásquez, A.M. (2018, November). Application of business intelligence in the quality management of higher education institutions. *In Journal of Physics: conference series* (vol. 1126, No. 1, p. 012053). IOP Publishing.

Salaki, R.J. and Ratnam, K.A. (2018). Agile Analytics: Applying in the Development of Data Warehouse for Business Intelligence System in Higher Education. In: Rocha, Á., Adeli, H., Reis, L.P., Costanzo, S. (eds) *Trends and Advances in Information Systems and Technologies. WorldCIST*'18 2018. *Advances in Intelligent Systems and Computing*, vol. 745. Springer, Cham. https://doi.org/10.1007/978-3-319-77703-0_101

Scholtz, B., Calitz, A. and Haupt, R. (2018). A business intelligence framework for sustainability information management in higher education. *International Journal of Sustainability in Higher Education*, 19(2), 266–290.

Shah, T.H. (2022). Big data analytics in higher education. Research Anthology on Big Data Analytics, Architectures, and Applications, 1275–1293. DOI: 10.4018/978-1-6684-3662-2.ch061. Retrieved from: https://www.igi-global.com/chapter/big-data-analytics-in-higher-education/291037 on 8/12/2024

Singh, H. and Miah, S.J. (2020). Smart education literature: A theoretical analysis. *Education and Information Technologies*, 25(4), 3299–3328.

Sorour, A. and Atkins, A.S. (2024). Big data challenge for monitoring quality in higher education institutions using business intelligence dashboards. *Journal of Electronic Science and Technology*, 22(1), 100233.

Williamson, B. (2015). Digital education governance: data visualization, predictive analytics, and 'real-time' policy instruments. *Journal of Education Policy*, 31(2), 123–141. https://doi.org/10.1080/02680939.2015.1035758

10 Can FoMO Enhance Continuous Intention to use ChatGPT?: Mediation of Innovative Capability of using Data Analytics

Poornima Jirli[1*], Anuja Shukla[2] and Arpana Kumari[3]

AI has been evolving quickly lately, and its applications span a wide range of fields, from academia to various facets of daily life. The chapter talks about the ongoing use of AI technologies in academic tasks, how ChatGPT may support teachers, and how it can help with tailored learning. We investigate the factors that influence students' continuous intention to use generative artificial intelligence (GenAI). In quantitative analysis, data was collected from 296 post-graduate management students trained on the usage of responsible AI. The results have shown that fear of missing out (FOMO) significantly affects continuous intention to use AI tools, besides increasing innovative capability to use data analytics, which again increases the likelihood of continued AI use in academic settings. It considers ethical issues and challenges related to the adoption of AI within educational systems, focusing on the threats of digital tools to academic integrity and data privacy. The findings underscore the significance of ethical frameworks and guidelines that ensure the responsible use of AI tools in academics. Practical suggestions are offered for management and academic institutions. To effectively and sustainably integrate AI, it is important to understand the relationship between FOMO, innovation, and continuous intention.

1. Introduction

With the rapid growth and development in artificial intelligence over the past few years, innovative tools have been introduced within different sectors, including academics (Kumari and Saharan, 2020). ChatGPT is an AI-powered tool taken by force from OpenAI. This innovative and powerful tool is considered one of the most versatile and influential tools of this era. ChatGPT was an innovative language model that found broad applications in academic settings; deep learning techniques have been used to construct human-like tests from the given inputs. (Brown et al., 2020).

[1] 0009-0000-3796-6974
[2] 0000-0002-3001-5962
0000-0001-8752-674X
* Corresponding author: Poornima.j85@gmail.com

DOI: 10.1201/9781003514282-10

AI-powered tools, like ChatGPT, will eventually enable a transformative revolution in management research and learning. Personal tutors, ChatGPT will be able to perform detailed questioning and homework assignments in many subjects. In such a capacity, ChatGPT allows students to comprehend the most elusive concepts with depth, thereby facilitating independent learning beyond the confines of a conventional classroom (Jones, 2022). ChatGPT allows for optimization of resources, thereby increasing productivity and benefiting the educators. ChatGPT can help with composing quiz questions and creating educational content. Feedback could also be provided through ChatGPT, and it could assist students in formulating the ideas for their assignments. In that respect, educators could focus on personalized instruction and mentorship. This would only enhance the quality of education, which is achieved with the AI tool ChatGPT (Smith and Lee, 2021).

It is a very important tool for research, ranging from literature reviews to data analysis. It could summarize relevant information from different sources and suggest a possible area of research effectively and efficiently. ChatGPT's capacity to foster creativity and critical thinking in researchers will speed up the research process and spur innovation (Wang et al., 2021). The biggest role of ChatGPT in academics shall be streamlining administrative functions. These include handling routine inquiries, course and program information, and other such administrative tasks. All these tasks would find their perfect fit with ChatGPT, thereby reducing the workload and raising operational efficiency (Taylor and Kumar, 2023). The use of an AI tool like ChatGPT is a two-edged sword because it carries immense benefits while also raising several critical concerns related to issues of equity and academic integrity. Integrating artificial intelligence (AI) like ChatGPT into education would require careful thought and a careful assessment of how technology can be used for good without sacrificing moral principles (Farrelly and Baker, 2023; Rana et al., 2024; Kumari and Singh, 2022).

AI tools, including ChatGPT, have a host of advantages in academics. Because of this, serious scrutiny of issues relating to data privacy, academic integrity, and general reliance will be warranted. Therefore, very strong guidelines and an ethical framework must be implemented to ensure that the use of ChatGPT complements learning rather than degrades it (Nguyen and Chang, 2023).

In addition, new technologies such as data analytics are gaining much importance within this evolving world of academics due to their offerings and hence lead from the front in shaping the world of education (Kumari et al., 2023; Sharma et al., 2021). The reason for this lies in the massive gain of technologies in academics because of FOMO. The feeling of FOMO makes students try the AI tool ChatGPT; hence, students not using ChatGPT feel that they are going to miss something, whereby those using the tools will ace their assignments and gain deep insights on hard subjects. FOMO does not spare even the field of education. Students not utilizing the technology may feel they are missing out since other students are using it to produce engaging lesson plans, intelligent exam questions, and real-time feedback. This is because they are driven to learn how to use ChatGPT, an AI tool, to avoid feeling behind in terms of advanced features. This could lead to improved teaching quality and more inventive capabilities.

FOMO motivates educators and students to innovate, and the usage and intention to use ChatGPT are further strengthened. In this respect, FOMO and the increasing use of AI tools are a bridge for their innovative capability; in academics, it's always the race and the need to be topmost. FOMO drives people not only to utilize AI tools like ChatGPT but also to create innovation and understand the engagement with such technologies more deeply.

Thus, the study aims to answer the following research questions:

RQ1: How does the fear of missing out (FOMO) influence the intention to use ChatGPT among students, educators, and researchers in academic settings?

RQ2: To what extent does the fear of missing out (FOMO) drive innovative capability to use data analytics in the academic use of ChatGPT?

RQ3: Does innovative capability to use data analytics mediate the relationship between the fear of missing out (FOMO) and the intention to use ChatGPT in academic environments?

The results from this study will be useful in formulating guidelines and best practices that can help apply ChatGPT ethically and responsibly within an academic setting. These shall be developed in a way that the benefits of this AI tool are maximized while lessening potential risks in its use and making sure it integrates into academia in a manner that is meant to safeguard academic integrity and equity.

2. Literature Review

There has been substantial growth in education because of AI applications. Moving further, one advanced version of natural language processing (NLP) was ChatGPT, which has the potential to process and transform various components of academics. This literature review, therefore, tries to develop an encompassing review of the role of ChatGPT in academically synthesizing the available literature and exploring various implications, advantages, and challenges associated with the integration of ChatGPT in academics (Pasca and Arcese, 2024).

One of the most talked-about topics in recent studies is ChatGPT's capacity to recognize different learning styles. Its ability to analyze the student's learning preferences based on learning history provides features of personalized tutoring, tailored content, and delivers according to the requirement. AI learning tools, such as ChatGPT, offer that viability where one does not have to continuously bear anxiety of lagging in academic tasks by continuous engagement and instantaneous feedback (Alshahrani, 2023). Different characteristics were varied per user, including age, educational background, and culture. (Kumari and Saharan, 2021). More importantly, there is improved understanding and knowledge since ChatGPT was able to make every experience unique for its users (Koyutürk et al., 2023).

Learning experiences were enhanced by the use of ChatGPT in the classroom, and institutional operations responded to the query of whether adopting AI technologies would have positive effects. According to Alhumaid et al. (2023), integration with ChatGPT will make possible personalization in the learning domain, offering dynamic assessment and touch points for online, mobile, and blended learning with authentic interactions. The usage of ChatGPT in education augmented and enhanced learning experiences and institutional functions, exemplifying gains from the adoption of AI technologies (Alhumaid et al., 2023). Through dynamic evaluation and efficient engagement with online, mobile, and blended learning environments, integration with ChatGPT could be implemented to customize the learning experience (Popescu, 2023). Integrating ChatGPT will allow one to construct effective learning experiences among students and educators.

2.1 Fear of Missing out (FOMO) and Continuous Intention to use

The use of the FOMO concept has been extensively researched in relation to social media and technology use. FOMO is the anxiety experienced upon the belief of one's self missing out on some of the quintessential experiences that others are enjoying. The concept has been extended to the field of education.

Students not using new technologies, such as ChatGPT, would feel FOMO from friends or colleagues using the technology. This feature, by supporting continuous engagement and immediate feedback, reduces anxiety related to being behind in academic activities. Hence, the integration of AI tools within blended learning environments will reduce FOMO (Alshahrani, 2023). The perceived compatibility, relative advantage, and trialability factors alone mostly govern the adoption of AI in education. Therefore, understanding the implication of FOMO in technology adoption becomes important (Alhumaid et al., 2023). Misusing tools like ChatGPT for specific AI-powered cheating and academic dishonesty might lead to several unintended consequences because of the psychological drive arising with FOMO (Lo, 2023). Any risks involved should be decided in a way such that, on the whole, AI tools can be utilized for good in education.

The Fear of Missing Out has recently been a popular concept in the context of technology adoption. It is generally defined as a "pervasive apprehension that others might be having rewarding experiences from which one is absent" (Przybylski et al., 2013). This FOMO forces people to always remain observant and active on the digital platforms and technologies lest they miss something (Dossey, 2021). In emerging research related to technology usage, FOMO is a direct connector between the use of new technologies. Individuals experiencing higher levels of FOMO are most liable to use social media sites frequently and adopt them since such anxiety about missing out on the social content and interaction force them to do so (Abel et al., 2016). In education, students adapt to new learning platforms such as ChatGPT, experiencing FOMO, to ensure that they do not fall behind their peers (Roberts and David, 2020).

FOMO acts as a strong motivator for technology adoption through which students can meet their needs related to seeking and sustaining relevance in both the social and academic realms. FOMO increases the frequency of using technology and strengthens the intention to integrate those technologies into daily routines

(Wegmann et al., 2022). Therefore, the literature study currently indicates that FOMO has a favorable effect on the intention to utilize new technologies, which lends some credence to the idea that FOMO is what motivates these individuals to embrace ChatGPT-like tools. Therefore, we hypothesize that:

H1: Fear of missing out (FOMO) positively impacts continuous intention to use AI in academic writing

2.2 Fear of Missing out (FOMO) and Innovative Capability to use Data Analytics

FOMO acts not only as a driver of the intention to use new technologies but also considerably contributes to the creation of innovative capability to use data analytics. The sheer idea of FOMO naturally motivates people to keep up with or ahead of their peers in a variety of spheres of life, including embracing new concepts and technologies (Przybylski et al., 2013). Apart from this psychological drive, it might provoke heightened creativity and a proactive stance toward the search for new solutions (Beyens et al., 2016). The linkage to technology is such that FOMO has been associated with a greater readiness to experiment with and adopt new tools, hence improving innovative capability (Li, 2021). FOMO leads people to keep searching and adopting newer technologies continuously to avoid exclusion, thereby developing innovative skills and approaches in the process (Gupta et al., 2022).

For example, students with FOMO in educational contexts may exhibit a higher desire to explore new ways of learning and new tools, such as ChatGPT, which could also enhance their creative problem-solving skills and academic performance (Li et al., 2021). According to other recent research, FOMO may foster a competitive environment where people constantly push themselves, encouraging creativity and innovation (Wegmann et al., 2022). Thus, the literature depicts how FOMO influences technology adoption and positively affects the ability to innovate among people.

H2: Fear of missing out (FOMO) positively impacts innovative capability.

2.3 Innovative Capability to use Data Analytics and Continuous Intention to use

Traditionally, educational innovation has been driven by early adopters of technology. ChatGPT is a new AI tool with opportunities for developing new learning experiences with more interactivity and personalization. Therefore, students who consider themselves technology early adopters are supposed to be more active users of the AI tool and use data analytics and ChatGPT. In most cases, such an attitude is encouraged by the students' interest in exploring new educational possibilities and finding ways to improve their learning process. Farrelly and Baker (2023) refer not only to fostering innovation in the blended learning system but also to the sustainability and scalability of educational models through enhanced content delivery with AI-powered tools like ChatGPT and automated administrative tasks. The impact of such innovation is corroborated in research that suggests significant enhancement in problem-solving skills and critical thinking abilities among students with the use of AI technologies (Alhumaid et al., 2023). The fact that AI tools like ChatGPT help students develop their critical thinking and problem-solving abilities, making learning outcomes more creative and efficient, further encourages early adopters (Popescu, 2023).

Innovative capability to use data analytics and intention to use new technologies are closely related, as observed in the literature. Innovative capability is defined here as the capacity of an individual to generate new ideas and put them into practice effectively, leading mostly to the adoption of new tools and practices (Amabile, 1988). Various studies prove that highly innovative-ability individuals like to adopt new technologies more often since innovators are curious enough to try new and further advanced solutions (Souto, 2015; Mishra and Shukla, 2023). Of course, the trend applies in educational technologies where highly innovative individuals will continue applying and using tools like ChatGPT for learning and teaching (Zhou and George, 2001). Innovative capability refers to one's level of creativity and is most of the time juxtaposed against higher tolerance for ambiguity and higher exploratory intention that necessarily translates to a higher intention to use untried new technologies (Hughes et al., 2020). Most individuals with a higher innovative capability realize very clearly the potential efficiencies that new technologies will bring to them, thereby reinforcing their intention to begin their use (Fang et al., 2021). Recent research has also pointed out that innovative individuals are those who tend to be the early adopters and thus also tend to influence others in their social and professional circle to use it too (Rogers, 2003).

H3: Innovative capability to use data analytics positively impacts continuous intention to use

2.4 Innovative Capability to use Data Analytics Mediates the Relationship between FOMO and Continuous Intention to use

Continuous intention refers to a user's intention to continue using technology over some time. It encompasses a set of factors including perceived usefulness, ease of use, and satisfaction derived from the technology. ChatGPT enhances the speed and quality of academic writing. Continuous intention to use may, therefore, arise because some of these experiences have been positive—ranging from coherence to relevance of the content generated—thereby reducing cognitive load on students. According to Alshahrani (2023), the integration of AI into blended learning environments, including advanced tools like ChatGPT, plays an important role in providing ongoing support to students and enhancing the intention to continue using these technologies in academic settings.

Recently, the interaction of FOMO, innovative capability, and the intention to use new technologies has drawn huge interest in research studies. The concept of mediation—innovative capability to use data analytics as an intermediary variable—allows a more subtle explanation of how FOMO influences the intention to use emerging technologies such as ChatGPT. Generally, FOMO is characterized by a feeling of apprehension for not being able to participate in some mentally rewarding experience (Przybylski et al., 2013). In most cases, FOMO drives people to explore and adopt new technologies, but the relationship is not always straightforward. Innovation ability mediates this anxiety-driven behavior into creating a real intention to use new tools (Li, 2021).

Innovative abilities for data analytics enable one to envision the possible benefits of new technology and find creative ways to embed it in work or study routines (Souto, 2015). In this way, effective innovative capacity bridges the gap between the initial state of FOMO and the actual decision to adopt new technology (Gupta et al., 2022). Individuals with high levels of innovative capability and FOMO are more likely to experiment with and commit to using new technologies compared to others (Hughes et al., 2020). Previous research findings have the power of innovative capability for FOMO in the context of technology adoption (Fang et al., 2021). Thus, the literature explains the need to study the mediating role of innovative capability between the relationship of FOMO and the intention to use new technologies, to bring about a sustainable usage of technology.

H4: Innovative capability mediates the relationship between FOMO and continuous intention to use.

2.5 Conceptual Model

In the conceptual model, Fig. 1, FOMO, Innovative Capability, and Continuous Intention (CI) are integrated through Self-Determination Theory (SDT) and Dual-Process Theory (DPT).

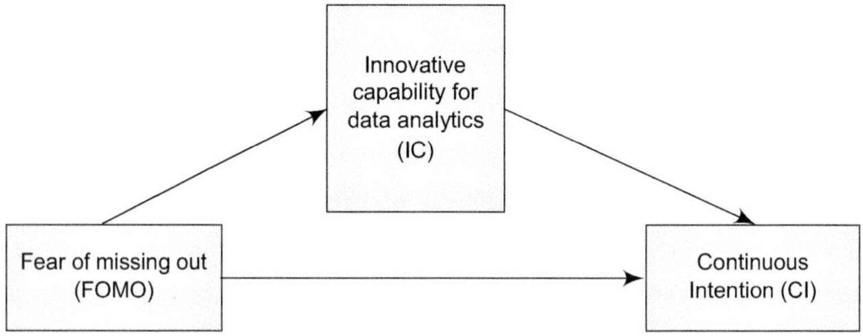

Fig. 1 Conceptual Model.
Source: Author's own

According to SDT, human motivation (Mishra et al., 2024) is due to three in-built psychological needs, namely, autonomy, competence, and relatedness (Deci and Ryan, 2000). FOMO represents a kind of psychological response in the framework as well, tied to the need for relatedness in terms of staying connected to avoid losing ground when something else could potentially give them increased standing or status. This desire for relatedness, however, adds to the fear of being left behind and motivates innovative behaviors in individuals and organizations want to be competitive and relevant (Przybylski et al., 2013). Dual-Process Theory offers the cognitive framework in which these motivational drivers.

These have their impact on decisions. The two types of cognitive modes are defined in theory as follows: System 1 is fast, automatic, and often emotion-driven, whereas System 2 is slower, deliberative, and analytical (Kahneman, 2011). In the model, FOMO relies predominantly on System 1 processing; that is, quick, emotionally driven decisions to capitalize on emerging opportunities. However, innovative capability moderates this process through the activation of System 2-that is, more thoughtful and strategic innovation and the fact that decisions are not just reactive but in alignment with long-term goals (Gronchi et al., 2024). That way, FOMO and innovative capability interact to shape the ability to generate continuous intention, which is important for a strategic advantage in a dynamic environment.

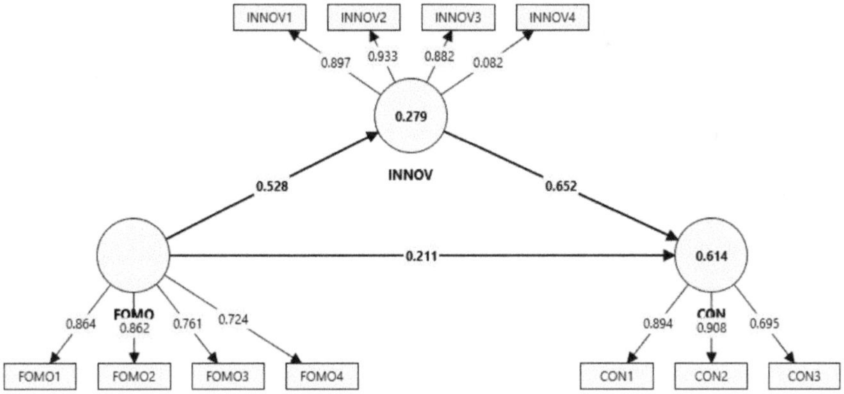

Fig. 2 PLS RUN.

Source: Authors' own

Note: FOMO = Fear of missing out, INNOV = Innovative capability for using data analytics, CON = Continuous intention

3. Research Methodology

The study used a mixed method approach. The study was done in two steps. Stage one, an assignment was shared with students of a PGDM institute. The submission was taken through Turnitin. Over 91% of students used AI as high as 100%. A training session was done to explain the responsible usage of AI. The second phase, a questionnaire was shared to survey the continuous intention of AI by students. The survey was completed by 312 students, incomplete responses were deleted to

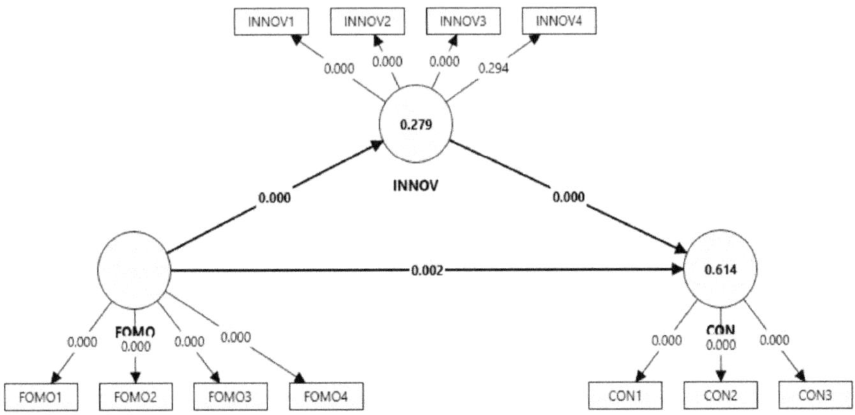

Fig. 3 Bootstrap RUN.
Source: Authors' own
Note: FOMO = Fear of missing out, INNOV = Innovative capability for using data analytics, CON = Continuous intention

get 296 usable responses. The items were adapted from the standard scale (refer Appendix 1).

4. Results

4.1 Descriptive Statistics

The demographic characteristics of the study sample are presented in Table 1. The majority of respondents (89.53%) are within the 28-38 age group (n = 265) with an average age of 24. A smaller proportion of the sample falls within the 38-48 age group (8.78%, n = 26), and the smallest segment consists of individuals above 48 years of age (1.69%, n = 5). This distribution suggests that the sample is predominantly young, with a significant majority being early to mid-career professionals. The gender composition of the sample shows a slight male predominance. Males constitute 55.41% of the respondents (n = 164), whereas females account for 44.59% (n = 132). This relatively balanced gender representation ensures that the findings are generally applicable across genders, though the slight male bias should be noted. Regarding educational attainment, a substantial proportion of respondents hold a postgraduate degree (59.46%, n = 176), indicating a high level of educational achievement within the sample. Respondents with an undergraduate degree represent 31.42% of the sample (n = 93). The remaining 9.12% (n = 27) are classified under 'Others,' which includes various non-traditional educational backgrounds. The high incidence of postgraduate education suggests that the respondents can have significant professional and academic expertise.

Table 1 Descriptive Statistics

S.No.	*Category*	*Frequency*	*Percentage*
Age Group	28–38	265	89.53
	38–48	26	8.78
	Above 48	5	1.69
Gender	Female	132	44.59
	Male	164	55.41
Education	Others	27	9.12
	PG	176	59.46
	UG	93	31.42
	Total	296	100%

Table 2 CFA

Item	*Outer Loading*	*Cronbach's alpha*	*Composite reliability*	*Average variance extracted*
CON1	0.894	0.783	0.875	0.702
CON2	0.908			
CON3	0.695			
FOMO1	0.864	0.825	0.880	0.648
FOMO2	0.862			
FOMO3	0.761			
FOMO4	0.724			
INNOV1	0.897	0.723	0.835	0.615
INNOV2	0.933			
INNOV3	0.882			
INNOV4	0.082			

Table 3 Fornell Larker criterion

	CON	*FOMO*	*INNOV*
CON	0.838		
FOMO	0.555	0.805	
INNOV	0.763	0.528	0.784

Table 4 HTMT

	CON	*FOMO*	*INNOV*
CON			
FOMO	0.693		
INNOV	0.938	0.606	

Table 5 Hypothesis Testing

	Path Coefficients	*Tstatistics*	*Pvalues*
H1: Fear of missing out (FOMO) positively impacts continuous intention to use AI.	0.211	3.143	0.002
H2: Fear of missing out positively impacts innovative capability.	0.528	10.171	0.000
H3: Innovative capability positively impacts continuous intention to use AI.	0.652	11.176	0.000
H4: Innovative capability mediates the relation between FOMO and continuous intention to use.	0.344	8.463	0.000

4.2 Measurement Model Assessment

The factor loadings were higher than 0.7, Discriminant validity was assed using HTMT and Fornell Larcker and HTMT criterion. The Fornell-Larcker criterion assesses discriminant validity by comparing the square root of the Average Variance Extracted (AVE) for each construct with the correlations between the constructs. Discriminant validity is confirmed if the square root of the AVE for a construct is greater than its correlations with any other construct. The square root of AVE (diagonal) is higher than its correlation (non-diagonal elements) indicating good discriminant validity. HTMT is another method to assess discriminant validity, particularly in PLS-SEM. It compares the ratio of the between-trait correlations to the within-trait correlations. Discriminant validity is confirmed if the HTMT values are below a threshold, commonly 0.90. The HTMT value ranges between 0.693 and 0.89, which is below the threshold of 0.90, indicating discriminant validity.

4.3 Structural Model Assessment

The hypothesis regarding Fear of Missing Out and Intention to Use was also supported (β = 0.211, p = 0.002). The hypothesis regarding Fear of Missing Out and Innovative capability was also supported (β = 0.528, p = 0.000). The hypothesis regarding innovative capability and Intention to use was also supported (β = 0.652, p = 0.000). The hypothesis regarding Innovative capability mediating the relationship between Fear of Missing Out and continuous intention to use was also supported (β = 0.344, p = 0.000).

The R-square values provided show how well each independent variable explains the variance in the dependent variable. With an R-square of 0.614, it provides a robust explanation of the dependent variable's variance. This suggests that the control variable is likely a significant predictor within the model. An R-square of 0.279 for the innovation variable (INNOV) shows that while this variable still contributes to explaining the dependent variable, its impact is not as strong as that of the control variable. There might be other factors not included in the model that could better explain the variance.

5. Discussion

The first hypothesis represents the positive impact of FOMO on intention to use, which was supported. This means that individuals who experience a higher level of Fear of Missing Out (FOMO) are more likely to have a greater intention to use a particular technology or product. The positive path coefficient ($\beta = 0.211$) indicates a positive relationship, and the *p*-value (0.002) confirms that this relationship is statistically significant. This implies that as FOMO increases, so does the intention to use the product. H2 reflects that FOMO positively impacts innovative capability was supported. This indicates that individuals who experience higher levels of FOMO are more likely to demonstrate higher innovative capability. The strong positive path coefficient ($\beta = 0.528$) shows a significant and robust relationship, while the *p*-value (0.000) indicates that this finding is highly significant. This suggests that FOMO can drive individuals to be more innovative for using data analytics, possibly as a means to avoid missing out on new opportunities. H3 shows Innovative capability for data analytics positively impacts Intention to Use. This hypothesis posits that individuals with higher innovative capability are more likely to have a stronger intention to adopt a product or technology. The very strong positive path coefficient ($\beta = 0.652$) indicates a substantial relationship, and the *p*-value (0.000) confirms its statistical significance. This suggests that individuals with greater innovative tendencies are more likely to embrace and adopt new technologies or products. H4 suggests innovative capability mediates the relationship between Fear of Missing Out and continuous intention to use. This mediation hypothesis suggests that Innovative capability serves as a mediator, explaining the relationship between FOMO and Intention to Use. The positive mediation path coefficient ($\beta = 0.344$) indicates a significant mediation effect, with the *p*-value (0.000) confirming its statistical significance. In other words, FOMO influences Innovative capability, which then impacts Intention to Use. Thus, one reason FOMO increases the intention to use a product is that it enhances innovative capability for using data analytics, which subsequently drives the intention to adopt and use the product.

6. Implications

6.1 Academic Implications

The present paper contributes strongly to theoretical contexts of AI tools adoption and usage, specifically ChatGPT in the educational system. The study found that Fear of Missing Out is strongly related to innovation as FOMO drives individuals to continuous innovation to stay in line with the ongoing requirements. The study contributes to the prevailing research on technology adoption and data analytics. The study brings ChatGPT as an AI tool in light and discusses that the adoption of this tool for innovative capacity is a significant step in an academic context. Previous studies (Przybylski et al., 2013; Li, 2021) have already proven that FOMO positively influences innovation and the intention to use AI tools, and this study solidifies that link. The study also demonstrated that the innovative capacity of a person mediates individuals' FOMO and their continuous intention. Previous studies have already been discussing the adoption of ChatGPT. This study adds value by exhibiting that if FOMO is an influential element that encourages the continuous intention for adoption of the ChatGPT tool, enhancement of innovative capability for using data analytics can help improving the curiosity within the education processes (Hughes et al., 2020; Gupta et al., 2022). These studies indicate that if educational systems can capture FOMO and encourage innovation, these systems may better adapt to AI, enhancing student learning experiences and engagement with technology (Farrelly and Baker, 2023).

6.2 *Practical Implications*

The study contributes immensely to the practical implications of the findings. The chapter recognizes the significance of FOMO in driving innovation and creativity amongst individuals. This also encourages fostering an environment where individuals' innovative capability can be fostered. Such opportunities, where individuals face fear of missing out, can drive them to adopt technology and tools that can support them for being productive. Therefore, this study reveals that innovative capability serves as a mediator that can streamline the individual's FOMO in enhancing their continuous intention. The chapter advocates that FOMO can encourage competitive behavior by enhancing the intention for AI adoption with the help of innovation. Usage of tools like ChatGPT can be encouraged in the organization, considering that, such AI tools can foster the innovative capability and manage FOMO for enhancing the effectiveness of the outcome.

7. Limitations and Future Scope

One of the primary limitations of the research study is the limited sample size. A small or non-representative sample can skew the results and may not accurately reflect the broader population. When studying phenomena like FOMO, which can vary significantly across different demographics, a larger and more diverse sample is necessary to obtain generalizable findings. Measuring FOMO can be inherently subjective, often relying on self-reported data through surveys or questionnaires. This subjectivity can introduce bias, as individuals may not accurately report their feelings or behaviors related to FOMO. Future research should consider using a combination of qualitative and quantitative methods to gain a more nuanced understanding. Further, factors such as market dynamics, organizational culture, and technological advancements can all influence these relationships. Failing to control for these variables may lead to misleading conclusions. Thus, future studies can investigate the role of these factors in the prevailing context.

Conclusion

Adoption of AI and technology has become mandatory for individuals and organizations in the fiercely competitive business sector. According to this study, ChatGPT may be a useful tool for people who have FOMO. FOMO is the realization that some crucial information is being overlooked, which may affect performance. This study advocates that fear of missing out influences individuals to enhance their competitiveness and innovativeness towards data analytics because they fear lagging in the lack of information, skills, or expertise. To be efficient and effective, it is important to have required competencies. AI tools enhance innovativeness and make the work faster and easier. This technology allows one to focus more on an innovative angle rather than repetitive and basic non-creative tasks. As a result, this study offered insightful information on how AI and technology may be adopted to improve creative capacity for data analytics that help manage FOMO.

Reference

Abel, J.P., Buff, C.L. and Burr, S.A. (2016). Social media and the fear of missing out: Scale development and assessment. *Journal of Business & Economics Research* (JBER), 14(1), 33–44.

Alhumaid, K., Al Naqbi, S., Elsori, D. and Al Mansoori, M. (2023). The adoption of artificial intelligence applications in education. *International Journal of Data and Network Science*, 7, pp.457–466.

Alshahrani, A., 2023. The impact of ChatGPT on blended learning: Current trends and future research directions. *International Journal of Data and Network Science*, 7, pp. 2029–2040.

Amabile, T.M. (1988). A model of creativity and innovation in organisations. *Research in Organizational Behavior*, 10(1), 123–167.

Beyens, I., Frison, E. and Eggermont, S. (2016). "I don't want to miss a thing": Adolescents' fear of missing out and its relationship to self-esteem, time spent on social media, and problematic smartphone use. *Cyberpsychology, Behavior, and Social Networking*, 19(8), 501–506.

Brown, T., Mann, B., Ryder, N., Subbiah, M., Kaplan, J., Dhariwal, P.... and Amodei, D. (2020). Language models are few-shot learners. *arXiv preprint arXiv:2005.14165*.

Deci, E.L. and Ryan, R.M. (2000). The "What" and "Why" of Goal Pursuits: Human Needs and the Self-Determination of Behavior. Psychological Inquiry, *11*(4), 227–268. https://doi.org/10.1207/S15327965PLI1104_01

Dossey, L. (2021). FOMO, digital dementia, and our dangerous experiment. *Explore*, 17(5), 341–347.

Fang, M., Zhang, Y. and Yan, J. (2021). How innovative behavior influences digital technology adoption: The mediating role of perceived value. *Journal of Innovation & Knowledge*, 6(3), 176–183.

Farrelly, M. and Baker, R. (2023). Ethical implications of AI in education. *Journal of Educational Ethics, 15*(3), 45–60.

Farrelly, T. and Baker, N. (2023). Generative Artificial Intelligence: Implications and Considerations for Higher Education Practice. *Education Sciences*, 13(1109). Available at: https://doi.org/10.3390/educsci13111109.

Gronchi, G., Gavazzi, G., Viggiano, M.P. and Giovannelli, F. (2024). Dual-Process Theory of Thought and Inhibitory Control: An ALE Meta-Analysis. Brain Sciences, *14*(1), 101.

Gupta, M. and Arora, N. (2022). Fear of missing out and its impact on technology adoption behavior: An exploratory study. *Technology in Society*, 70, 101996.

Hughes, C., Lee, S. and Tian, L. (2020). Creativity in education: Innovations in teaching methods and classroom practices. *Journal of Creative Behavior*, 54(3), 672–689.

Jones, H. (2022). The role of AI in self-directed learning: A case study of ChatGPT. *International Journal of Educational Technology, 34*(1), 12–25.

Koyutürk, C., Yavari, M., Theophilou, E., Bursic, S., Donabauer, G., Telari, A., Testa, A., Boiano, R., Gabbiadini, A., Hernández-Leo, D., Ruskov, M. and Ognibene, D. (2023). Developing Effective Educational Chatbots with ChatGPT prompts: Insights from Preliminary Tests in a Case Study on Social Media Literacy. Preprint.

Krämer, W. Kahneman, D. (2011): Thinking, Fast and Slow. Stat Papers 55, 915 (2014). https://doi.org/10.1007/s00362-013-0533-y

Kumari, A. and Saharan, T. (2021). Organisational culture as a stimulant to knowledge management practices: an empirical analysis on Indian real estate companies. *International Journal of Knowledge and Learning, 14*(4), 360–384.

Kumari, A. and Singh, A.K. (2022). Technology Adoption for Facilitating Knowledge Management Practices in Firms. *Emerging Technologies in Data Mining and Information Security: Proceedings of IEMIS 2022, Volume 1*, 117–129.

Kumari, A. and Saharan, T. (2020, March). Knowledge management as a catalyst to sustainable development in banking industry. In *2020 International Conference on Computer Science, Engineering and Applications (ICCSEA)* (pp. 1–6). IEEE.

Kumari, A., Khan, M. and Lakshmi, N. (2023). Assessing antecedents of individual readiness to adopt knowledge management in higher educational institutions. *Cogent Business & Management, 10*(2), 2238393.

Li, C. (2021). How does fear of missing out influence entrepreneurial creativity? The mediating roles of intrinsic motivation and knowledge sharing. *Journal of Business Research*, 134, 353–363.

Li, Y., Wang, S. and Zhao, L. (2021). Exploring the effect of fear of missing out on learning engagement and academic performance among university students: A mediating role of self-efficacy. *Education and Information Technologies*, 26, 7683–7698.

Lo, C.K. (2023). What Is the Impact of ChatGPT on Education? A Rapid Review of the Literature. *Educ. Sci., 13*(4), p. 410. https://doi.org/10.3390/educsci13040410.

Mishra, A., Malik, N. and Shukla, A. (2024). Decoding individual motivations and responses to misinformation: insights from thematic analysis. *Journal of Research in Interactive Marketing.*

Mishra, A. and Shukla, A. (2023). Gyan fresh: digital transformation of dairy business with resilience and technology innovation. *FIIB Business Review, 12*(1), 20–30.

Nguyen, A. and Chang, H. (2023). Balancing AI integration in higher education: Benefits and challenges. *AI and Education Journal, 11*(2), 90–105.

Pasca, M.G. and Arcese, G. (2024). ChatGPT between opportunities and challenges: an empirical study in Italy. *The TQM Journal*, (ahead-of-print).

Popescu, A. (2023). AI's Secret Weapon in Education: ChatGPT - The Future of Personalized Learning. *Bulletin of the Transilvania University of Braşov, Series V: Economic Sciences*, 16(65) No. 2. https://doi.org/10.31926/but.es.2023.16.65.2.5.

Przybylski, A.K., Murayama, K., DeHaan, C.R. and Gladwell, V. (2013). Motivational, emotional, and behavioural correlates of fear of missing out. *Computers in Human Behavior*, 29(4), 1841–1848.

Rana, G., Sharma, R. and Parashar, B. (2024). Application of Artificial Intelligence for the Success of Supply Chain Operations in the Age of Data Analytics. In *2023 International Conference on Smart Devices (ICSD)* (pp. 1–6). IEEE.

Roberts, J.A. and David, M.E. (2020). The social media party: Fear of missing out (FOMO), social media intensity, connection, and well-being. *International Journal of Human-Computer Interaction*, 36(11), 1026–1032.

Rogers, E. M. (2003). Diffusion of Innovations (5th ed.). Free Press.

Sharma, R., Saini, A.K. and Rana, G. (2021). Big data analytics and businesses in industry 4.0. *Design Engineering*, 238–252.

Smith, L. and Lee, J. (2021). AI-driven pedagogy: Enhancing educator productivity with ChatGPT. *Educational Research Review, 28*, 101–115.

Souto, J.E. (2015). Business model innovation and business concept innovation as the context of incremental innovation and radical innovation. *Tourism Management*, 51, 142–155.

Taylor, R. and Kumar, S. (2023). Automating academia: The role of AI in administrative efficiency. *Journal of Higher Education Management, 20*(4), 67–82.

Wang, X., Liu, Y. and Zhao, Z. (2021). Accelerating academic research with AI: The case of ChatGPT. *Research Technology Management, 64*(6), 22–30.

Wegmann, E., Stodt, B. and Brand, M. (2022). The tendency to fear missing out and internet-use expectancies contribute to symptoms of internet-communication disorder. *Addictive Behaviors Reports*, 16, 100251.

Zhou, J. and George, J. M. (2001). When job dissatisfaction leads to creativity: Encouraging the expression of voice. *Academy of Management Journal*, 44(4), 682–696.

Appendix

Fear of Missing Out (FOMO)

1. I fear missing out on important information if I don't use ChatGPT.
2. I feel left out when I see my peers using ChatGPT and I am not.
3. I worry that not using ChatGPT will negatively impact my academic performance.
4. I feel compelled to use ChatGPT because I see others using it.
5. I check ChatGPT frequently to avoid missing out on any valuable information.

Innovation (Inn)

1. I am always looking for innovative tools to enhance my learning.
2. I consider myself an early adopter of new technologies for education.
3. Using ChatGPT is a novel experience for me.
4. I prefer using new technologies like ChatGPT over traditional methods.
5. I am interested in using innovative solutions like ChatGPT for my studies.

Continuous intention to use AI for academic writing (CI)

1. I am willing to use AI/GPT for my assignments.
2. I will feel happy to interact with AI devices for academic writing.
3. I am likely to interact with AI devices.

11 Smart Tourism Supply Chains: Integrating AI, Advanced Analytics, and Robotics for Enhanced Customer Experiences

Devvandya Mishra[1*] and Lokesh Vijayvargy[2]

In the last two decades, tourism has become one of the most competitive and high-growth sectors in the world, which has, in turn, contributed much to the economic and social development involved. The United Nations World Tourism Organisation and the World Travel and Tourism Council estimate that tourism has significantly added value to global gross domestic product, estimated at 9.9 trillion U.S. dollars in 2023. In terms of 2024, estimates are around 11.1 trillion U.S. dollars. The purpose of this chapter is to identify the integration of smart supply chain technologies in the tourism sector into AI-based predictive analytics, machine learning, and automation for improving the operational dimension in tourism, namely, efficiency, effectiveness, personalization, and sustainability. It brings forward the importance of a smart supply chain to enhance logistics, demand forecasting, customer experience, and operational cost, which helps the industry grow and remain competitive. Further, it talks about the impact of such technological advancement strategies on the tourism supply chain in stimulating innovation, improving global competitiveness, and addressing issues related to sustainability and resilience in the tourism system. This book chapter outlines the rising importance of artificial intelligence and advanced analytical techniques as significant factors in determining the future economic importance of tourism to the global economy.

1. Introduction

The tourism industry has emerged as one of the world's most competitive and rapidly growing service sectors, as confirmed by the United Nations World Tourism Organization and the World Travel and Tourism Council through their statistical research, significantly contributing to socioeconomic development (UN Tourism, 2024). In 2023, the travel and tourism industry directly contributed to the gross domestic product, which was approximately 9.9 trillion U.S. dollars, according to Fig. 2. Figure 1 shows that it was predicted to reach an estimated 11.1 trillion U.S.

[1] 0009-0001-0098-069X
0000-0002-7032-4962
* Corresponding author: devvandya.mishra.fpm23j@jaipuria.ac.in

DOI: 10.1201/9781003514282-11

dollars in 2024 (Statista, 2024). The World Travel and Tourism Council estimates that by 2026, the value will increase by USD 10,986.5 billion, representing 10.8% of the world GDP (Travel and Tourism Economic Impact 2024: Global Trends, 2024).

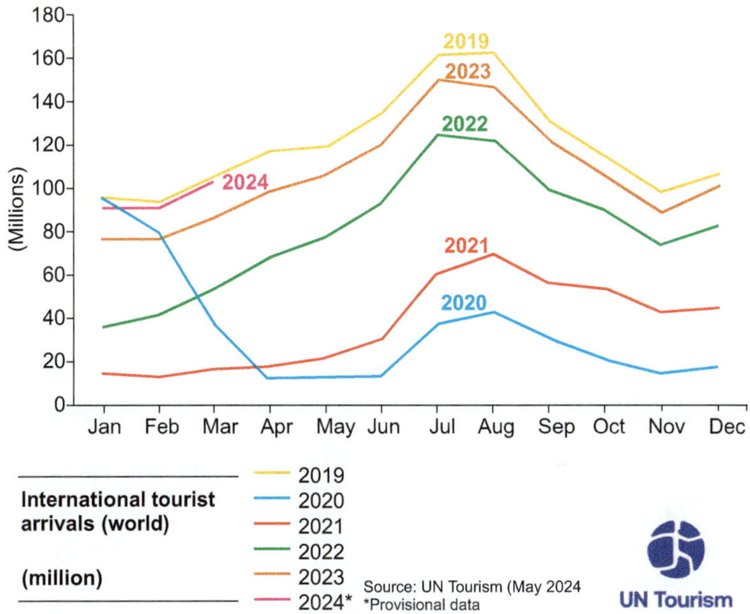

Fig. 1 United Nations Tourism.
Source: https://www.unwto.org/about-us

Products and services: Usually, many organizations are involved in the preparations to offer a variety of services to the clients. The dynamic and multifaceted industry depends on the smooth coordination of various organizations, including travel companies, airlines, transport agencies, hotels, and local attractions. These interconnected components are collectively known as the tourism supply chain. The supply chain forms the backbone of industries in the service sector, ensuring the seamless flow of information, goods, and services across various stakeholders. A supply chain contains interconnected processes that involve procurement, production, distribution, and customer engagement. The evolution of technology has revolutionized the supply chain, transitioning them from traditional, linear systems to innovative, hybrid, dynamic networks. Smart supply chains leverage technologies such as artificial intelligence, advanced analytics, robotics, machine learning, and Metaverse to improve efficiency, responsiveness, and resilience. This transformation has significant implications for the tourism supply chain, where customer satisfaction and experience are paramount.

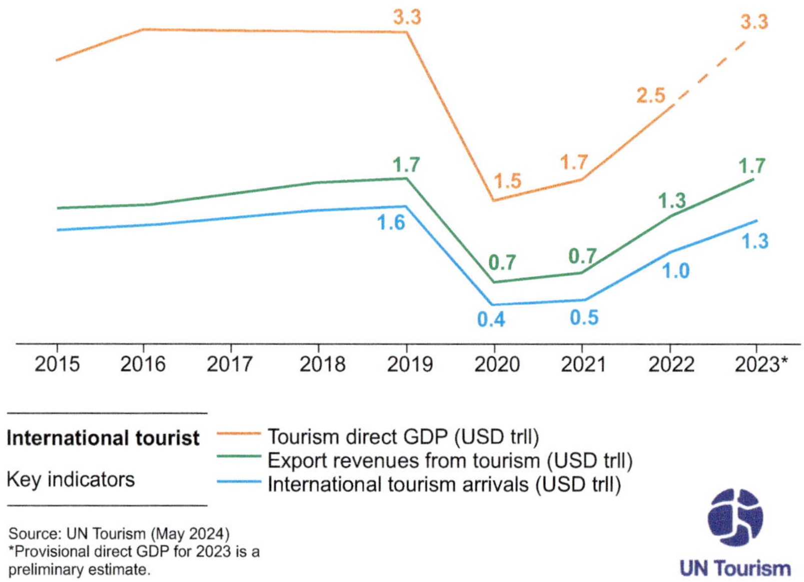

Fig. 2 Contribution of travel and tourism in world GDP.
Source: https://www.unwto.org

Background: The supply chain in traditional industries and manufacturing sectors primarily focuses on optimizing costs, improving efficiency, and ensuring the smooth and fast delivery of goods from the supplier to the end consumers. The traditional supply chain is linear, comprising interconnected processes such as procurement, production, inventory management, transportation, and distribution. The key objectives revolve around reducing lead time, minimizing waste, and maintaining cost-effectiveness while meeting market demands. The rapid advancement of technology has revolutionized supply chain management, transforming it from traditional, reactive systems to proactive and intelligent networks. Advancements such as artificial intelligence, the Internet of Things, Advanced analytics, and robotics have transformed supply chain operations, enabling real-time data sharing, predictive insights, and process automation. As the global market becomes increasingly interconnected, the supply chain must address challenges such as rising customer expectations, environmental concerns, and volatile market conditions. This revolution from a traditional supply chain to a smart supply chain, technology-driven ecosystem showcases the paradigm shift in how goods and services are produced, distributed, and consumed. The integration of advanced technology has become essential for industries and service sectors to remain competitive and resilient in today's dynamic business environment.

The tourism supply chain is different from the manufacturing supply chain. Here, the complexity and focus on customer-centricity. A wide range of interconnected services, such as accommodation, transportation, food, beverage, and local experiences, characterizes this supply chain. Each service provider operates a node in the supply chain, requiring effective coordination to deliver a seamless customer experience and satisfaction. Historically, the tourism supply chain has faced various challenges in operating the services. Inefficiencies are major challenges, such as reliance on manual processes and outdated systems. Fragmentation is another issue, because of the involvement of multiple stakeholders and independent parties. Last, there is a high susceptibility to external disruption like pandemics, geopolitical tensions, and economic fluctuations. The advent of smart technologies such as Artificial Intelligence, Robotics, and Advanced analytics in the tourism supply chain will overcome real-time decision-making, predictive insights, and automation, which, no doubt, collectively enhance both operational efficiency and customer satisfaction.

The application of smart technology in the tourism supply chain has vast implications. For example, AI-powered tools can predict customer preferences and seasonal trends, and advanced analytics allow for the creation of hyper-personalized travel experiences based on individual preferences. Smart technology helps to optimize resource utilization, reduce waste, and minimize environmental footprints. This book chapter's main objective is to explore and investigate the transformative potential of smart technologies and advanced analytics in the tourism supply chain, focusing on integrating artificial intelligence, advanced analytics, and robotics.

1.1 Objectives of the Study

The objectives of the study were as follows:

I. To explore the transformative potential of smart technology in the tourism supply chain, emphasizing the integration of AI, advanced analytics, and robotics. It seeks to comprehensively understand how these technologies are reshaping the tourism supply chain while highlighting the benefits and challenges associated with their adoption.
II. Empowering stakeholders to enhance operational efficiencies and customer satisfaction by offering actionable insights and strategies.
III. Additionally, it will examine the real-world applications and future trends shaping the tourism industry, contributing to the broader discourse on building a sustainable and customer-centric tourism ecosystem.

1.2 Organization of Chapter

This chapter has been divided into six sections. The first section consists of the introduction, the objective, and the organization of the chapter. The second section consists of a methodology of this book chapter, which explains the research design, data resources, and article selection criteria. The third section covers a detailed explanation of the tourism supply chain and its components, and it discusses the challenges of the tourism supply chain. The fourth section discusses AI and advanced analytics in the tourism supply chain and other aspects, advantages, challenges, future trends, etc. A case study is discussed in the fifth section of this chapter. A conclusion follows this in the sixth section.

2. Methodology

The methodology employed in this theoretical exploration is the integration of artificial intelligence (AI) and advanced analytics in the tourism industry supply chain. The intent is to provide a comprehensive framework to understand the importance of artificial intelligence and advanced analytics in the tourism industry supply chain system. The chapter employs a descriptive and exploratory research design to examine the transformative role of artificial intelligence and advanced analytics in optimizing tourism supply chains. This chapter follows a conceptual approach. The conceptual studies do not offer the data, instead, they offer an integration of previous literature, apart from the qualitative approach used to analyse the secondary data collected from the academic literature, various industry reports, and various case studies. With the help of a comprehensive literature review, a theoretical foundation is established. All the relevant studies, industry frameworks, and reports were analyzed to understand the existing applications and potential of AI and advanced analytics in the tourism supply chains. The key themes include demand forecasting, customer services, sustainability, and inventory management. The data collection process involved a thorough and rigorous literature review to identify and synthesize existing knowledge and the applications of artificial intelligence in the tourism supply chain. Relevant studies, industry frameworks, and reports were identified using a structured search strategy across scholarly databases such as Scopus, Google Scholar, Research Gate, and Web of Science, supplemented by industry-specific sources like McKinsey reports, etc.

The research philosophy underpinning this study aligns with interpretivism, which emphasizes understanding the implications of potential artificial intelligence and analytics by integrating diverse insights from theoretical and practical perspectives.

The article selection process followed a set of inclusion and exclusion criteria. Articles published in peer-reviewed journals, conference proceedings, FT-50, and other reputed industry publications from the last date of (2015–2024) were prioritized. Key search terms include "Supply Chain Management" and "Artificial Intelligence," "Advanced Analytics in Supply Chain," "Sustainability in Tourism Supply Chain" and "Demand forecasting with AI." Selected studies were critically reviewed to extract the themes such as demand forecasting, customer services, Customer satisfaction, and inventory management from a theoretical framework. By integrating these critical processes and perspectives, the chapter establishes a robust conceptual foundation for understanding the role of artificial intelligence and advanced analytics in optimizing the tourism supply chain.

3. Tourism Supply Chain Management

The Tourism Supply Chain is a network of businesses and multiple services that deliver seamless travel experiences. Balanced cooperation among various business partners in the tourism supply chain plays a significant role in a product's success in the tourist market. When the product succeeds in the market, this collaboration helps optimize business operations. In 1975, the United Nations World Tourism Organization (UNWTO) published a report on the tourism industry's distribution channel.

The tourism supply chain is a network of tourism organizations that engage in various types of activities, these activities encompass the provision of diverse components of tourism products and services, such as accommodation, flights, and transportation also it involves primarily all marketing and distribution activities in the chain. The entire network involves a broad range of entities from public and privates sectors, all contributing to the creation, sale, and promotion of the tourism products and services (Zhang, Song, and Huang, 2009; Hermawan, Pangestuti, Artono, and Supriono, 2023).

The tourism supply chain refers to a network connected to various tourism businesses engaged in different activities. This activity involves managing the customer and delivering various types of services, just as accommodation, food, transportation, railways, and also the sale of diverse tourism products (González-Torres, T., and Pelechano-Barahona, 2021; Song, 2011). The tourism supply chain involves several components, not just accommodations, transport, and excursions, but also handicrafts items, waste disposal, cafés, restaurants, and infrastructure that support tourism in the destination (Tapper, Fon, Schwartz, Kornilaki, and Carey, 2004). Each element within the supply chain plays a significant role in ensuring customer satisfaction and the smooth functioning of the industry. The success of the tourism supply chain mainly relies on the efficient coordination and collaboration between these entities (Zhang and Song, 2018). For Example, airlines and hotels must align their capacities to match and fulfill expected demand during the peak travel seasons or vacation time. The close connection and interdependencies highlight the need for a well-organized and flexible supply chain that can easily handle changes and be ready for unexpected challenges. The most essential element of this entire process is especially the suitable selection of partners in the supply chain, which focuses on the preparation of a competitive tourism product. The correct contractor selection process is very complex and complicated, requiring a high and strong commitment to the criteria for evaluating potential partners. This is an extremely important issue, especially in the case of tourism services, whose production occurs in the presence of the customer (the Tourist) and they are ready to be consumed at the moment of production (Szpilko, 2017). A strong supply chain benefits tourists by providing smooth and seamless experiences and helps the entire tourism ecosystem grow and thrive.

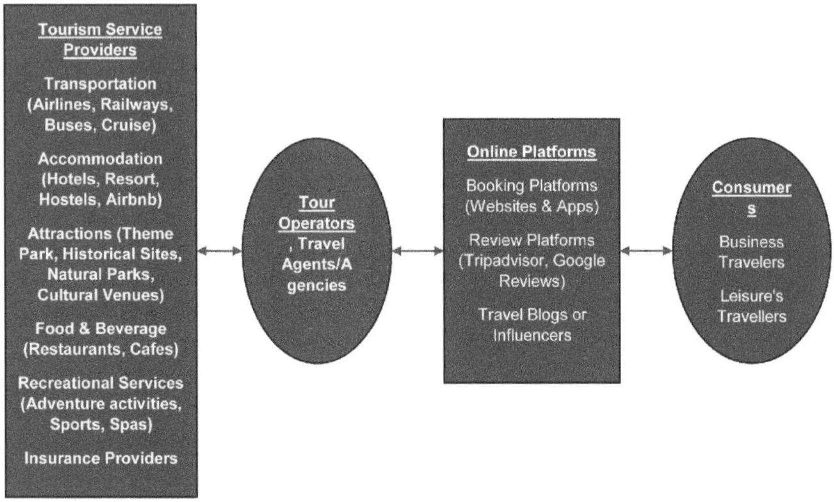

Fig. 3 Tourism Supply Chain Management
Source: Authors Self-Created

Compared to the traditional industrial supply chain in manufacturing, the tourism supply chain possesses unique and advanced characteristics, such as complexity, high risk, and challenges in quality control (Ling, 2015). Other characteristics include automation, real-time processing, and data-driven decision-making.

The complexity arises in the tourism supply chain due to the involvement of diverse stakeholders, such as transportation providers, hotel service providers, local attractions, travel agencies, and government agencies, as well, all of which need to work in a system and in harmony to deliver a seamless experience to the tourist. Each stakeholder operates within its unique operational framework, objective, goals, and constraints, yet they must all align their effort to create a seamless and cohesive experience for the tourist. Simplifying the procedure and improving the overall efficiency of the tourism supply chain in the best possible way necessitate effective communication and information flow.

High Risk is another very significant feature of the tourism supply chain. Various factors such as seasonality, demand fluctuations, uneven economic conditions, unstable market conditions, and some external factors such as political instability, geopolitics issues, or global pandemics can significantly and directly impact the tourism sector's stability, progress, and profitability (Song, 2011).

The third and most crucial is upholding quality control, a difficult undertaking in the tourism supply chain. The subjective nature of customer (tourist) experience and higher dependencies on human interactions make it difficult and unpredictable to ensure consistent service standards across the board. Unlike physical products/goods where quality can be measured and monitored in a set of parameters, tourism services rely on customer perception, likes, and choices, making quality assurance a dynamic and ongoing process.

The smart supply chain is dependent on the automated system and process to minimize wastages, reduce human intervention, reduce errors, ensure maximum output, and enhance efficiency and effectiveness. For example, self-driving vehicles, automated warehouses, and robotics process automation (RPA) (Wu, Yue, Jin, and Yen, 2016; Tussyadiah, 2020).

The smart supply chain enables continuous monitoring and real-time updates allowing the supply chain to respond instantly to disruptions such as supply shortages, fluctuations in demand, and weather forecasting (Zamani, Smyth, Gupta, and Dennehy, 2022; Tsolakis, Schumacher, Dora, and Kumar, 2023). Real-time data collection and analysis enables informed decision-making at every stage of the supply chain. It ensures better forecasting, inventory management, and dynamic pricing strategies.

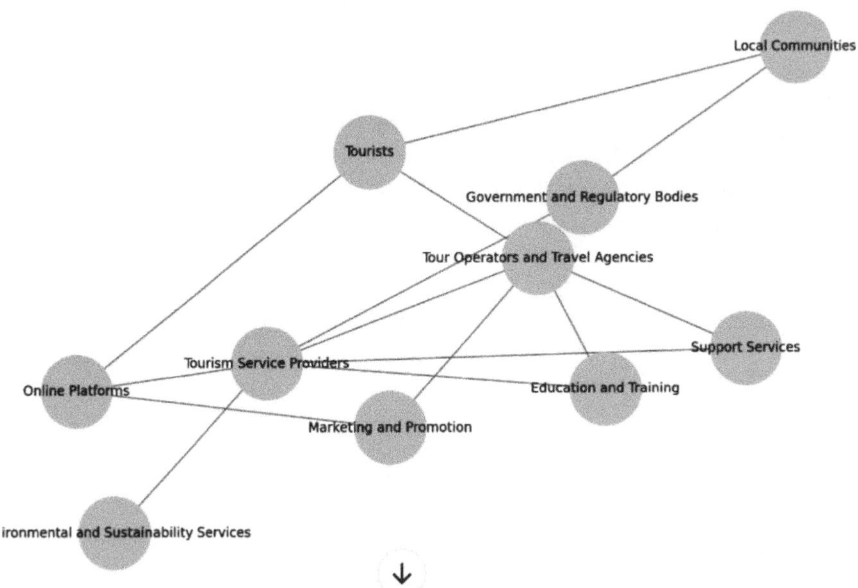

Fig. 4 Components of tourism supply chains.
Source: Authors Self-Created

The tourism supply chain is a complex network that involves several key components, each element playing a significant role in delivering tourism products and services to the end users or tourists (Song, 2011). Figure 4 shows the important components of tourism supply chain, i.e., Environmental and sustainability services, online platforms, marketing and promotions, tourist service providers, support services, government regulatory bodies, and local communities. The multifaceted ecosystem involves numerous stakeholders, ranging from service providers to intermediaries, who collaborate to create seamless and satisfying travel experiences.

The primary entities are tourism service providers, which offer essential services such as transportation (air, railway, bus, taxi, car, etc), accommodation (hotels, resorts, Airbnb), attractions (Natural sites, historical places, theme parks, mixed sites, etc), and food or beverages (Cafes, Restaurants, local food, etc.). These providers supply the core elements of the tourist experience and satisfaction (Ghaderi, Hatamifar, and Khalilzadeh, 2018; George Q. Huang, 2012).

Another crucial component is tour operators and agencies, which act as intermediaries in the tourism supply chain. They offer organized travel packaged tours, guided experiences, and itineraries; they work closely with the service providers to ensure a seamless travel experience for the tourist and ensure maximum customer satisfaction (Guo and He, 2012). Their ability to negotiate and coordinate effectively ensures tourists benefit from reliable services, optimized schedules, and sometimes exclusive experiences. Apart from that, they can take additional responsibility, such as managing contingencies, resolving issues, and adapting to any last-minute change, ensuring a stress-free and enjoyable journey for the tourist.

Along with them, various online platforms such as booking websites and apps (MakeMyTrip, booking.com, Expedia, Skyscanner, Culture Trip, hotels.com, Momondo, Agoda, AllTrails, Hostelworld, etc), other review platforms (Trip Advisor), and travel blogs/influencers are vital in making tourism products accessible to the end consumers or tourist. At the heart of the tourism supply chain are the tourists, who are the end consumers, who drive demand.

By creating rules and guidelines that protect privacy, ensure safety, promote ethical practices, and support sustainable tourism in this industry, all regulatory bodies and government organizations also contribute significantly. They also focus on proper infrastructure development and marketing campaigns to promote tourism on world platforms, for instance, the Indian government launched various schemes to promote tourism such as the Swadesh Darshan scheme, Prashad Scheme, Dekho Apna Desh, etc. (Ministry of Tourism, 2024). Ethical guidelines are developed by all these authorities to prevent exploitation and promote fair trade practices across all stakeholders in the tourism supply chain. By supporting innovation and public-private partnerships, these entities help create a thriving, resilient, and inclusive tourism ecosystem, that benefits all other stakeholders (Chen, 2009).

Local communities are a fundamental, integral, and indispensable part of the tourism supply chain, providing authentic cultural, historical, and regional experiences that benefit economically from tourism activities in this area. These communities offer a bundle of services and activities, ranging from traditional crafts, development of local business, cuisine, and performance to guided tours and homestay, which not only enhance the tourist experience but also provide a significant advantage to the local area people. By participating in and benefiting from tourism, local communities gain opportunities for income generation, employment, and the development of local business, fostering economic growth and resilience. It is essential for the safety and preservation of local and historical heritage, monuments, sites, and sustainable management of natural resources.

In the tourism supply chain, marketing and promotions efforts are carried out by various advertisement agencies, government entities, destination marketing organizations, and influencers who help promote tourist places, destinations, and products to a national and international audience. Their effort plays a significant role in shaping the tourist demand, creating awareness, generating interest, and influencing travel decisions among potential tourists. Their initiative includes organizing roadshows, hosting travel fairs, or creating comprehensive promotional material showcasing the cultural, natural, and historical assets of a region. Some government marketing cases such as the Chalo India Campaign, Incredible India Camping, promote on the world travel market in London, World Tourism Day celebrations, etc. (Ministry of Tourism, 2024).

The tourism industry's success mainly relies on the skills, expertise, and professional workforce. To address these challenges posed by rapid advancement in technology, and higher customer expectations, various entities provide education and training to the workforce to gain new skills. They offer periodic training programs and sessions for hospitality professional and all local communities, helping to improve their service standards and promote sustainable practices apart from fostering inclusivity by involving all local communities in sustainable tourism practices (Tussyadiah, 2020).

3.1 The Interface of Technology, Process, and People in the Tourism Supply Chain

The global tourism supply chain has been transformed by technological advancement. The tourism business model has three critical components: Technology, People, and Processes, ensuring a seamless and efficient system.

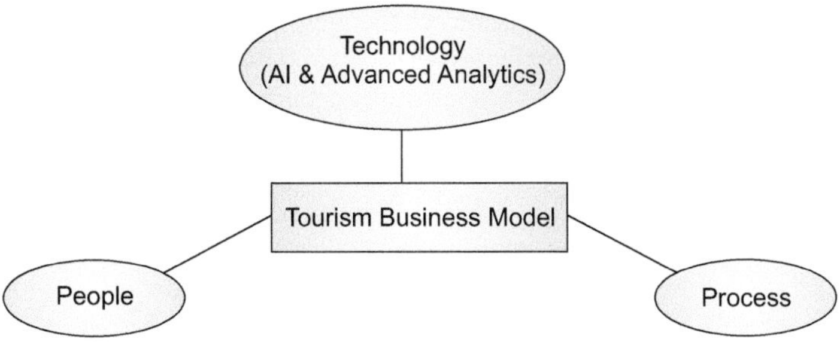

Fig. 5 Tourism Business Model.
Source: Authors Self-created

Technology, particularly artificial intelligence (AI) and advanced analytics, plays a major role in enabling predictive analytics, demand forecasting, enhancing decision-making, and customizing the services. These technological advances in the supply chain make sure to streamline operations, optimize the utilization of resources, and finally contribute to improved service quality. The advancement of technology independently triggered growth for the industry, and customers are the biggest beneficiaries (Tussyadiah, 2020).

The second component of this model, people, showcases the indispensable human element in the tourism supply chain. This includes all stakeholders, such as employees, managers, customers, service providers, local communities, and the government. The collaboration of these diverse groups, their creativity, and their focus on customer-centric practices are essential for delivering value-added experiences or services and ensuring sustainable tourism practices. They also remain a driver of culture exchange, economic growth, and environmental stewardship.

The last component of this model, encompassing the workflow, logistics, payment mechanism, booking system, and feedback loop, highlights the operational backbone of the system. A classic example, airlines nowadays can forecast passenger loads to optimize flight schedules and fuel usage. Aggregators deliver the services at an affordable price to their customers because technology has opened up new business opportunities. The interplay between all these three elements represents a dynamic and classic model. It is difficult to imagine a supply chain without the interface of technology, which is now more like a routine and close to being a universal truth (Palang and Tippayawong, 2019). No doubt, the interface between technology, process, and people has benefited all the stakeholders. This model underscores the significance of integrating advanced technology, human expertise, and a well-defined and structured process to create a sustainable, efficient, and customer-focused tourism ecosystem (Tussyadiah, 2020; Soratana, Landis, Fu, and Suto, 2021).

3.2 Challenges in the Tourism Supply Chain

The tourism supply chain faces various challenges that directly affect its efficiency and effectiveness. One of the major issues is demand fluctuations or higher demand uncertainty, it is hard to predict because travel patterns can shift unexpectedly due to various other reasons such as economic conditions, global events, evolving travel trends, natural disasters, political instability, pandemics, and seasonality adds another layer of complexity (Zhang, Song, and Huang, 2009; Song, 2011).

Apart from that, another major problem is the smoothness of information flow is related to the quality of service. Due to a lack of adequate and effective information, and information asymmetry, the service providers cannot timely identify the real-time needs of the tourist, which seriously affects the accuracy of service and also affects the tourist trust (Tsolakis, Schumacher, Dora, and Kumar, 2023). There are numerous stakeholders involved in the tourism supply chain and their goals, interests, and behaviours are also different. In practice, lack of effective cooperation and coordination between all stakeholders often leads to problems such as service interpretation, and inconsistent service quality, which not only affect the tourist experience, trust, and satisfaction level but also damage the brand image and market competitiveness of the organization to a certain extent and showcase a negative image in the market (Cheunkamon, Jomnonkwao, and Ratanavaraha, 2022).

The tourism supply chain frequently faces a variety of risks, including demand swings, political concerns, and some natural risks. The existing risk management mechanism is not well established and operated, and there is a lack of an early warning system, and a lack of response measures for all potential risks, which makes it difficult for enterprises to perform effectively in adverse conditions, which may lead to service interruption and loss of customers and profits.

Another major challenge is that modern information technologies such as big data analytics, Artificial Intelligence, the Internet of things, robotics applications, and advanced analytics have broad application prospects in the tourism supply chain. However, many organizations still adhere to traditional operating models and cannot utilize advanced technology. Due to this, they are unable to provide adequate services to the customers and lose customer trust (Cheunkamon, Jomnonkwao, and Ratanavaraha, 2022). Seriously limiting their innovation ability and market competitiveness. Each element in the tourism supply chain has different standards and parameters, which can lead to uneven service quality, and varying degrees of impact on tourist experiences, and may also damage the enterprise's reputation, image, and goodwill. In addition, these different parameters also make cooperation and coordination between enterprises difficult.

4. AI and Advanced Analytics in Tourism Supply Chains

Integrating Artificial intelligence and advanced analytics into the tourism supply chain has significantly transformed the industry's ability to manage its functions, operations, forecasting, and enhance customer satisfaction. AI and advanced analytics ensure supply chain optimization, allowing for value addition in service delivery, increasing efficacy and efficiency, and facilitating decision-making across several activities.

Artificial intelligence is vital for streamlining various activities and work in the tourism supply chain. AI-powered tools and techniques streamline and rationalize supply chain operations by automating tasks such as demand forecasting, efficient resource allocation, reducing operational cost, inventory management, and minimizing waste or risk (Cannas, Ciano, Saltalamacchia, and Secchi, 2024; Zamani, Smyth, Gupta, and Dennehy, 2022). For instance, automated booking systems facilitate faster reservation and check-ins, minimizing lead times, and enhancing customer experience and satisfaction (Soratana, Landis, Fu, and Suto, 2021). Machine learning algorithms and artificial intelligence allow dynamic pricing strategies to allow airlines, hotels, and tour operators, and agents adjust their prices based on real-time data, which is based on customer experience, behavior, market competition, trends, seasonality, maximizing profit while offering customized and personalized recommendations based on customers data, likes and dislikes, purchase history which leads to increase customer satisfaction and provide a seamless experience (Tsolakis, Schumacher, Dora, and Kumar, 2023). For example, chatbots and virtual assistants have made it possible for travel agencies to offer round-the-clock customer support and answer inquiries using real-time data, which increases client satisfaction and fosters trust (Kim, So, Shin, and Li, 2024). Sustainability is another aspect where AI has made a crucial contribution to the tourism supply chains. It contributes to carbon footprint reductions by identifying various sustainable practices, such as waste management, energy-efficient transportation, enhancing operational efficiency, and lodging options, optimizing transportation routes to minimize fuel consumption, reducing delivery time, and promoting the use of locally sourced and sustainable products. Various tools like smart fleet management, real-time tracking systems, and route optimization enhance the efficiency, effectiveness, and reliability of transportation services and improve overall customer experience (Rane, Paramesha, Choudhary, and Rane, 2024). Advanced analytics includes data analytics capabilities, improvisation, anticipation, and enhanced capability, also enhancing supply chain resilience.

Advanced analytics techniques are one of the techniques that revolutionize the optimization of tourism supply chains enabling the business to make smarter decisions, enhance customer experiences, and streamline operations activities (Adeleye, Oyeyemi, Asuzu, Awonuga, and Bello, 2024; Zhao and Hou, 2021). Additionally, it plays a role in improving transparency and sustainability. For instance: hotels and airlines can optimize their energy consumption and waste management process by leveraging data to track usage patterns and make informed decisions about sustainability initiatives. This will ensure optimum utilization of resources, cost reduction, and improved service level.

Advanced analytics in the tourism supply chain converts large and complex data into useful information, based on this insight companies can analyze market trends, and demand forecasting with the help of predictive analytics, make competitive strategies, set benchmarks against competitors, and work more efficiently.

Big data analytics and Real-time analytics are emerging as significant drivers for the implementation of systematic and effective tourism supply chain practices in the industry (Cheng, et al., 2020). In the tourism supply chain, big data analytics enables dynamic pricing strategies, inventory optimization, and price demand forecasting. For instance, hotels can charge their room rents in real time based on booking trends and competitors' rates. Together, big data and real-time analytics empower the tourism supply chain most effectively, apart from that, all stakeholders can move from reactive to proactive management. They can easily anticipate tourist needs, reduce risk, and implement adequate strategies to enhance operational efficiency and sustainability.

4.1 Benefits of Artificial Intelligence and Advanced Analytics in the Tourism Supply Chain

In response to all challenges, artificial intelligence and advanced analytics are transforming supply chain management in the tourism industry. Modern technologies enable more accurate demand forecasts, demand and supply management, efficient resource allocation, inventory management, and real-time problem-solving (Zhang, Song, and Huang, 2009). Various AI-powered tools analyze large volumes of data to predict upcoming trends, enhance decision-making capacity, appropriate forecasting, and optimize the utilization of resources.

Furthermore, tourism organizations may react more rapidly, enhance coordination amongst stakeholders, and build a more robust and flexible supply chain with the aid of sophisticated analytical tools (Lee and Fernando, 2015). With the aid of an AI-driven system that examines past data and finds patterns, AI can forecast future trends, including the busiest travel seasons, the most popular destinations, and possible interruptions.

AI-powered chatbots and virtual assistants can provide 24/7 customer support, answer queries, ensure tourists get instant assistance, personalized recommendations, and resolve their issues promptly. Advanced analytics can easily identify potential equipment failures before they occur in the supply chain, which helps minimize downtime and reduce maintenance costs (Wamba, Queiroz, Guthrie, and Braganza, 2022). By leveraging advanced analytics and predictive modeling, AI systems can monitor and analyze vast amounts of data from various sources, such as operational metrics, service logs, etc. By integrating AI and advanced analytics solutions in the tourism supply chain, providing solutions across customer services and operational management, due to the tourism supply chain becomes more efficient, resilient, and customer-centric, meeting the demands of modern travellers with ensuring the long-term viability (Wamba, Queiroz, Guthrie, and Braganza, 2022).

AI can analyze the largest dataset to identify patterns of fraudulent activity, such as fake booking, reservation, or identity theft. This will help protect enterprises and customers from financial losses. By leveraging the advanced analytics system in the tourism supply chain enterprises can gain a competitive edge by offering a high quality service to the customers, optimizing the operations, and making data-driven decisions, apart from those advanced analytics empowers the tourism supply chain to increase revenue by optimizing pricing strategies, personalized marketing and also an open door for upselling/cross-selling opportunities for the enterprises. By embracing these innovations, the tourism industry can overcome challenges and limitations, provide a seamless customer experience, and ensure sustainable growth (Soratana, Landis, Fu, and Suto, 2021).

Smart supply chain refers to a technology-driven approach to managing entire supply chain operations that emphasizes automation, decision-making based on data-driven, and a focus on real-time processing (Wu, Yue, Jin, and Yen, 2016). A smart supply chain is a system that seamlessly integrates its stakeholders, partners, and processes and autonomously adapts to changes in the environment. The main key features of a smart supply chain include adaptability, integration, intelligence, and self-optimization (Zhang, Yang, and Yang, 2022). A smart supply chain includes all advanced features and technologies, including the Internet of Things (IoT), Artificial intelligence, machine learning, and intelligent infrastructure, especially in intelligent or advanced infrastructure such as automated warehouses, to enable efficient handling. These intelligent systems adapt to operational changes, ensuring agility and scalability.

4.2 The Significance of AI and Advanced Analytics in Tourism Supply Chains

Artificial intelligence and advanced analytics play a crucial role in the modern tourism supply chain by improving service efficiency, reducing waiting time, enabling fast data-driven decision-making, and reducing the complaint rates of customers (Wamba, Queiroz, Guthrie, and Braganza, 2022). With the intensification of competition in the tourism market, the core competitiveness of an enterprise tends to provide personalized and differentiated services to the customers.

Through high-tech supply chain management, enterprises can better integrate all internal and external resources and actively and quickly respond to market changes, and tourists can experience customized tourism experiences. Under differentiated services, enterprises can stand out in highly competitive market competition.

With the help of AI and advanced analytics in the supply chain enterprises reduce their operational cost and also improve economic benefits through a centralized procurement system, process optimization, and other means; On the other hand, supply chain management promotes cooperation, and resource sharing among various enterprises, promotes industrial clustering development and enhances the competitiveness of the entire industry (Zamani, Smyth, Gupta, and Dennehy, 2022). The most important criterion for measuring the quality of tourism services is to figure out tourist satisfaction. From the perspective of the supply chain, tourism service management can provide a high-quality tourist experience throughout the entire journey, striving to provide very high-quality tourist services in various aspects such as hotel booking, transportation, ticket confirmation, feedback, etc.

A strong supply chain will ensure the achievement of long-term goals of organizations apart from building customer trust and enhancing satisfaction levels. All tourism businesses should respond to the global economy's sustainable development by giving environmental preservation, resource conservation, and social responsibility enough thought (Mandal, 2018). Tourism services management from the perspective of the supply chain enables all enterprises to achieve green procurement, carbon emission reduction, and energy conservation, promote cooperation with all the local communities and public enterprises, and provide tourists with more pleasing travel experiences. The long-term growth of the businesses is guaranteed by this sustainable tourist development approach, which also helps to preserve the local economy and culture (Soratana, Landis, Fu, and Suto, 2021).

4.3 Challenges in the Implementation of AI and Advanced Analytics in the Tourism Supply Chain

The implementation of AI and advanced analytics in the tourism supply chain comes with several challenges, because this industry deals with a vast amount of the crucial and sensitive data of customers, including their personal details, payment data, travel history, preferences, likes or dislikes. This reliance on data exposes businesses to a significant risk. One of the biggest challenges is data integration and quality because the tourism industry relies on numerous data sources such as weather patterns, new trends in tourism, customer preference, flight availability, and booking capacity, which must be seamlessly integrated for effective artificial intelligence applications (Nassar, Salah, Rehman, & Svetinovic, 2020; Tsolakis, Schumacher, Dora, and Kumar, 2023).

Apart from that, data privacy and security remain one of the major concerns because tourism data involves sensitive customer information, personal details, etc (Tsolakis, Schumacher, Dora, and Kumar, 2023). Mishandling or leakage of such sensitive information not only imposes a risk to customer trust but also leads to potential regulatory and legal challenges for tourism organizations. For customers, a data breach can result in identity theft, financial fraud, or unauthorized access to personal accounts, causing distress and inconvenience. For tourism enterprises, such incidents erode the trust of customers, which is a cornerstone of success in this highly competitive industry. A damaged reputation can lead to customer attrition, negative publicity, a decline in market share, loss of customer trust, a bad image in the market. Beyond these immediate reputational risks, data breaches can expose the tourism organization to substantial regulatory and legal challenges.

The third and major challenge is the high cost associated with implementing AI and Advanced analytics in the tourism supply chain, because it includes technology acquisitions, staff training, and high-tech infrastructure development, which's why for small enterprises it's hard to install and execute (Sutono, 2019). For many small and medium-sized enterprises (SMEs) within the tourism sector, these financial demands present substantial barriers, making it difficult to adapt and execute advanced solutions.

Last but not least, there is a shortage of qualified workers to handle these cutting-edge technologies. Resistance to change is also a common problem, as many tourism enterprises may fear the loss of jobs, displacements, or struggle to understand the advantages of high-tech artificial intelligence and advanced technology. In addition, some tourism enterprises struggle to fully understand the tangible benefits AI and Advanced analytics can bring. For many, perceiving complexity and cost of implementation outweigh the potential advantages, such as improved efficiency, enhanced customer experiences, and data-driven decision making. The lack of awareness or reluctance to explore the possibilities of technological transformation further slows the pace of adoption.

Another major challenge is that ethical concerns also arise, particularly with the risk of AI models perpetuating biases. AI systems are only as good as they are trained on and if the training data reflects historical biases or inequalities, these biases can be embedded into the AI algorithms. For example, AI systems used in the tourism sector supply chain for pricing, customer recommendations, or employee selection and recruitment could unintentionally favour certain groups over others, leading to discriminatory outcomes or results. Moreover, the lack of diversity in terms of developing AI models can exacerbate the risk of bias. If AI engineers, data scientists, or product developers come from narrow demographics, they may inadvertently overlook the needs, preferences, or challenges faced by the other groups, leading to exclusionary practices in AI-powered tools and services.

5. Future Trends in Smart Supply Chain for Tourism

The future of smart supply chains in the tourism industry is shaped by rapid technological advancement, shifting customer expectations, and growing demand for reference and sustainability. One of the most significant trends in the smart supply chain will be the use of artificial intelligence and machine learning for predictive analytics (V. Harish et al., 2023). With the help of this advanced technology, tourism businesses can focus on the demand, optimize pricing strategy and improve inventory management (Wu, Yue, Jin, and Yen, 2016; Tussyadiah, 2020).

AI and advanced analytics tools will help in more accurate planning and provide more personalized services to customers. Blockchain technology will also help improve supplier relationships, track the provenance of services, and ensure that ethical and sustainability standards are met throughout the supply chain (Deepak Eldho Babu, 2018; Suleiman, 2023). As the tourism industry continues to embrace automation, Robotic Process Automation will improve administrative tasks, such as invoicing, reservation, and customer service inquiries. By automating all the routine processes, businesses can reduce operational costs, minimize human error, defects, and wastages, and enhance overall productivity and efficiency (Tussyadiah, 2020).

The advanced analytics adopt energy-efficient transportation methods, reducing lead time and sourcing local sustainable products. It will be useful to track sustainability metrics, ensuring enterprises meet their environmental goals and providing transparent reporting to the customers. The future of the tourism supply chain will also be increasingly personalized and enhance its performance (Zhang and Tavitiyaman, 2022).

AI and data analytics will enable hyper-localized service offers to the customers. This means that accommodations and activities will be prepared based on personal profiles, resulting in a more customized experience. Integrating augmented reality (AR) and Virtual Reality (VR) technologies in the tourism sector will reshape how consumers interact with the supply chain. After implementing this technology, tourists can explore the destination virtually before bookings, while AR could enhance the real-world experience by providing interactive, context-sensitive information during the travel (Tussyadiah, 2020; Kim, So, Shin, and Li, 2024).

The concept of collaborative supply chain and network supply chain is also another best option in the tourism sector, wherein stakeholders such as hotels, airlines, tour operators, and local businesses work together to create seamless and integrated travel experiences (Zhang and Song, 2018). The main objective of these collaborations is to enhance customer experience, improve overall efficiency, and optimize operations. By fostering cooperation, stakeholders can share their information, and resources, and align strategies (Soratana, Landis, Fu, and Suto, 2021). Artificial intelligence and advanced analytics play a significant role in facilitating these collaborative efforts. These advanced technologies will create new touchpoints in the supply chains.

6. Case Study: Enhancing Customer Experience in the Age of Robotics and Automation

In this age of great technological progress, the service sector has undergone a lot of change due to robotics and automation, especially in the hospitality domain. However, though these technologies bring efficiency and reduce the cost of operations, they are sometimes disappointing for customers who enjoy personal contact or have problems with technological systems. This case study analyzes opinions from Prasun Choudhary, founder of DailyBetter and erstwhile President of OYO International, regarding the balance between automation and customer satisfaction in the hospitality sector.

A hotel chain incorporated AI-based kiosks and robot room service to improve customer experience and streamline processes. Though a few customers appreciated the efficiency, many were disappointed as they felt they could not get personalized interaction and were unable to understand automated systems, and special requests were not met.

It would raise an essential question regarding "how will the hospitality industry balance all those benefits of robotics and automation while minimizing customer frustration as well as ensuring satisfaction."

Prasun Choudhary said, "The biggest issue is that most people do not know how their customers behave, which impacts the service design. Technology is an enabler and not a replacement for genuine customer care. Irritation often arises when automation is done without considering the different tastes of customers. As service providers, we need to design experiences that are aligned with the traits and expectations of our customers. This would mean offering a hybrid model, automating where it adds value, yet making sure human touchpoints are accessible when needed. Flexibility and empathy must guide our approach."

Strategies for Success

Some strategies that may reduce customer dissatisfaction with automation are based on Choudhary's observations.

This calls for an attempt to segment consumer preferences first before serving the diversified customer needs. This means creating a detailed customer profile that will determine which group prefers automated services over those that value human interaction. From there, the identified customer groups will create ways of service delivery from a business. This would include installing self-service kiosks for those who are tech savvy and traditional receptions for the guests who wish to interact in person; this would ensure all the guests are catered to and hence improve general satisfaction and loyalty.

This approach is based on building a harmonious integration of human interaction and automation in offering exceptional customer experience. The use of robots and artificial intelligence should be applied in mundane, low-value processes like check-ins, room deliveries, or standard queries that can easily be done and completed. Meanwhile, the human aspect should be employed only in complex or personalized situations wherein empathy and inventiveness are crucial. This balance will be achieved if the staff are trained to complement automation by honing skills in emotional intelligence and personalized service, thus making them provide meaningful and memorable interactions that technology alone cannot achieve.

There should be mechanisms through which the automation improves satisfaction above all by effective feedback processes. Feedback systems that provide information straight and directly can immediately perceive the problems the customers face with automated services; therefore, quicker solution finding and minimal frustration result. This information can be amended correctly where faults may be seen to improve on this generally. Employee training should be guided by these insights so that staff members are better equipped to help consumers who cannot be automated.

Continuous development is necessary to keep these automated systems relevant and effective. These technologies ought to be updated and tested periodically to spot any possible flaws or hiccups, improve usability, and adapt to the shifting needs of consumers. In doing so, the quality of services that can be provided will be met and eventually overridden by automation in the long run.

The systems will demand constant updates and stringent testing to avoid malfunction and thus ensure smooth running. Thus, the whole system becomes strong in terms of reliability and, as a spinoff benefit, enriches the user experience also. This will help the business remain abreast of the current and emergent technologies in addition to the changing expectations of customers. Organizations can provide their clients with sophisticated services by always updating and upgrading their automated systems, thus successfully maintaining their competitive advantage in the market.

Prasun Choudhary mentioned that introducing robotics and automation into the hospitality industry would require careful planning about the various characteristics of customers. Since expectations, preferences, and comfort levels of guests are all divergent, a customer-centric approach is required. This harmonious integration promises not only efficiency in the operations but also the delivery of positive treatment without frustration to all the guests.

Even though technology in robotics and automation changes the face of the hospitality industry towards efficiency, precision, and even cost-effectiveness, among other virtues, these give rise to new problems, specifically when technology fails in delivering personified service. In this line, for example, the guest would feel a very low sense of the service rendered if he arrives when there is no human interface at the kiosk to allow self-check in. The same case applies across organizations. After all, technology should support humans without replacing them.

Choudhary believes that technology should be an enabler rather than a replacement. Automation fits best to tasks that involve repetition and routine, like room service, check-in procedures, and data management tasks where efficiency and precision are paramount. However, for complex or emotionally nuanced interactions, such as addressing complaints, accommodating special requests, or fostering memorable experiences, the human element is still indispensable.

Flexibility is the most important quality to embrace since it adapts to the shifting needs of the hospitality industry. Businesses have to stay flexible and keep reviewing the technological abilities to improve them in keeping pace with what is new and what a customer needs. Sympathy is equally important to be integrated into the growingly mechanized world and help customers feel acknowledged and understood. The other point for successful integration is an ongoing learning process. Train the employees in both advanced technologies and using them with empathetic, personalized services. Organizations need to support research and development to monitor technological changes; for instance, AI-based personalization, or robotic systems that behave more like human beings can be adopted.

Adding robotics and automation is a strategic and significant step regarding the shape that the service experience takes in light of the changes occurring in the hospitality sector. With flexibility, empathy, and continued learning, such an organization will succeed in offering service excellence, where technology and human contact work together. Such an organization will be defined in terms of both operational effectiveness and a touch that is warm and personable, which is synonymous with excellence in hospitality.

In addition to providing better customer service, these firms that can successfully integrate and negotiate this will emerge as leaders in a customer-centric, technology-driven environment.

Conclusion

In conclusion, artificial intelligence and advanced analytics will bring revolutionary changes in the tourism supply chain, making processes more effective, efficient, customer and environmentally friendly. Integrating these advanced technologies and advanced analytics such as artificial intelligence, AR and VR, and robotic process automation (RPA) will reshape how tourism businesses operate and serve customers. AI and machine learning will help businesses predict demand more accurately, set better pricing strategies, and be helpful in strategic planning. These tools and techniques will improve transparency, and trust within the tourism supply chain (Tsolakis, Schumacher, Dora, and Kumar, 2023). With the help of these technologies, the tourism supply chain works very smoothly, apart from that, tracking of service origin will be possible, ensure ethical sourcing, and confirm that sustainability standards are being met. This is significant because these days travellers prefer eco-friendly and responsible travel operations. Sustainability will be a key focus of future tourism supply chains. Advanced analytics will help enterprises adopt greener transportation options and reduce delivery times (Gruchmann, Topp, and Seeler, 2022). AR and VR technologies enable people to explore the destinations virtually before booking, gaining confidence in their choices. Despite the challenges, the rewards are immense: a more efficient, transparent, and customer-centric tourism industry that meets the needs of travellers while preserving the planet for future generations. The combination of smart technology and advanced analytics will not only make the tourism industry more efficient but also provide travellers with unique, memorable, and personalized experiences, setting a new standard for the way we explore the world. As the tourism industry continues to grow, the adoption of artificial intelligence and advanced analytics in the tourism supply chain will not only define competitive advantages apart from but also a set of excellences in an increasingly interconnected and dynamic world. Businesses that adapt and innovate today will lead the way in creating an unforgettable experience for travellers tomorrow.

References

Adeleye, R.A., Oyeyemi, O.P., Asuzu, O.F., Awonuga, K.F. and Bello, B.G. (2024). Advanced Analytics in Supply Chain Resilience: A Comparative Review of African and USA Practices. *International Journal of Management & Entrepreneurship Research, 6*(2), 2664–3588. doi:https://doi.org/10.51594/ijmer.v6i2.771

Cannas, V.G., Ciano, M.P., Saltalamacchia, M. and Secchi, R. (2024). Artificial intelligence in supply chain and operations management: a multiplecase study research. *International Journal of Production Research, 62*(9), 3333–3360. doi:https://doi.org/10.1080/00207543.2023.2232050

Chen, D. (2009). Innovation of Tourism Supply Chain Management. *IEEE.* doi:10.1109/ICMeCG.2009.79

Cheng, T.C., Kamble, S.S., Belhadi, A., Ndubisi, N.O., Lai, K.-h., and Kharat, M.G. (2020). Linkages between big data analytics, circular economy, sustainable supply chain flexibility, and sustainable performance in manufacturing firms. *International Journal of Production Research, 60*(22), 6908–6922. doi:https://doi.org/10.1080/00207543.2021.1906971

Cheunkamon, E., Jomnonkwao, S., and Ratanavaraha, V. (2022). Impacts of Tourist Loyalty on Service Providers: Examining the Role of the Service Quality of Tourism Supply Chains, Tourism Logistics, Commitment, Satisfaction, and Trust. *Journal of Quality Assurance in Hospitality & Tourism, 23*(6), 1397–1429. doi:https://doi.org/10.1080/1528008X.2021.1995564

Deepak Eldho Babu, A.K. (2018). Sustainability practices in tourism supply chain: Importance performance analysis. *Benchmarking: An International Journal, 25*(4), 1148–1170. doi:https://doi.org/10.1108/BIJ-06-2016-0084

George Q., Huang, H.S. (2012). *A comparative analysis of quantity and price competitions in tourism supply chain networks for package holidays.* Routledge. Retrieved from https://www.taylorfrancis.com/chapters/edit/10.4324/9781315878065-2/comparative-analysis-quantity-price-competitions-tourism-supply-chain-networks-package-holidays-george-huang-haiyan-song-xinyan-zhang

Ghaderi, Z., Hatamifar, P. and Khalilzadeh, J. (2018). Analysis of tourist satisfaction in tourism supply chain management. *An International Journal of Tourism and Hospitality Research, 29*(3), 433–444. doi:https://doi.org/10.1080/13032917.2018.1439074

González-Torres, T., R.-S. J. and Pelechano-Barahona (2021). Managing relationships in the Tourism Supply Chain to overcome epidemic outbreaks: The case of COVID-19 and the hospitality industry in Spain. *International Journal of Hospitality Management*, 721–728. doi:10.5267/j.uscm.2022.5.003

Gruchmann, T., Topp, M. and Seeler, S. (2022). Sustainable supply chain management in tourism: a systematic literature review. *Supply Chain Forum: An International Journal, 23*(4), 329–346. doi:https://doi.org/10.1080/16258312.2022.2085504

Guo, X., and He, L. (2012). Tourism Supply-Chain Coordination: The Cooperation between Tourism Hotel and Tour Operator. *Tourism Economics, 18*(6), 1361–1376. doi: https://doi.org/10.5367/te.2012.0179

Harish, V., Sharma, R., Saini, R.K. and Negi, D. "Intelligent Supply Chain Orchestration: A Framework for Seamless Integration of Industry 4.0 Technologies," 2023 Seventh International Conference on Image Information Processing (ICIIP), Solan, India, 2023, pp. 265–270, doi: 10.1109/ICIIP61524.2023.10537750

Hermawan, R., Pangestuti, E., Artono, G. and Supriono, S. (2023). The tourism supply chain management: Case Study on Mount Kelud, Kediri regency. *International Journal of Green Tourism Research and Applications, 5*(2). doi:https://doi.org/10.31940/ijogtra.v5i2.108–113

Kim, H., So, K.K., Shin, S. and Li, J. (2024). Artificial Intelligence in Hospitality and Tourism: Insights from Industry Practices, Research Literature, and Expert Opinions. *Journal of Hospitality & Tourism Research.* doi:https://doi.org/10.1177/10963480241229235

Lee, H.K. and Fernando, Y. (2015). The antecedents and outcomes of the medical tourism supply chain. *Tourism Management, 46*, 148–157. doi:https://doi.org/10.1016/j.tourman.2014.06.014

Ling, R. (2015). A review of Tourism Supply Chain based on the Perspective of Sustainable Development. *Journal of Economics and Sustainable Development, 6*(22). Retrieved from https://core.ac.uk/download/pdf/234647319.pdf

Mandal, S. (2018). Exploring the influence of big data analytics management capabilities on sustainable tourism supply chain performance: the moderating role of technology orientation. *Journal of Travel & Tourism Marketing, 35*(8), 1104–1118. doi:https://doi.org/10.1080/10548408.2018.1476302

Ministry of Tourism. (2024). Retrieved from tourism.gov.in: https://tourism.gov.in/

Nassar, M., Salah, K., Rehman, M.H. and Svetinovic, D. (2020). Blockchain for explainable and trustworthy artificial intelligence. *Data Mining and Knowledge Discovery, 10*(1). doi: https://doi.org/10.1002/widm.1340

Palang, D. and Tippayawong, K.Y. (2019). Performance evaluation of tourism supply chain management: the case of Thailand. *Business Process Management, 25*(6). doi:https://doi.org/10.1108/BPMJ-05-2017-0124

Rane, N., Paramesha, M., Choudhary, S. and Rane, J. (2024). Artificial Intelligence for Enhancing Resilience. *Social Science Research Network.* doi:10.2139/ssrn.4831911

Song, H. (2011). *Tourism Supply Chain Management.* Routledge. doi:https://doi.org/10.4324/9780203804391

Soratana, K., Landis, A.E., F.J. and Suto, H. (2021). *Supply Chain Management of Tourism Towards Sustainability.* Cham; Springer. Retrieved from https://link.springer.com/book/10.1007/978-3-030-58225-8

Statista. (2024). Retrieved from www.statista.com: https://www.statista.com/statistics/233223/travel-and-tourism-total-economic-contribution-worldwide/

Suleiman, M.A. (2023). The impact of tourism supply chain on sustainable performance in sub-Saharan Africa: evidence from Tanzania. *Management of Environmental Quality, 34*(2), 492–510. doi:https://doi.org/10.1108/MEQ-01-2022-0007

Sutono, A. (2019). Supply chain management: implementation issues and research opportunities in tourism industry. *Uncertain Supply Chain Management*, 427–438. doi:10.5267/j.uscm.2018.12.004

Szpilko, D. (2017). Tourism Supply Chain-Overview of Selected Literature. *Procedia Engineering, 182*, 687–693. doi:10.1016/j.proeng.2017.03.180

Tapper, R., Fon, X., Schwartz, K., Kornilaki, M. and Carey, B. (2004). *Tourism supply chains. Report of a Desk Research Project for the Travel Foundation.* Leeds Metropolitan University and Environment Business & Development Group.

Travel & Tourism Economic Impact 2024: Global Trends. Research Hub. WTTC. Retrieved from https://researchhub.wttc.org/product/economic-impact-report-global-trends

Tsolakis, N., Schumacher, R., Dora, M. and Kumar, M. (2023). Artificial intelligence and blockchain implementation in supply chains: a pathway to sustainability and data monetisation? *Annals of Operations Research, 327*(1), 157–210. doi:https://doi.org/10.1007/s10479-022-04785-2

Tussyadiah, I. (2020). A review of research into automation in tourism: Launching the Annals of Tourism Research Curated Collection on Artificial Intelligence and Robotics in Tourism. *Annals of Tourism Research, 81*. doi:https://doi.org/10.1016/j.annals.2020.102883

UN Tourism. (2024). Retrieved from UNWTO: https://www.unwto.org/tourism-data/global-and-regional-tourism-performance

Wamba, S.F., Queiroz, M.M., Guthrie, C. and Braganza, A. (2022). Industry experiences of artificial intelligence (AI): benefits and challenges in operations and supply chain management. *Production Planning & Control, 33*(16), 1493–1497. doi:https://doi.org/10.1080/09537287.2021.1882695

Wu, L., Yue, X., Jin, A. and Yen, D.C. (2016). Smart supply chain management: a review and implications for future research. *The International Journal of Logistics Management, 27*(2), 395–417. doi: 10.1108/IJLM-02-2014-0035

Zamani, E.D., Smyth, C., Gupta, S. and Dennehy, D. (2022). Artificial intelligence and big data analytics for supply chain resilience: a systematic literature review. *Annals of Operations Research, 327*, 605–632. doi:https://doi.org/10.1007/s10479-022-04983-y

Zhang, G., Yang, Y. and Yang, G. (2022). Smart supply chain management in Industry 4.0: the review, research agenda and strategies in North America. *Annals of Operations Research*, 1075–1117. doi:10.1007/s10479-022-04689-1

Zhang, X. and Song, H. (2018). An integrative framework for collaborative forecasting in tourism supply chains. *International Journal of Tourism Research, 20*(2), 158–171. doi: https://doi.org/10.1002/jtr.2168

Zhang, X. and Tavitiyaman, P. (2022). e-Supply Chain Management in Tourism Destinations. *Handbook of e-Tourism*, 1289–1309. Retrieved from https://link.springer.com/referencework/10.1007/978-3-030-48652-5

Zhang, X., Song, H. and Huang, G.Q. (2009). Tourism supply chain management: A new research agenda. *Tourism Management, 30*(3), 345–358. doi:https://doi.org/10.1016/j.tourman.2008.12.010

Zhao, X. and Hou, J. (2021). Applying the Theory of Constraints Principles to Tourism Supply Chain Management. *Journal of Hospitality & Tourism Research, 46*(2). doi:https://doi.org/10.1177/109634802199679l

12 Emerging Trends and the Future of Business Analytics

Bhakti Parashar[1*], Amrita Chaurasia[2], Vipul Jain[3], Sanjay Sharma[4] and Shivani Agarwal[5]

Artificial intelligence (AI) and machine learning (ML) have transformed business analytics in today's data-driven environment, bringing new trends, technologies, and capabilities. Focusing on areas such as machine learning, predictive analytics, natural language processing (NLP), text analytics, computer vision, and robotic process automation (RPA), this research examines the AI-based technologies needed to advance business intelligence (BI). By integrating these solutions, organizations can improve operations, gain valuable insights, and make data-driven decisions. Finding patterns and predicting trends in big data is accomplished through machine learning and predictive analytics. Natural Language Processing (NLP) and text analytics help unlock the potential of redundant data by facilitating user insights, service automation, and better decision making. This expands business intelligence to include visual data processing. Also, RPA and AI are used to automate processes to increase productivity and efficiency. Big data and IoT can also help businesses adapt to changing business conditions. Advanced design techniques enhance human-computer interaction and enable intelligent data interpretation. This article explores the future potential of AI-ML NLP to reinvent analytics and decision making by highlighting the latest developments and impacts of AI-ML NLP on BI.

1. Introduction

The business intelligence industry has grown rapidly in recent years, enabling business users to gain greater insights from operational data, particularly from the business world. A well-known example is data mining in e-commerce, which has developed into a major application of data mining (Kohavi, R., and Provost, F. 2001; Kohavi, R., et al., 2000). By combining clickstream data from their websites

[1] 0000-0002-2561-117X
[2] 0000-0003-3618-6954
[3] 0000-0002-0833-9354
[4] 0000-0001-7625-5091
[5] 0000-0002-3205-552X
* Corresponding author: sharma.bhakti@gmail.com

DOI: 10.1201/9781003514282-12

with demographic and behavioral data, businesses are creating data sets that far exceed the size and complexity of data collections even a few years ago (Kimball, R., and Merz, R. 2000). To extract useful information about website activity and visitor characteristics from these large databases, a combination of analytics techniques and human experience is required. As millions of clickstream files are created daily and combined into a database containing hundreds of attributes, the need for technology to find important patterns is increasing.

1.1 Artificial Intelligence and Machine Learning: The Backbone of Modern Business Analytics

Machine learning (ML) and artificial intelligence (AI) have revolutionized industries such as banking, healthcare, manufacturing, retail, supply chain, logistics, and electronic products, which have received positive reviews due to rising prices (Dwivedi et al., 2021 Schmidt, 2020). In today's fast-paced and competitive global economy, data-driven decision making powered by AI and machine learning has become imperative (Davenport, 2018). This decision-making process is well supported by economic analysis (Schmitt, 2023). Business analytics is an interdisciplinary field that combines management science, operations science, statistics, information systems, and machine learning (Sharda, Delen, and Turban, 2018). Prescriptive analytics, predictive analytics, and descriptive analytics are the three main categories that are often divided (Delen and Ram, 2018).

While predictive models are important for decision making, they can only be used by workers who have the skills necessary to effectively use the measured resources (Grover, Chiang, Liang, and Zhang, 2018). Although the movement to expand the scope of educational science data is growing, there is still a need for experts who can analyze the data and draw conclusions (Clayton and Clopton, 2019) (Kar, Kar, and Gupta, 2021). Interacting with machine learning algorithms poses significant challenges because a great deal of knowledge and experience is required to properly design and modify these models (Schmitt, 2023).

1.2 Technological Advancements Paving the Way for the Future

Technology has improved people's lives and transformed many sectors. For example, artificial intelligence (AI) has become the basis for innovation in many sectors, including healthcare and transportation, by increasing automation, improving accurate decisions, and streamlining processes (Smith, 2021). Similarly, renewable energy technologies such as solar and wind are changing the electricity generation sector by reducing dependence on fossil fuels and reducing climate change (Johnson and Lee, 2020). The creation of 5G networks that will make connections faster and more reliable is another important change in the development of the Internet of Things (IoT). This connectivity can create smart cities, and data analysis can be used to effectively manage the process (Brown, 2019).

Additionally, effective treatments such as CRISPR gene editing have the potential to alter medicines and treat genetic diseases (Taylor et al., 2022).

Despite this progress, legal issues, cybersecurity risks, and access to technology remain equally important. These issues need to be resolved if progress is to be made. Collaboration between scientists, business, and politics can ensure that technological innovation paves the way for a sustainable and inclusive future.

2. The Role of AI in Decision-Making

Artificial Intelligence (AI) technology has improved business operations, improved supply chains, and enhanced customer relationships by analyzing historical and real-time data (Brynjolfsson and McAfee, 2017). The government uses AI for urban planning, policing, and efficient allocation of resources. For example, AI is needed in traffic management to improve traffic flow, reduce pollution, and reduce accidents (Goodman,2020; Parashar et al., 2023; Sharma et al., 2019). Although AI can analyze unprecedented data, the purpose of its conclusions depends on the data it displays. Therefore, ethical issues such as impartiality, accountability, and transparency are important in AI decision-making (O'Neil, 2016).

All things considered, AI's analytical and computational capabilities are revolutionizing decision-making, helping individuals and businesses make efficient, smart, and successful choices in various sectors. Figure 1 shows the capabilities of AI in business intelligence.

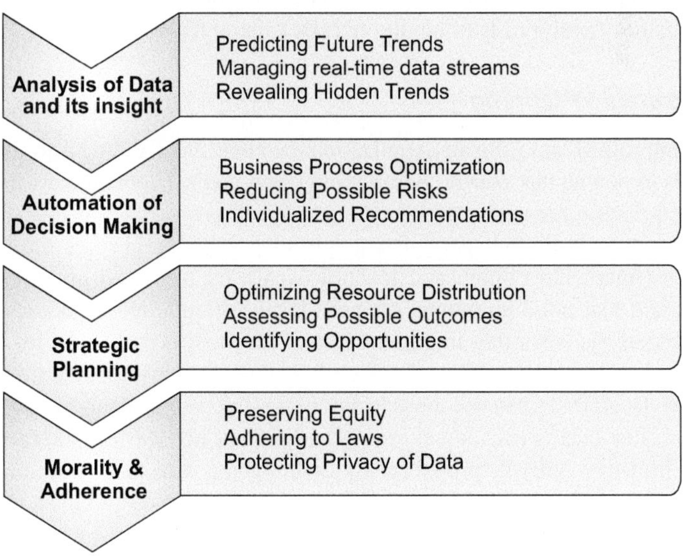

Fig. 1 Role of AI in Business Analytics.

2.1 Analysis of Data and its Insight

Data analytics and insights are critical to business decisions today as they enable organizations to manage complex and unpredictable situations. One important topic is predictive modeling, which uses historical data and machine learning models to predict future patterns. For example, companies use predictive models to predict customer behavior, monitor product levels, and measure market performance for better profitability (Shmueli et al., 2017). Real-time data processing allows businesses to respond quickly to real-time data, improving decision-making.

2.2 Automation of Decision Making

As AI-driven decision making becomes more important in today's fast-paced and competitive business world, interest in machine learning (ML) is increasing. However, the demand for analytics experts far exceeds what is currently available. To fill this gap, attempts have been made to make machine learning techniques more accessible to beginners. As a solution, automated machine learning (AutoML) reduces the need for expertise by providing efficient, user-friendly tools for activities such as model selection and exchange hyperparameters. Data-driven AI/ML-driven decision-making is becoming increasingly important in the business world (Davenport, 2018). Today's decision-making approach is through business analysis (Schmitt, 2023). It is an interdisciplinary approach that incorporates ideas from management science, operations research, statistics, information systems, and machine learning (Sharda, Delen, and Turban, 2018). Prescriptive analytics, predictive analytics, and descriptive analytics are the three main categories into which business analytics is often divided (Delen and Ram, 2018).

2.3 Strategic Planning

It is the process of choosing an organizational course and deciding how to allocate resources to pursue that course. Setting long-term goals, choosing how to achieve them, and determining what resources are needed are all part of this. Resource allocation is an important part of strategic planning that ensures the available resources (financial, human, and technological) are allocated as efficiently as possible and that good intentions are used (Hitt, Ireland, and Hoskisson, 2017). Scenario analysis is another important part of the process that identifies different scenarios and how they affect the organization's strategy to determine future outcomes. Businesses can use these technologies to change situations and prepare for unforeseen events (Schoemaker, 1995). Last but not least, market expansion aims to find new growth prospects by diversifying into new product lines or breaking into new geographic markets (Porter, 1990). These strategic components guarantee that a company maintains its competitive edge while setting itself up for future expansion.

2.4 Morality and Adherence

Truthfulness and compliance, especially in matters of justice, law, and privacy, are essential to the functioning of society and organizations. Although ethics is concerned with these ideas, especially in law and ethics, ethics is concerned with concepts that help people know the difference between good and bad. Social justice requires that everyone be treated fairly and equally in all relationships; this requires businesses and individuals to treat everyone equally, regardless of their background. This effort helps maintain harmony and peace (Rawls, 1971). Another important aspect of morality is obeying the law. Laws are social structures designed to ensure justice and order. Following them prevents people from getting hurt and helps promote health (Hart, 1961). Justice prevents injustice and thus maintains social order.

In today's digital environment, data privacy has become important. Respecting people's privacy rights and protecting them from harm is an important aspect of personal data ethics. Organizations have moral and legal obligations to protect consumer privacy (Westin, 1967).

3. Emerging Trends in Business Analytics

The fact that we come across terms such as "Business Intelligence", "Business Analysis", "Artificial Intelligence", "Machine Learning", "Deep Learning", "Internet of Things", "Big Data" and "ChatGPT" in the critical analysis word shows that the communication network of technology has changed the jobs. The principles of business intelligence (BI) and business analytics (BA) enable data-based decision making, and artificial intelligence (AI) and its subsets such as machine learning (ML) and deep learning develop these models (DL). This technology provides advanced algorithms for model prediction and data analysis. By integrating intelligence into BI platforms, businesses can gain deeper insights and make better decisions, thus increasing their data potential. Predictive analytics can predict trends, customer behaviour, and market dynamics and rely on machine learning.

As the reliance on computer-aided technology increases, researchers are focusing on better communication technologies that can transform human interactions and understand people's natural and human languages (Rana et al., 2019). The amount of inappropriate information that is considered useless continues to increase due to the problem of information dissemination in all industries, including business, healthcare, and education. To improve natural language understanding and processing, NLP is a powerful tool that can be used with technologies such as machine learning, artificial intelligence, and machine learning to benefit human-computer interaction and benefit humans. Computer chat is possible. Analyse and create large, unstructured, unprofitable forms and documents in various sectors. Significant results will be created, leading to improved decision-making processes and thus effective work.

The IoT is well paired with Big Data Analytics, creating a transformative force that is changing the commercial and societal landscapes due to its unparalleled connection and record-breaking data creation. Millions of connected gadgets have released tons of data, which gives organizations ready to apply cutting-edge analytics a lot of room to exploit. Figure 2 below shows the most recent developments in business analytics.

Fig 2 Recent development in Business Analytics.

3.1 Artificial Intelligence (AI) and Machine Learning (ML)

Artificial Intelligence (AI) and machine learning (ML) have become popular in the business world in recent years due to their increasing use in this field. Today, they are increasingly used in many sectors, including marketing, supply chain management, fraud detection, and customer support. Although using such technologies requires some human control, it also creates many problems and issues for companies, including fraud and data protection. This article discusses the potential and challenges of AI and machine learning in business. AI and machine learning have many uses in the business world. AI-powered chatbots can assist customers around the clock, eliminating the need for human service representatives. Machine learning (ML) can be used in supply chain management to predict demand and thus improve product levels in the supply chain. AI can be used in business to tailor information to specific customers.

3.2 Natural Language Processing (NLP) and Conversational Analytics

Recent developments in NLU and NLP have led to a resurgence of interest in natural language processing. Through natural communication, non-professional workers can easily access and request information. Early systems moved away from keyword searches and focused on simple fact-finding queries, but over time, both input and SQL queries have become more complex. The use of interactive systems for data analysis has also recently gained traction, allowing users to access data automatically and instantly.

3.3 Edge Analytics

The goal of edge technology is to place cloud resources (specifically, computing, storage, and communication) closer to the edge or smart devices that create and consume data. Two new concepts in edge technology—edge computing and edge analytics—are the result of being incorporated and applied to edge devices. Edge analytics use different methods or algorithms to analyse data from edge devices. With the introduction of edge analytics, edge devices have now become a full-fledged commercial product. So far, edge analytics has not supported the use of analytical methods. Low power, small memory footprint, limited resources, and other limitations of end devices make it impossible to implement sophisticated analytical techniques.

3.4 *Internet of Things (IoT) Analytics*

With the introduction of Industry 4.0 into our lives, IoT has become more widespread and has begun to grow in many areas. From devices that can send data to objects on any network, the Internet of Things allows people and everything to stay connected anytime, anywhere. As a result, a smart environment (including smart devices that transmit data over a smart network) emerges. As mentioned earlier, data is sent along with basic information and converted into information that can be used in business decisions. Although IoT is growing rapidly today, large enterprises are also interested in business analytics.

3.5 *Blockchain for Data Integrity*

The new way to create and store financial information and business transactions is through distributed ledger and blockchain technology. This innovative collaboration between these technologies opens new ways to analyse data and identify changes. Major financial institutions have invested heavily in developing new blockchain services related to this technology and are still at the forefront. However, most business schools have yet to incorporate blockchain concepts into their business and marketing programs.

3.6 *Immersive Data Visualization*

Over the past few years, augmented reality (AR) and virtual reality (VR) technologies have made great strides in terms of hardware capabilities and ease of use, which has increased the use of these tools in many places. In an integrated scenario, the complexity of managing multiple profile views and tracking workflows increases. Once Immersive Insights provides data scientists with access to their data to perform analytics based on content usage, data scientists need to track the data source of the screens and metrics that impact the solution. At the same time, there needs to be a central place where users can discuss whether to use certain methods, or even how to collect data to fulfill the agreement, or complete instructions for other methods.

3.7 Explainable AI (XAI)

Artificial intelligence (XAI) helps end users understand a model by using a set of visual representations to explain and interpret its inner workings and how it was achieved. In recent years, many research papers have featured the effectiveness of the model and the XAI method, but their impact on human-machine teamwork has received little attention. Further expansion is one of the goals of XAI, which will lead to more powerful machine learning models.

3.8 Cloud and Hybrid Analytics

For businesses looking to improve their IT infrastructure and achieve data-driven success, hybrid cloud computing is a great idea. It creates an ecosystem that combines the management of private spaces with the transformation of the public climate. However, this integration also brings new challenges in information security, management, and analysis. One of the most important aspects of this ecosystem is hybrid cloud analytics, which allows companies to gain insights from diverse data, increase operational efficiency, and develop a competitive advantage. This chapter reviews best practices, discusses how Veeam solutions can provide seamless data and analytics in a hybrid environment, and examines how hybrid cloud analytics can transform data management.

3.9 Automated and Self-Service Analytics

Business users with no IT experience can use self-service analytics software to analyze data and identify business opportunities. These technologies represent advanced analytics. They enable users without statistics or background knowledge to extract valuable insights from business data. Since using these computer tools in their businesses, many companies have realized the value of self-service software.

3.10 Data Privacy and Ethical Analytics

Big data has been a great resource for many organizations, promising improved operations and new business opportunities. However, big data has increased access to sensitive information that when processed can directly jeopardize the privacy of individuals and violate data protection laws. As a result, data controllers and data processors may be imposed tough penalties for non-compliance that can result in even bankruptcy. The massive collection and processing of personal data in big data analytics has raised serious privacy issues, linked to large-scale electronic surveillance, profiling, and disclosure of private data. To exploit the benefits of analytics without entering the private sphere of the individual, it is necessary to set limits on big data processing and integrate data protection safeguards into the analytics value chain. ENISA, with the current report, supports this approach and the position that the challenges of technology (for big data) should be addressed by the opportunities of technology (for privacy). When using the analytics process, companies should consider the risks it poses to individuals' information privacy and develop responsible measures to accompany its use. This set of ethical standards asks companies to embrace accountable practices that mirror particular risks in the specific use of the analytics process.

3.11 Digital Twins in Analytics

Businesses are leveraging the business potential of various technologies, such as digital twins, to create innovation and technology management. A digital twin is a representation of the real world, such as a system, process, or product. Since digital twins are connected to real-time information, the connection between the twin's information and the physical twin differs from other related concepts. Solutions for digital twins are developing rapidly, and using them is becoming easier and cheaper every day with imagination. Most digital twin studies are conducted in different areas of engineering science. However, the connection between the concept and innovation must be understood to use the digital twin concept.

3.12 Big Data Integration with AI

Due to recent developments in sensor networks and Internet of Things (IoT) technology, a lot of data is being collected. Examining such large data requires more efficient methods with accurate analysis. Artificial intelligence (AI) methods such as machine learning and evolutionary algorithms in big data analytics can produce more accurate, faster, and measurable solutions. Due to the rapid advancement of technology, digital information is growing rapidly (Klein, 2017). Therefore, a lot of data is being produced from different sources such as networks, smartphones, sensors, etc. Big data refers to large amounts of data that cannot be stored and processed using traditional relational data and analysis techniques. New tools and analyses must be developed to extract patterns from large data sets. Big data is created from many sources and inputs. Therefore, to use rapidly changing data, newly developed analytical tools must be able to analyse the connections between them.

4. Future Technologies in Business Analytics

Here are some of the top future predictions in business analytics that can affect organizations in the long run:

- Business intelligence (BI) tools are becoming increasingly popular and will become increasingly independent. According to Allied Market Research, the market for self-service BI tools is anticipated to reach $14.19 billion by 2026.
- Future cloud data use will rise in tandem with the growth of business data.
- Decision intelligence will continue to grow, giving businesses the ability to extract insights from their current data.

5. Challenges

Notwithstanding its ongoing expansion, business analytics is not without its difficulties:

- Despite evidence of analytics' usefulness, some businesses are unwilling to integrate it into their workflow.
- The volume of unstructured data generated today is still unfathomable despite numerous solutions.
- Developing a comprehensive analytics solution for business operations remains a challenge for organizations.
- Data processing and quality are still relevant challenges.
- Stakeholders are opposing the integration of business analytics and modifications to the current workflow.
- Data silos make it difficult for organizations to deliver data points to stakeholders.

Conclusion

In the future, AI will become more important in decision-making. The amount of data produced by many businesses will increase, requiring analytical tools to quickly examine data and make recommendations. Artificial intelligence (AI) technology, including machine learning and predictive analytics, will enhance organizations' ability to make data-driven decisions, allowing them to compete accurately and consistently in an ever-changing market. In addition, AI will inevitably support greater collaboration in decision-making by combining machine intelligence with human emotion and intelligence. With the help of this combination, businesses will be able to make the most of human intelligence and decision-making to make better, more informed decisions. Ethics will influence the future development of cognitive skills and decision-making. If organizations want to gain the trust of all stakeholders, they must prioritize accountability, honesty, and transparency. If AI is used in decision-making, businesses can realize their potential and follow social norms. Manage risk fairly and responsibly.

References

Brown, J. (2019). *The rise of smart cities: 5G and IoT*. Technology Review.

Brynjolfsson, E. and McAfee, A. (2017). Machine, platform, crowd: Harnessing our digital future. W. W. Norton & Company.

Clayton, P.R. and Clopton, J. (2019). Business curriculum redesign: Integrating data analytics. *Journal of Education for Business*, *94*(1), 57–63.

Davenport, T.H. (2018). From analytics to artificial intelligence. *Journal of Business Analytics*, *1*(2), 73–80.

Delen, D. and Ram, S. (2018). Research challenges and opportunities in business analytics. *Journal of Business Analytics*, *1*(1), 2–12.

Dwivedi, Y.K., Hughes, L., Ismagilova, E., Aarts, G., Coombs, C., Crick, T.,... and Williams, M.D. (2021). Artificial Intelligence (AI): Multidisciplinary perspectives on emerging challenges, opportunities, and agenda for research, practice and policy. *International Journal of Information management*, *57*, 101994.

Goodman, B. (2020). Ethical challenges in the age of artificial intelligence. *AI & Society*, 35(2), 333–345.

Grover, V., Chiang, R.H., Liang, T.P. and Zhang, D. (2018). Creating strategic business value from big data analytics: A research framework. *Journal of Management Information Systems*, *35*(2), 388–423.

Hart, H.L.A. (1961). *The concept of law*. Oxford University Press.

Hitt, M.A., Ireland, R.D. and Hoskisson, R.E. (2017). Strategic management: Concepts and cases: Competitive advantage and globalization. *Cengage Learning*.

Johnson, R. and Lee, M. (2020). Renewable energy solutions for a sustainable future. *Journal of Energy Innovation*.

Kar, S., Kar, A.K. and Gupta, M.P. (2021). Modeling drivers and barriers of artificial intelligence adoption: Insights from a strategic management perspective. *Intelligent Systems in Accounting, Finance and Management*, *28*(4), 217–238.

Kimball, R. and Merz, R. (2000). The data webhouse toolkit: Building the Web-enabled data warehouse. *Industrial Management & Data Systems*, *100*(8), 406–408.

Klein, S. (2017). *IoT Solutions in Microsoft's Azure IoT Suite* (pp. 273–289). Berkeley, CA: Apress.

Kohavi, R. and Provost, F. (2001). Applications of data mining to electronic commerce. In *Applications of data mining to electronic commerce* (pp. 5–10). Boston, MA: Springer US.

Kohavi, R., Brodley, C.E., Frasca, B., Mason, L. and Zheng, Z. (2000). KDD-Cup 2000 organizers' report: Peeling the onion. *Acm Sigkdd Explorations Newsletter*, *2*(2), 86–93.

O'Neil, C. (2016). Weapons of math destruction: How big data increases inequality and threatens democracy. Crown.

Parashar, B., Sharma, R., Rana, G., and Balaji, R.D. (2023). Foundation concepts for industry 4.0. In *New Horizons for Industry 4.0 in Modern Business* (pp. 51–68). Cham: Springer International Publishing.

Porter, M.E. (1990). *Competitive strategy: Techniques for analyzing industries and competitors*. Free Press.

Rana, G., Sharma, R. and Goel, A.K. (2019). Unraveling the power of talent analytics: Implications for enhancing business performance. *Business governance and society: Analyzing shifts, conflicts, and challenges*, 29–41.

Rawls, J. (1971). *A theory of justice*. Harvard University Press.

Schmitt, M. (2023). Deep learning in business analytics: A clash of expectations and reality. *International Journal of Information Management Data Insights*, *3*(1), 100146.

Schoemaker, P.J.H. (1995). Scenario planning: A tool for strategic thinking. *Strategic Management Journal*, 16(3), 117–137. https://doi.org/10.1002/smj.4250160305.

Sharma, R., Singh, S.P. and Rana, G. (2019). Employer branding analytics and retention strategies for sustainable growth of organizations. *Understanding the role of business analytics: Some applications*, 189–205.

Sharda, R., Delen, D. and Turban, E. (2018). *Business intelligence, analytics, and data science: A managerial perspective*. Pearson.

Shmueli, G., Patel, N.R. and Bruce, P.C. (2017). *Data mining for business analytics: Concepts, techniques, and applications*. Wiley.

Smith, A. (2021). Artificial intelligence and the future of work. *AI Research Journal*.

Taylor, S. (2022). The psychology of pandemics. *Annual review of clinical psychology*, *18*(1), 581–609.

Westin, A.F. (1967). *Privacy and freedom. Atheneum.*

13 Analysis and Forecasting Consumer Price Index of Indian Economy via Predictive Modeling Algorithms

Manoprabha M[1*], Mansurali A[2] and Swamynathan Ramakrishnan[3]

One of the most crucial macroeconomic indicators for assessing inflation is the Consumer Price Index (CPI), which influences several financial markets, productivity, and economic variables, and a nation's ability to forecast changes in the CPI is critical to its economic growth. An increase in CPI may lead to a decline in the market. India's CPI increased by 0.3% and stood at 185.80 in February 2024. Thus, it is essential to estimate the CPI variation precisely. This chapter aims to compare predictive regression algorithms such as Random Forest, Decision Tree, and Support Vector Models in predicting the CPI of India and uses classical regression to examine the relation between the economic indicators like crude oil rate, gold rate, and Federal fund effective rate (FFER) with CPI. Predictive models have been validated using robust metrics including MAE, RMSE, MSE, and R2, from which random forest regression is measured as the best algorithm for predicting CPI. Additionally, from Multiple Linear Regression (MLR) results, changes in the FFER and gold rate have a greater impact on India's CPI.

1. Introduction

Financial market investors and the Federal Reserve frequently use the CPI to estimate inflation and adjust monetary policy over time. CPI figures are frequently used as deflators in national accounts, as well as a macroeconomic indicator of inflation and a tool for governments and central banks to manage inflation and monitor price stability. The CPI is also used to index employee dearness allowances in response to price increases. For this reason, the CPI is regarded as one of the most significant economic indicators. Two necessary elements for creating CPI numbers are pricing data that is gathered regularly and weighting diagrams, which show consumption patterns. The information is based on the base year 2010 and 2012 which is divided into 6 groups like food and bevarages, pan, tobacco and intoxicants, clothing and footwear, housing, fuel and light and other miscellaneous

[1] 0000-0002-2578-9891
[2] 0000-0002-4326-8360
[3] 0000-0002-1818-0471
* Corresponding author: manoprabha@cutn.ac.in

DOI: 10.1201/9781003514282-13

which has been collected every month as per a weekly schedule from selected 1114 urban markets in 310 towns and 1181 villages by the statistical unit covering the entire population of the country. From which the All-India Consumer Price Index for Rural and Urban areas are calculated. Every month on the 12th, the Central Statistical Office releases the dataset. The CPI is a Laspeyres index which has been calculated throughout the country. The Indian CPI, which has been computed over the years, indicates a pattern of rising values, which could raise the nation's cost of inflation. Unevenly growing prices in an inflationary environment ultimately cause some customers' purchasing power to decline, and this depreciation of real income is the single highest cost of inflation. Over time, inflation may also erode the purchasing power of fixed interest rate payers and receivers. Thus, CPI is considered the better predictor of future inflation, therefore accurate prediction of CPI changes is important for a country's economic development.

1.1 Objective of the Study

1. To conduct a background study on the prediction of CPI using the classical approach and Machine learning algorithms.
2. To study the relationship between the most important indicators like the world gold rate, the crude oil price, and FFER with CPI.
3. Predicting the CPI for India using Machine learning regression algorithms, such as random forest regressor, support vector regressor, and decision tree regressor.
4. To find the best predictive algorithm using the performance metrics like MAE, MSE, RMSE, and R^2.

Organization of the Chapter

The chapter is organized as follows:

1. **Introduction:** Overview, background, scope, problem definition, technical information, objectives, and chapter organization.
2. **Related works:** Analysis of existing research on prediction of economic indicators using machine learning algorithms.
3. **Materials and Methods:** Detailed explanation of the research methods used in the chapter, and about the CPI dataset.
4. **Experimentation, Results and Analysis:** About the Performance metrics used to evaluate the predictive algorithm and the results obtained.
5. **Conclusion:** Summary of the overall idea of the study, the analysis carried out and the insights from the results.

2. Related Works

The consumer price index is the most watched economic barometer in the world. In recent years, the public and media have paid increasing attention to the monthly CPI announcements (Blinder et al., 1980). Accurately measuring CPI and its rate of change, inflation is an economic issue. CPI also overstates the change in cost of living (Boskin et al., 1998). NSSO & CSO, which works under Ministry of Statistics and Program Implementation (MOSPI) has generally relied on its sample surveys to collect the price and expenditure information necessary to produce the CPI (Konny et al., 2019). The purpose of the Chained Consumer Price Index is to approximate the cost of living index more accurately than the current CPI measure for all urban consumers (Cage et al., 2003). CPI is immune to one of the potentially important biases inherent in the measure of inflation weighting bias. The Dynamic factor model is used to compute the inflation element in a broad cross section of consumer price changes (Bryan and Cecchetti, 1993). Empirical results show there is a unidirectional causality between producer price index and consumer price indices in Finland and France and bidirectional causality between two indices in Germany (Akcay, 2011). The two main purposes of measuring CPI are to calculate real wages and to measure inflation (Edvinsson and Soderberg, 2011). The Schultze panel recommends that the Bureau of Labor Statistics should endeavor to introduce new and improved goods into the consumer price index. The CPI is measured based on the Laspeyres formula, while the Personal Consumption Expenditures (PCE) is measured based on the Fisher Ideal formula, the difference is referred to as the formula effect. The formula impact accounts for about half of the 0.4 percentage point discrepancy in growth rates between the PCE and the CPI (Broby, 2022) reviewed the use of predictive analytics in finance. The study comprehensively covers classification, regression, clustering, and time series models in the analysis of financial time series and cross-sectional data. The MLR, SVR, ARDL and MARS were employed in forecasting the US CPI value, in which the MARS algorithm generated higher accuracy. These results support the US government in overseeing economic policies, sectors, social security and boosting national economic development (Nguyen et al., 2023). In our dynamic economy, new items come into the consumer marketplace continuously. The difference between the price of the old version of the item observed the previous time and the price of the substitute version this time may present a pure price change that occurred between the two observations (Armknecht et al., 1997). In the United States, there are two primary measures of the prices paid by consumers for goods and services. One is the CPI and the other is the Personal Consumption Expenditures (PCE) prepared by the Bureau of Economic Analysis (BEA). Based on investigation, CPI has been significantly affected by using an inappropriate measure of cost of housing, with the rise in housing costs, the CPI being overstated (Dougherty and Van Order, 1982). In his analysis of the causes and remedies for bias in the CPI, Hausman (2003) found that the cost of living index, new commodities, and services, all had a major impact on the CPI.

A rate of increase in the average level of prices for goods and services relative to the decline in buying power of the currency is known as inflation (Cogoljevic et al., 2018) found that there is still a gap in the analysis of the components influencing inflation and attempted to use MLR to ascertain the effects of the exchange rate, CPI, monetary aggregates, and discount rate on inflation. One method that makes it possible to examine correlations between more than two variables, a scenario that most often arises in macroeconomic research, is multiple regression. The least squares method is the most well-known technique for multiple regression estimation. We select the sample regression function and reduce the sum of squared residual values, just like in the two-variable regression (Anghelache et al., 2014).

(Hansen, 2018) conducted a study in which he addressed the quantification of unstructured data and how to recover information helpfully for economists, emphasizing the applications of machine learning algorithms for economic research and policymaking. The resources required to properly take advantage of the potential of rich new digital data sources are provided by machine learning (Stock and Watson, 1999) forecasts inflation at the 12-month horizon. Inflation forecasts produced by the Phillips curve have been more accurate than forecasts based on other macroeconomic variables, including interest rates, money and commodity prices. People will benefit immensely from stock price prediction in understanding where and how to invest to reduce the danger of losing money. Two methods are used by the suggested system to operate: regression and classification. Machine learning regression analysis uses this information to forecast a company's closing stock price, while classification uses it to forecast whether closing stock prices will rise or fall the next day (Ravikumar and Saraf, 2020). The random forest formulation and examine prediction performance on simulated and real-world datasets, where maximum large trees do overfit, combining the knowledge from boosting methods. These investigations show that extra tuning can be used to control tree size by restricting the number of splits and/or the size of nodes that are permitted to divide. However, by utilizing bigger (than default) primary tuning parameter values, good performance for random forests can still be achieved even in these settings (Segal, 2004). (Maulud and Abdulazeez, 2020) had done a review study on linear regression, comprehensive in machine learning, discussed numerous studies on polynomial and linear regression conducted by various researchers, comparing the results using the most effective method for maximizing accuracy and prediction.

A comparative research between ARIMA and ANN was conducted by (Purbasari et al, 2020) to forecast the value of the Indonesian CPI between January 2009 and December 2017. The results showed that ANN provides a smaller MSE value, making it the preferred method for CPI forecasting. A statistical technique called quantile regression neural network uses an artificial neural network to capture nonlinear patterns in data and models non-homogeneous variance. The Consumer Price Index is actual data that is purported to have these properties, CPI (Rantini et al., 2018). The traditional approach to forecast CPI is the Autoregressive Integrated Moving Average (ARIMA) model. However, a major flaw in the ARIMA model is that it ignores inter-regional economic interaction, which has a considerable impact on CPI, and only takes into account the time effect. The Generalized Space Time Autoregressive Integrated (GSTARI) model is designed to study the CPI data of four major Chinese cities to evaluate and validate the forecast accuracy of the GSTARI model with the standard ARIMA model. These tests reveal that the

GSTARI model has a 7%–38% better prediction accuracy than the ARIMA model for the majority of cities (Dong et al., 2020).

3. Materials and Methods

3.1 *Materials*

3.1.1 *Dataset*

India is one of the founding members associated with the International Monetary Fund (IMF), which assists the flow of statistics between the Government of India and the Reserve Bank of India. The Consumer price index and the inflation rate of India from January 2014 to January 2024 have been collected from IMF data using the link https://data.imf.org/. This data consists of the CPI value of different commodities like food and beverages, tobacco and intoxicants, clothing and footwear, housing, fuel and light, consumer food price index, and other items over a year, and the price value in Rural, Urban, and overall index. In addition, the world gold price, Crude oil price, and Federal Fund Effective Rate (FFER) have also been collected from the sources such as Statista, MOSPI, and IMF to study the impact of these variables on CPI and Inflation Rate.

3.2 *Methods*

3.2.1 *Multiple Linear Regression*

To develop the relationship between the dependent and independent variables and reflect on the WQI value, a statistical approach called multiple linear regression is employed. The general form of the MLR model is,

$$Y = \beta_0 + \beta_1 X_1 + \beta_2 X_2 + \ldots\ldots + \beta_n X_n + \varepsilon \tag{1}$$

where Y is the response variable, $X_1, X_2, \ldots, X_n$ are the independent variables, β_0 is the intercept, and $\beta_1, \beta_2, \ldots, \beta_n$ are the regression coefficients. And finally, ε is the error term that is not explained by the independent variables.

3.2.2 *Decision Tree regressor*

The decision tree is used to fit a sine curve with additional noisy observations. As a result, it learns local linear regressions approximating the sine curve. If the maximum depth of the tree is set too high, the decision trees learn too fine details of the training data and learn from the noise, i.e., they overfit.

A decision tree regressor predicts a continuous target variable by dividing the feature variables into small zones, each with one prediction. We will begin with one continuous variable and then two continuous variables. We will not use categorical variables since, for decision trees, continuous variables are ultimately treated like categorical data when the splits are made.

3.2.3 *Random Forest Regressor*

Random forest regression is an invaluable tool in data science. It enables us to make accurate predictions and analyse complex datasets with the help of a powerful machine-learning algorithm. A random forest regression model combines multiple decision trees to create a single model. Each tree in the forest builds from a different subset of the data and makes its independent prediction. The final prediction for input is based on the average or weighted average of all the individual trees' predictions.

3.2.4 *Support Vector Regressor*

SVR is a type of support vector machine (SVM) that is used for regression tasks. It tries to find a function that best predicts the continuous output value for a given input value. It tries to find the hyperplane in a high dimensional space that maximally separates different classes or output values. SVR can use different types of kernels, which are functions that determine the similarity between input vectors. A linear kernel is a simple dot product between two input vectors, while a non-linear kernel is a more complex function that can capture more intricate patterns in the data. The choice of kernel depends on the data's characteristics and the task's complexity (Comito and Pizzuti, 2022).

4. Experimentation, Results and Analysis

4.1 Experimental Setup

The Python programming language is used for data preprocessing, statistical analysis, and forecasting analysis.

4.2 Performance Metrics

Forecasting models' performance is assessed by using four metrics: MAE, RMSE, MSE and R^2. Lower values of these metrics indicate better performance of the model.

4.2.1 *Mean Absolute Error (MAE)*

MAE measures the mean of the errors between actual value and forecasted value. It can be mathematically expressed as,

$$M.A.E = \frac{1}{n}\sum_{i=1}^{n} |\hat{Y}_i - Y_i| \tag{2}$$

Here, n is the number of observations, $\hat{Y}_i$ is the forecasted value and Y_i is the actual value.

4.2.2 Root Mean Squared Error (RMSE)

RMSE is similar to MAE but it penalises large errors more heavily because of the squaring of errors. It is expressed mathematically as,

$$R.M.S.E = \sqrt{\frac{1}{n}\sum_{i=1}^{n}(\hat{Y}_i - Y_i)^2} \tag{3}$$

Here, n is the number of observations, $\hat{Y}_i$ is the forecasted value and Y_i is the actual value.

4.2.3 Mean Squared Error (MSE)

MSE of an estimator measures the average squared difference between the estimated values and the actual value. MSE is a risk function, corresponding to the expected value of the squared error loss. If a vector of n predictions is generated from a sample of n data points on all variables, and Y is the vector of observed values of the variable being predicted, with $\hat{Y}$ being the predicted values, then the within sample MSE of the predictor is computed as

$$M.S.E. = \frac{1}{n}\sum_{i=1}^{n}(Y_i - \hat{Y}_i)^2 \tag{4}$$

4.2.4 R-Square

R^2 or the coefficient of determination is a statistical measure in a regression model that determines the proportion of variance in the dependent variable that can be explained by the independent variable. R^2 can take any values between 0 and 1. A low R^2 is generally a bad sign for predictive models.

$$R^2 = \frac{\text{Sum of squares due to regression}}{\text{Total sum of squares}} \tag{5}$$

The sum of squares due to regression measures how well the regression model represents the data used for modelling. The total sum of squares measures the variation in the observed data.

4.3 Results and Analysis

The characteristics of all the variables have been presented in Table 1, which shows the average value and Standard deviation of CPI, Inflation rate, Crude oil rate, Gold rate, and FFER are 145.8, 5.17,63.7 USD,1512.09 USD, 1.2974 USD, and 20.8, 1.157, 20.04, 308.13 and 1.588 respectively. The minimum and the maximum value of all the variables have also been estimated. The maximum CPI value has reached 186.3 and the minimum value is 113.6, same as the inflation rate was 8.6 at its peak and lowered up to 1.46.

Table 1 Characteristics of all the attributes.

Summary	*CPI*	*Inflation Rate*	*Crude oil rate*	*Gold Rate*	*FFER*
Mean	145.8	5.17	63.7	1512.09	1.2974
S.D	20.8	1.157	20.04	308.13	1.5888
Minimum	113.6	1.460	18.84	1068.90	0.0400
Maximum	186.3	8.6	114.67	2168.58	5.3300

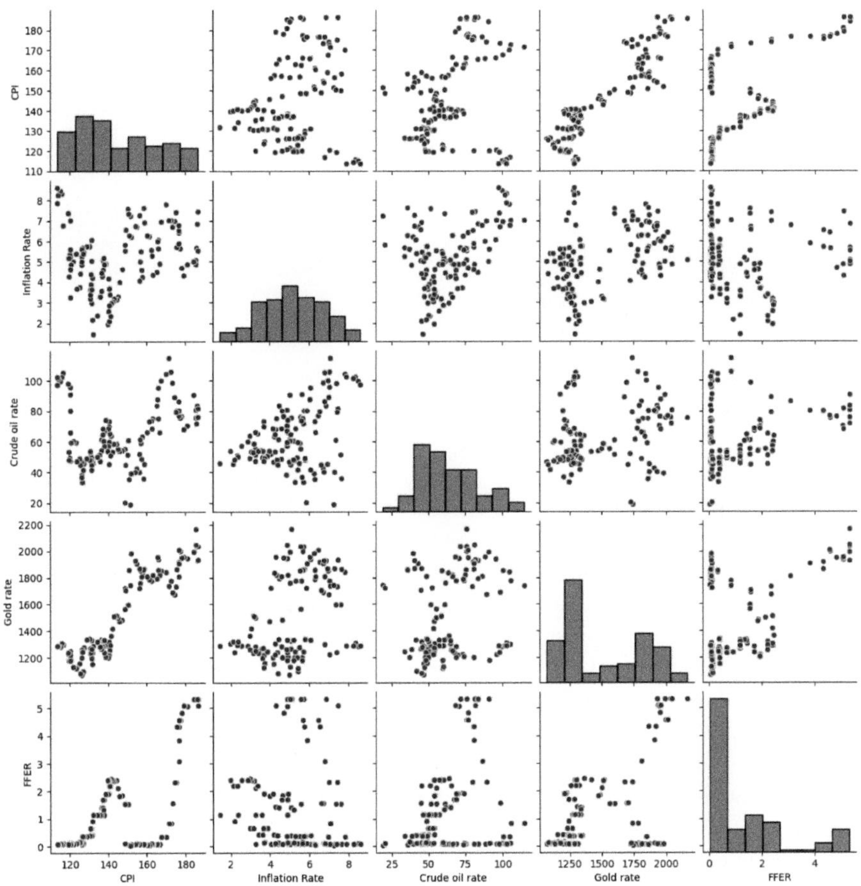

Fig. 1 Matrix scatter plot.

Figure 1 visualizes the spread of data of all the variables included in our study. And Fig. 2 represents the relation between CPI and the Inflation rate.

Figure 3 is beneficial to picture the CPI for diverse commodities like food and beverages, tobacco and intoxicants, clothing and footwear, housing, fuel and light,

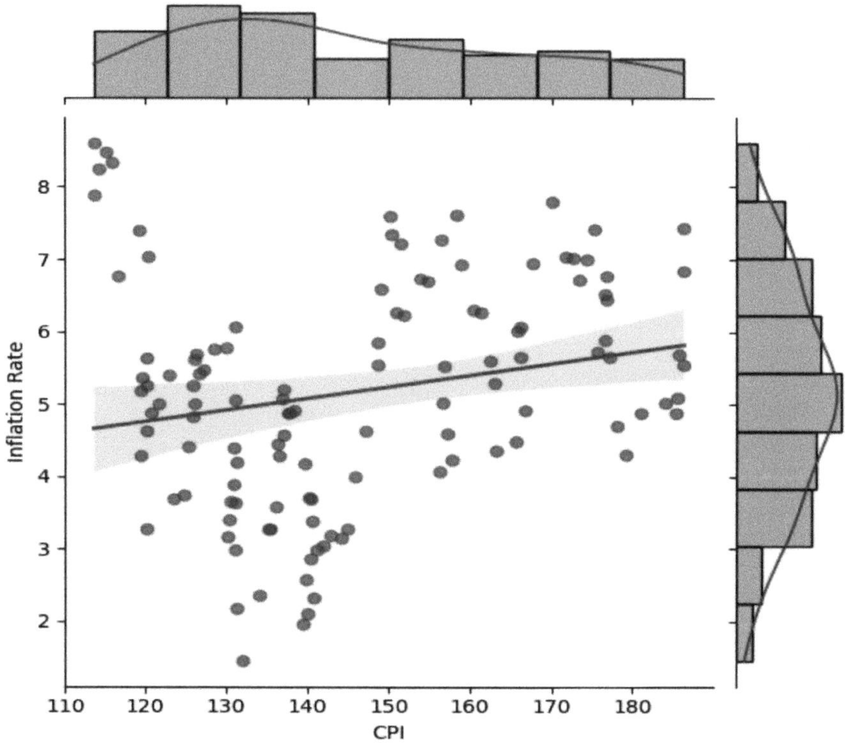

Fig 2 Diagrammatic representation of the relation between CPI value and Inflation rate.

consumer food price index and other items and the value conforming to different subdivisions like rural and urban separately and overall general index with their range of distribution has been shown.

The correlation matrix Fig. 4 shows the association between each variable. The range of correlation is between −1 and 1. The shade of green in the diagram represents positive correlation, the darkest shade represents perfect correlation and the shade of brown represents negative correlation. From Fig. 4, it is clear that the Gold rate has a 0.91 correlation with CPI and predominantly all the variables showed positive correlation with the consumer price index.

The Multiple Linear Regression Model is used to understand the relationship between the predictor and response variables and to forecast the CPI using an equation. The input parameters connected to the economic parameters are analyzed utilizing the MLR for CPI modeling. We excluded the highly correlated parameter before developing the model using the remaining parameters, as demonstrated by the Pearson correlation matrix, and have constructed a single equation related to MLR using the beta coefficient.

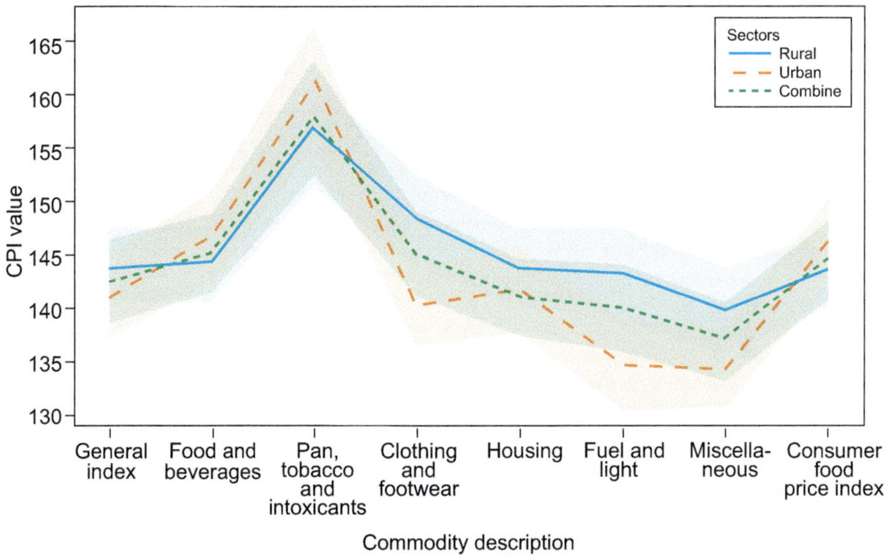

Fig. 3 Plot representing the CPI value of different commodities based on sectors.

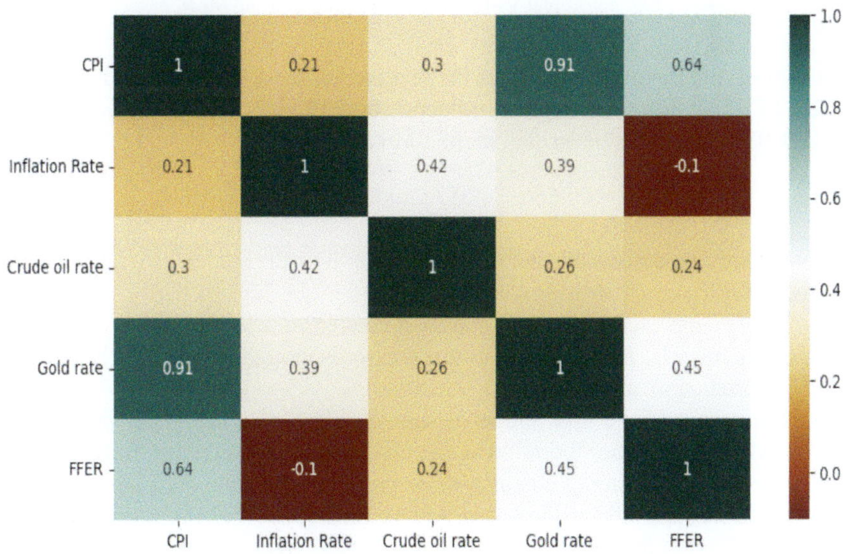

Fig. 4 Pearson's Coefficient Correlation matrix.

MLR has been used to establish the relation between CPI and the independent variables, and is shown in Table 2. After variable selection, the gold rate and FFER are regarded as the best variables to investigate regarding CPI out of the three independent variables—crude oil rate, gold rate, and FFER—for CPI prediction.

Table 2 Multiple Linear Regression results for CPI.

Model	*Beta*	*Std.Error*	*Std. Beta*	*t*	*Sig.*	*Lower*	*Upper*
Intercept	60.566	3.149	-	19.235	0.000	54.331	66.802
Gold rate	0.053	0.002	0.788	24.382	0.000	0.049	0.057
FFER	3.719	0.423	0.284	8.792	0.000	2.881	4.556

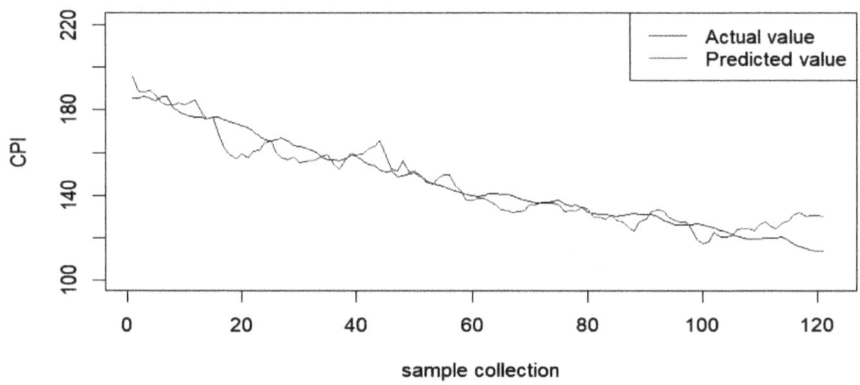

Fig. 5 Actual Vs Predicted value of CPI using MLR.

In Table 2, the Beta value indicates the relative influence of the entered variables. Here FFER indicates the highest influence on CPI (3.719) and the Gold rate as 0.053. The model can be written in the form of

$$\text{CPI} = 60.566 + 0.053 \text{ Gold rate} + 3.719 \text{ FFER}$$

Figure 5 represents the plot which indicates how the model fits with the actual value of CPI.

Figure 6 represents the actual and the predicted value of CPI by using different regression models. The random forest regressor is thought to have the best match for CPI prediction, as illustrated by the figure.

The output of the final model is shown in Table 3. The lower the RMSE, MSE, MAE, and higher the R^2 score are considered to be a better fit. Root mean square error of the model contains three predictors and the dependent variable CPI is observed, similarly the Mean Square Error was (0.0397) noted and Mean Absolute Error of the final MLR model is (0.1684) was observed. And the R^2 value of Random Forest regressor is estimated as 0.9636, which is the highest among all the models. Therefore, Random Forest is considered as the best model for predicting the consumer price index. And the plot shows the actual and the fitted value of CPI using different predictive algorithms.

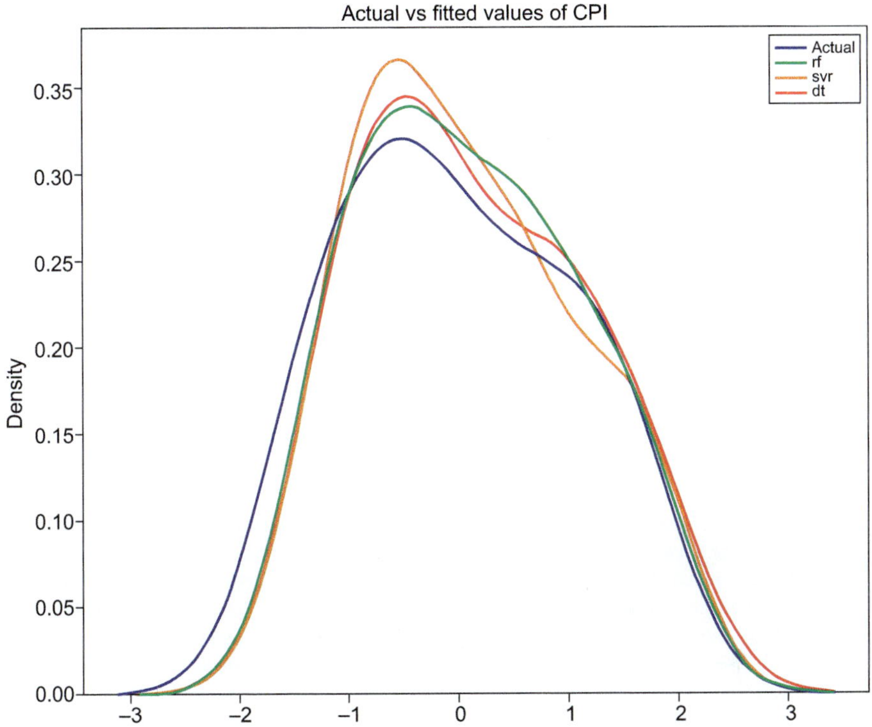

Fig. 6 Actual Vs Predicted value of different regressor models.

Table 3 Evaluation Metrics of Predictive models.

Predictive models	*Metrics*			
	MAE	*MSE*	*RMSE*	R^2
Decision Tree Regressor	0.2052	0.0626	0.2502	0.9374
Random Forest Regressor	0.1687	0.0397	0.1992	0.9636
Support Vector Regressor	0.1684	0.0513	0.2264	0.9487

Conclusion

Using monthly data gathered from January 2014 to January 2024, the Decision Tree Regression Algorithm, Random Forest Regression Algorithm, and Support Vector Regression Algorithm were used in this chapter to forecast the CPI of India. From the results, Random Forest Regression has the highest R^2 value of 0.9636, which indicates the best fit. It has the lowest MSE and RMSE value compared to other algorithms. Based on a comprehensive analysis, Random Forest is regarded as the most effective model for forecasting India's CPI. The MLR technique has been used to study the relationship between the dependent variables CPI and Inflation rate and the independent variables Crude oil rate, gold rate, and FFER. From the MLR results it is clear that the gold rate and FFER play a vital role in predicting the CPI value. There are certain limitations in CPI that the consumer goods considered do not represent all productions or consumption in the economy. Thus, the basket of goods for calculating CPI is not updated frequently. The study's future goals include incorporating all relevant factors to forecast the cumulative CPI, identifying the most accurate generalized model for doing so, and identifying patterns in these economic indicators that will aid in the country's improved development.

References

Akcay, S. (2011). The causal relationship between producer price index and consumer price index: Empirical evidence from selected European countries. *International Journal of Economics and Finance*, *3*(6), 227–232.

Anghelache, C., Anghel, M.G., Prodan, L., Sacala, C. and Popovici, M. (2014). Multiple linear regression model used in economic analyses. *Romanian Statistical Review Supplement*, *62*(10), 120–127.

Armknecht, P.A., Lane, W.F. and Stewart, K.J. (1997). New products and the US consumer price index. *The Economics of New Goods*, *58*, 375–391.

Blinder, A.S., Triplett, J.E., Denison, E. and Pechman, J. (1980). The consumer price index and the measurement of recent inflation. *Brookings Papers on Economic Activity*, *1980*(2), 539–573.

Boskin, M.J., Dulberger, E.R., Gordon, R.J., Griliches, Z. and Jorgenson, D.W. (1998). Consumer prices, the consumer price index, and the cost of living. *Journal of Economic Perspectives*, *12*(1), 3–26.

Broby, D. (2022). The use of predictive analytics in finance. *The Journal of Finance and Data Science*, *8*, 145–161.

Bryan, M.F. and Cecchetti, S.G. (1993). The consumer price index as a measure of inflation.

Cage, R., Greenlees, J. and Jackman, P. (2003, May). Introducing the chained consumer price index. In *International Working Group on Price Indices (Ottawa Group): Proceedings of the Seventh Meeting* (pp. 213–246). Paris: INSEE.

Cogoljević, D., Gavrilović, M., Roganović, M., Matić, I. and Piljan, I. (2018). Analyzing of consumer price index influence on inflation by multiple linear regression. *Physica A: Statistical Mechanics and its Applications*, *505*, 941–944.

Comito, C. and Pizzuti, C. (2022). Artificial intelligence for forecasting and diagnosing COVID-19 pandemic: A focused review. *Artificial Intelligence in Medicine*, *128*, 102286.

Dong, H., Guo, X., Reichgelt, H. and Hu, R. (2020). Predictive power of ARIMA models in forecasting equity returns: a sliding window method. *Journal of Asset Management*, *21*(6), 549–566. https://doi.org/10.1057/s41260-020-00184-z

Dougherty, A. and Van Order, R. (1982). Inflation, housing costs, and the consumer price index. *The American Economic Review*, *72*(1), 154–164.

Edvinsson, R. and Söderberg, J. (2011). A consumer price index for Sweden, 1290–2008. *Review of Income and Wealth*, *57*(2), 270–292.

Hansen, S. (2018). Machine Learning for Economics and Policy. *Social and Economics Studies*, *5*, 369–397.

Hausman, J. (2003). Sources of bias and solutions to bias in the consumer price index. *Journal of Economic Perspectives*, *17*(1), 23–44.

https://cpi.mospi.gov.in/

Konny, C.G., Williams, B.K. and Friedman, D.M. (2019). Big data in the US consumer price index: Experiences and plans. *Big Data for 21st Century Economic Statistics*.

Maulud, D. and Abdulazeez, A.M. (2020). A review on linear regression comprehensive in machine learning. *Journal of Applied Science and Technology Trends*, *1*(2), 140–147.

Nguyen, T.T., Nguyen, H.G., Lee, J.Y., Wang, Y.L. and Tsai, C.S. (2023). The consumer price index prediction using machine learning approaches: Evidence from the United States. *Heliyon*, *9*(10).

Purbasari, I.Y., Anggraeny, F.T. and Ardiningrum, N.A. (2020). Time-series Modeling for Consumer Price Index Forecasting using Comparison Analysis of AutoRegressive Integrated Moving Average and Artificial Neural Network. In *Proceedings of the International Conference on Culture Heritage, Education, Sustainable Tourism, and Innovation Technologies (CESIT 2020)* (pp. 599–604).

Rantini, D., Octavanny, M.A.D., Kruba, R., Kuswanto, H. and Fithriasari, K. (2018). Quantile Regression Neural Network Model for Forecasting Consumer Price Index in Indonesia. *Inferensi, 1*(1), 39–43.

Ravikumar, S. and Saraf, P. (2020, June). Prediction of stock prices using machine learning (regression, classification) Algorithms. In *2020 International Conference for Emerging Technology (INCET)* (pp. 1–5). IEEE.

Segal, M.R. (2004). Machine learning benchmarks and random forest regression.

Stock, J.H. and Watson, M.W. (1999). Forecasting inflation. *Journal of Monetary Economics*, *44*(2), 293–335.

14 The Meme Revolution: Integrating Cultural Phenomenon and Analytics in Branding Strategy

Ashima Agrawal[1*] and Ramzan Sama[2]

Marketers face challenges in engaging millennials and Generation Z (Gen Z) due to their short attention spans and aversion to traditional advertising. Memes, as concise, humorous, and culturally relevant content, have become a powerful tool for capturing their interest. These visually appealing and shareable snippets cater to younger audiences, enabling brands to establish deeper connections, boost engagement, and enhance brand awareness. Memes offer a unique blend of humor and relatability, helping brands humanize their image and build community within their target audience. When executed skilfully, memes can drive viral momentum and strengthen brand affinity. However, the humor inherent in memes is a double-edged sword. Missteps in tone, timing, or cultural context can lead to backlash, damage brand credibility, and alienate audiences. This chapter explores the evolution of memes as a promotional strategy with the help of content analysis, highlighting their benefits and potential pitfalls. It delves into how brands effectively use memes to align with audience preferences and maintain authenticity, ensuring campaigns resonate with cultural and social nuances. In the dynamic digital marketing landscape, brands must remain flexible and culturally sensitive. By thoughtfully leveraging memes, they can enhance lasting relevance, inspire engagement, and build meaningful connections with younger generations.

1. Introduction

Humans have always been naturally social, seeking interaction and connection with one another (Flanagan, 2020). Historically, socialization occurred in physical settings such as markets, salons, gardens, and other communal areas. These settings encouraged personal interactions and the exchange of ideas. However, as time has passed and technology has advanced, the nature of socialization has changed dramatically, moving from physical spaces to digital platforms. Among these digital platforms, social media has emerged as one of the most popular modes

[1] 0000-0003-4656-2140
0000-0001-5743-8175
* Corresponding author: ashima2897@gmail.com

DOI: 10.1201/9781003514282-14

of interaction, particularly among younger generations (Forsgren and Byström, 2018; Brinkman, Gabriel, and Paravati, 2020). Millennials (born 1981–1996) and Generation Z (Gen Z) (born 1997–2012) are particularly drawn to these platforms (Beresford Research, 2024). According to recent research by McKinsey (2023), 35% of Gen Z users spend more than two hours daily on social media, while 24% of millennials do the same. These statistics highlight how crucial social media is in the daily lives of these age groups.

For these generations, social media is more than just a means of communication; it is also a platform for self-expression, community building, and information sharing (Vries, Peluso, Romani, Leeflang, and Marcati, 2017). The ease of use, combined with its immersive nature, makes social media an essential part of their daily routine. It functions as a virtual marketplace for ideas, a contemporary salon for discussions, and a digital garden for cultivating connections (Kitchen, 2023). This shift from physical to digital socialization emphasizes the profound impact of technology on human interaction. Nonetheless, millennials and Gen Z spend a significant amount of time on social media, demonstrating its growing importance in shaping their identities and experiences in a digitally driven world (Smith, 2024).

Marketers have taken note of the significant attention that millennials and Gen Z devote to social media platforms. Recognizing that their target audience spends a significant amount of time on these platforms, marketers have identified social media as prime advertising real estate (Razzaq, Shao, and Quach, 2023). The logic is straightforward: if the audience is already present, why not meet them where they are? This realization has transformed social media into a thriving arena for brands, with fierce competition for consumer attention. To capitalize on this opportunity, brands have developed social media-specific strategies and content. This has resulted in the creation of dedicated social media campaigns that are visually appealing, engaging, and tailored to the platform's specific features. However, because social media is now overrun with branded material, brands use it to establish their presence and compete for consumer attention (Jung and Heo, 2021).

Social media began as a digital space offering users a chance to connect, share pictures, and socialize, but brands have turned this space into a battleground. Almost every brand now has a social media presence, flooding platforms with advertisements, promotions, and campaigns. As a result, the dynamics have shifted. Consumers have become more selective in response to the sheer volume of content available. Users on social media regularly block, ignore, or scroll past branded posts without hesitation, just as they do with YouTube ads (Waqas, Hamzah, and Salleh, 2021). This shift in behaviour has created a new challenge for brands: how to stand out in a sea of content while capturing meaningful attention. Traditional approaches are no longer enough. To overcome this challenge, brands must innovate by creating authentic, relatable, and value-driven content that speaks to their target audience. The key is storytelling, making genuine connections, and engaging customers in ways that go beyond simple promotion, transforming them into active participants rather than passive viewers.

Humor has long been used to draw people in and encourage interaction. A long-standing idea in marketing, humor has been extensively employed in conventional marketing channels such as newspapers, magazines, and billboards. Humor has also been used in several brand marketing initiatives and shared on social media. Humor as an advertising strategy gained momentum after the 1920s, first appearing in outdoor posters and radio, and subsequently on television by the 1950s (Weinberger and Gulas, 2021). Studies indicate that humor in advertising has a greater impact on consumer attitudes than many other persuasive techniques. Humor as a marketing tactic has received ample attention in marketing research. Humor plays a key role in capturing attention, which is widely regarded as the first step in effective communication (Chan, 2011).

A recent study by Kantar (2024) reveals that approximately 34% of advertisements incorporate humor. When broken down by advertising medium, humor is found in 37% of television ads, 28% of print and outdoor ads, 12% of website advertisements, 24% of Facebook and Instagram ads, and 30% of YouTube ads (Kantar, 2024). The marketing literature provides enough evidence of the effectiveness of amusing commercials (Chan, 2011). Researchers demonstrate that humorous advertisements elicit positive reactions from consumers toward the ads, the products featured, and the brand being promoted (Redondo, 2012). Additionally, they improve consumers' memory of low-involvement products. Humor has long been a popular and effective marketing tactic, consistently used by brands to engage audiences, build connections, and leave lasting impressions. While humor is a universal concept, its interpretation is deeply rooted in cultural meanings. The effectiveness of a humorous message largely depends on the cultural background of its audience, as different societies have unique perceptions of what is funny. Humor captivates audiences and improves their mood, leading to favorable brand outcomes. When humor is incorporated into a message, it tends to engage the audience more deeply, making it harder for them to ignore the content (Pelsmacker, Geuens, and Anckaert, 2002). With humor as a core and marketers' interest deeply rooted in creating something innovative and engaging, memes have emerged as a suitable category.

During the COVID-19 pandemic, as global lockdowns confined people indoors, boredom and anxiety about the widespread disease became common. To cope, many turned to social media platforms, spending increased amounts of time online (Priyadarshini, et al., 2022). Memes, with their light-hearted and humorous nature, became a popular way for individuals to reduce stress and uplift their spirits. During this period, the sharing of memes surged exponentially, with Instagram reporting over a million memes shared daily (Instagram, 2020). This explosion of memes caught the attention of marketers, highlighting their potential as a powerful communication tool.

Memes are like digital sparks, rapidly spreading ideas, trends, and concepts in formats such as images and videos, often paired with catchy phrases. They are inherently shareable, modifiable, and relatable (Malodia et al., 2022; Shifman, 2013). Their ability to distil complex messages into simple, engaging formats makes them not only entertaining but also a potent medium for communication (Nieubuurt, 2021). Memes often incorporate humor, sarcasm, and irony, making them highly engaging and relatable (Brubaker et al., 2018). Their influence is undeniable as a survey revealed that millennials view 20–30 memes daily, with 75% of 13–36 year-olds sharing them (YPulse, 2019). These statistics underscore the significant role memes play in social media, advertising, and marketing, making them an influential tool for modern communication.

1.2 Objectives of the Study

This chapter delves into key objectives critical in today's digital landscape. It aims to explore the effectiveness of memes as a tool for engaging millennials and Gen Z, focusing on their potential to enhance brand awareness and build deeper connections with younger audiences. Additionally, the chapter examines the challenges and risks inherent in meme marketing, shedding light on potential pitfalls such as missteps in tone, timing, or cultural relevance that could impact brand credibility and audience perception. By addressing these aspects, the study provides a comprehensive understanding of the opportunities and complexities of meme-based marketing in the digital era.

2. Literature Review

2.1 Evolution of Memes

In his book "The Selfish Gene," published in 1976, Richard Dawkins first used the term "meme." Similar to how genes transfer biological information, he used it to define an idea, habit, or style that spreads from person to person within a culture (Benaim, 2018; Taecharungroj and Nueangjamnong, 2015). According to Taecharungroj and Nueangjamnong (2015), the word itself comes from the Greek word "mimema," which means something that can be imitated. In the beginning, memes were thought to be cultural components that might be imitated and spread. This approach was based on the theory of cultural evolution, which holds that memes compete for survival and attention in a manner akin to biological evolution (Benaim, 2018). Catchphrases, melodies, and cultural customs that proliferated through social interactions were early instances of memes. A meme can be as basic as a gesture, symbol, or word that gains a life of its own within a community via common understanding and repetition. These recurrent symbols eventually influenced cultural discussions by bringing individuals together through comedy, shared knowledge, or social commentary. The development of memes underwent a dramatic shift with the advent of the internet in the latter half of the 20th century. Memes started to evolve with the emergence of social media and online discussion boards, mostly as image macros, GIFs, and videos. Platforms like Reddit and 4chan emerged around this time, allowing for the quick sharing and remixing of content (Brubaker, Church, Hansen, Pelham, and Ostler, 2018). Memes are known for their humor and ability to convey complex ideas with simplicity (Brubaker, Church, Hansen, Pelham, and Ostler, 2018). Since then, these memes have changed more and more every day. Previous popular memes, such as LoLCats and Rickrolling, have evolved into more sophisticated forms due to technological improvements. Memes are widely available in various formats nowadays, and they typically draw inspiration from pop culture events, politics, sports, current events, and movies.

2.2 Memes in Marketing

In the age of widespread social media use and digitization, memes have emerged as the youth's language. They have become the unorthodox tool of Internet marketing because of their innate tendency to be humorous, relatable, and viral (Yang, 2022). Marketers have taken notice of this and begun to take advantage of them. Brands have entered the meme culture, using memes on social media sites like Instagram, X (formerly Twitter), and Meta (formerly Facebook) to spread their messaging. Brands use witty memes on their social media pages, which attract the attention of users and engage with them. Several core components work together to convey the idea and message effectively:

- **Content Appeal:** Content is the heart of any advertising (Kusumasondjaja, 2018). Just like other traditional and digital advertising avenues, meme marketing is also influenced by the content of the meme. Brands have to strike a balance between the marketing message and a fair share of humor that the audience can relate to and does not feel overbragged advertising. Gen Z likes advertisements that are not too loud and subtle (Vardeman, 2024). Brands focus on a variety of appeals like emotional, informational, entertainment, and creativity while making memes.

  1. ***Emotional appeal:*** It is a persuasive communication technique that evokes an emotional response to a brand's message by using elements like drama, mood, music, and other strategies designed to trigger emotions (Wang, Morey, and Srivastava, 2014; Yoo and MacInnis, 2005). This appeal seeks to convey a message to its intended audience by emphasizing intangible qualities and subjective perceptions that cannot be easily verified objectively (Liu and Stout, 1987). Rather than highlighting a product's physical or functional features, this approach focuses on establishing emotional connections with customers. It captivates the audience by evoking specific feelings, reactions, or desires, frequently addressing deeper psychological, social, or emotional needs (Chou and Lien, 2010). The appeal's goal is to shape how consumers perceive the brand by eliciting emotional responses that can significantly influence their purchasing decisions (Zhang, Sun, Liu, and Knight, 2014). Emotions can be expressed either positively (happiness, satisfaction, love, hope) or negatively (anger, sadness, and anxiety) (Kusumasondjaja, 2018). Memes that evoke these kinds of emotions are used by brands, and audiences are drawn to these brand sites on social media because they can identify with them. Since emotions allow people

to connect, audiences frequently share these memes within their social networks, which increases their relevance.

2. ***Informational Appeal:*** To enable consumers to make well-informed purchasing decisions based on the product's purportedly improved functioning, the informational appeal seeks to give them concise, accurate, and pertinent knowledge about the promoted brand (Laskey, Day, and Crask, 1989). This strategy emphasizes presenting the brand's advantages logically and providing verifiable proof to back up its assertions (Zhang, Sun, Liu, and Knight, 2014). Informational appeals highlight the brand's measurable worth by emphasizing its objective attributes, such as performance, cost-effectiveness, popularity, convenience, or health benefits. Furthermore, it emphasizes the product's robustness and other useful features, establishing the brand as a sensible and trustworthy option (Dix and Marchegiani, 2013). By using this technique, informational appeal aims to convince customers by building faith in the brand's abilities and demonstrating that it is desirable. Brands also infuse memes with information about the brand's new product or service offerings. This enables brands to convey their ideas in a way that can reach consumers without skipping or ignoring the content.
3. ***Entertainment Appeal:*** Advertisements that use entertaining, pleasurable, and refreshing components to draw in and hold the attention of customers are said to have entertainment appeal (Kang, 2005; Gavilanes, Flatten, and Brettel, 2018). According to Joa et al. (2018), advertising entertainment is the ability of an advertisement to make consumers happy and content. Entertainment is a major component of memes. They are naturally amusing, which attracts attention and improves people's moods. People can now connect through memes. This attraction increases the use of memes.
4. ***Creative Appeal:*** Creativity is defined as "the art of establishing new and meaningful relationships between previously unrelated things in a manner that is relevant, believable, and in good taste, but which somehow presents the product in a fresh new light" (El-Murad and West, 2004). This approach sparks creativity and brings a fresh perspective, helping to capture attention that might otherwise be overlooked. Brands now have separate profiles for Chief Meme Officers in their organizations, where a dedicated person oversees such posts. An entire team of young people with zeal and a creative mindset who understand the pulse of youth create memes that surprise one another and form relationships that one could not have imagined. Brands are capitalizing on this trend. Many brands, including McDonald's, Wendy's, and Netflix, are succeeding with meme marketing strategies on social media platforms, particularly Instagram (Agrawal, et al., 2024).

- **Cultural Reference:** Memes are often created using popular templates that draw inspiration from cinema, politics, pop culture, sports, or current events. These references are iconic and deeply resonate with audiences. Brands leverage this effectively by incorporating widely recognized templates into their communication strategies. By cleverly twisting these familiar formats to align with their messaging, they connect instantly with their audience and add an element of surprise (Kim and Kim, 2024). This creative approach enhances engagement and makes the meme more appealing and memorable.

- **Caption:** Memes are frequently accompanied by text, which can be written above or below the image. These memes are known as image macro memes. These texts are usually one-liners that support the image, are humorous or contain satire (Yang, 2022). They make memes more understandable and point in the right direction. There is a good chance that just by using image memes will be misinterpreted, which can get brands in trouble. To avoid falling into such pitfalls, brands typically use image macro templates.
- **Visualization:** Visuals heavily influence social media platforms because they make anything more appealing than simple audio. Visualization is important for making memes appealing to the audience. The message's color scheme, as well as the creative and subtle placement of the brand logo, is not overly loud but is still visible and causes one to recall the brand (Brubaker, Church, Hansen, Pelham, and Ostler, 2018). Choosing colors that are easy to recall is one of the smart ways brands use meme visuals. Iconic brands such as Zomato (Food Delivery Application) use red color memes frequently because their brand color is red, whereas their main competitor Swiggy uses orange color memes the most. After all, orange is its brand color. Other visuals like brand logos, brand ambassador images, or taglines are at times used in memes, but recently they have taken a backseat.
- **Social Media Platform:** There are several social media platforms, including Meta (formerly Facebook), Instagram, X (formerly Twitter), LinkedIn, and Snapchat. Each platform has its nuances and serves a unique purpose. Meta is about sharing only text status updates, check-ins, photos, audio, and video. It currently has 3.03 billion monthly active users and is most popular among people aged 25 to 34, accounting for 29.4% (Kemp, 2024). Instagram has 2 billion monthly active users and is heavily used by the age group of 18-24 years (30.8%) and allows users to share more visuals in the form of pictures, videos, and reels, which have become quite popular (Statista, 2024). X is used for microblogging, in which users can share short text updates with a character limit and engage in quick real-time conversations. It has 666 million active users per month and is most popular among people aged 25 to 34, accounting for 29.63% (Duarte, 2024). LinkedIn is a professional networking site and the most widely used social media platform for B2B marketing. Because of its business focus, it's an excellent way for businesses to generate leads, share news, and keep up with others in their field. It has 310 million monthly active users, the majority of whom are between the ages of 25 and 34 (60%) (Shewale, 2024). Snapchat is the most popular platform among Gen Z, who prefer to share short snaps and videos from their daily lives to provide a glimpse into their lives. It has 750 million monthly active users, with the 18–24 demographic accounting for 38.1% of total usage. Memes are created and shared similarly on each platform because they serve different purposes and

age groups. Job-related memes are popular on LinkedIn, whereas Instagram serves various purposes. So, for a meme to be successful, brand managers need to be vigilant enough to make different memes for each platform so that they have meaning and can be shared easily, as one is likely to share something when they can easily connect with it and they believe that their social circle would also comprehend it easily.

Meme marketing has emerged as a significant area of research, driven by its growing adoption in brand strategies. The following table presents key insights from recent studies, shedding light on the effectiveness and impact of meme marketing in contemporary marketing practices.

3. Research Methodology

The study employs content analysis as its primary research methodology, complemented by a comprehensive literature review to establish a foundational understanding of memes. Content analysis, a systematic approach to examining large datasets of videos, images, and text, is instrumental in identifying patterns, trends, and frequencies, enabling researchers to extract meaningful insights (Stemler, 2001). This methodology has seen widespread application in various fields, including social media, television advertising, and digital communication, proving invaluable for simplifying complex data through organized categories and methodological coding systems (Weber, 1990).

Building on this foundation, the study analyzes a robust dataset of 954 memes collected from diverse social media platforms including Instagram, X, Meta, and LinkedIn. These memes represent a mix of both brand-generated content and user-generated content, offering a holistic view of meme culture in digital marketing. This approach allows for a nuanced understanding of the role memes play in engaging audiences, enhancing brand awareness, and building cultural relevance. Additionally, the literature review complements the content analysis by providing theoretical insights into the evolution of memes and their growing significance in communication and branding. Together, these methodologies create a well-rounded framework to explore the effectiveness and challenges of meme-based marketing strategies in the dynamic digital landscape.

4. Results and Discussion

After analysing and interpreting the content of brand-generated and user-generated memes across various platforms, several advantages and challenges have been identified. This comprehensive examination highlights memes' potential to enhance audience engagement and brand visibility while also uncovering risks such as misaligned messaging, cultural insensitivity, and potential backlash.

4.1 Pros and Cons of Meme Marketing

Like any other strategy, there are two sides to it. Meme marketing can be difficult because humor has traditionally been regarded as a double-edged sword. Things

Table 1 Extant literature on meme marketing

S. No.	*Authors*	*Objective of the Study*	*Theory Used*	*Number of Memes Analyzed/ Survey Respondents Considered*	*Major Findings*	*Branding Impact on Consumer Behavior Through Memes*
1.	(Kim & Kim, 2024)	To investigate the intricate dynamics of meme marketing and its profound impact on brand engagement and consumer perception in the digital ecosystem.	Uses and gratification theory, social influence theory	3,770 reviews and reactions were collected related to meme marketing activities, after which the data was cleaned.	Meme marketing increases brand engagement, virality, and cultural relevance, with authenticity, strategic alignment, and omni-channel distribution playing important roles in its effectiveness and long-term impact.	Brand engagement.
2.	(Agrawal, et al., 2024)	This study looks into and compares the content strategies used in brand-generated and user-generated memes on Instagram.	Uses and gratification (emotion, information, entertainment, creativity)	813 memes analyzed.	The study discovered that brand-generated memes use emotional, entertainment, and creative appeals with affiliative humor and silliness, whereas user-generated memes use informational, emotional, entertainment, and creative appeals with aggressive humor and sarcasm.	Not specified.

Contd.

Table 1 *Contd.*

S. No.	*Authors*	*Objective of the Study*	*Theory Used*	*Number of Memes Analyzed/ Survey Respondents Considered*	*Major Findings*	*Branding Impact on Consumer Behavior Through Memes*
3.	(Rathi & Jain, Impact of meme marketing on consumer purchase intention: Examining the mediating role of consumer engagement, 2024)	The study's goal is to investigate the relationship between meme marketing activities and consumer purchase intentions, using consumer engagement as a mediator.	Uses and gratification theory (entertainment, interaction, trendiness, informative, and e-WOM)	452 Indian social media users	Meme marketing activities have a positive impact on consumer engagement and purchase intentions. Consumer engagement helps to mediate the relationship.	Consumer purchase intention
4.	(Razzaq, Shao, & Quach, Meme marketing effectiveness: A moderated-mediation model, 2024)	This study looks into how different types of meme marketing (brand prominent vs. brand subtle) affect consumer attitudes and engagement, as well as how brand knowledge, meme literacy, and linguistic characteristics influence these effects.	Associative network theory, transportation theory	853 participants	The study concludes that brand-prominent meme marketing results in more positive consumer attitudes and engagement than brand-subtle memes. Furthermore, narrative transportation mediates this effect, with brand knowledge, meme literacy, and linguistic characteristics all influencing the efficacy of meme marketing campaigns.	Attitude toward advertisement engagement (likes, comments, and tags)

Table 1 *Contd.*

S. No.	*Authors*	*Objective of the Study*	*Theory Used*	*Number of Memes Analyzed/ Survey Respondents Considered*	*Major Findings*	*Branding Impact on Consumer Behavior Through Memes*
5.	(Vardeman, 2024)	This study investigates how Generation Z (digital native cohorts) perceive and evaluate brands' use of internet memes as advertisements in digital marketing contexts.	Not specified	Six semi-structured focus group studies with 3–10 undergraduate students from Western United States University.	According to the study, memetic advertising works well for non-serious brands with a consistent social media presence, but serious brands struggle to connect. Humor, cultural narratives, and social connections are essential for engaging Gen-Z audiences through memes.	Not specified
6.	(Shen, Lee, & Lin, 2023)	The study's research objective is to investigate how informational cues in brand memes influence consumer attitudes and brand relationships, with a focus on their potential as a communication tool.	Cue utilization theory	595 respondents.	The main findings show that cues such as comprehensibility and novelty positively influence cognitive and affective attitudes, whereas sarcasm can reduce the effectiveness of brand memes in consumer engagement.	Brand advocacy, brand intimacy and self-brand connection

Contd.

Table 1 *Contd.*

S. No.	*Authors*	*Objective of the Study*	*Theory Used*	*Number of Memes Analyzed/ Survey Respondents Considered*	*Major Findings*	*Branding Impact on Consumer Behavior Through Memes*
7.	(Razzaq, Shao, & Quach, Towards an understanding of meme marketing: conceptualisation and empirical evidence, 2023)	The study's research objective is to define meme marketing, establish its foundational elements, and investigate its communicative intentions through a taxonomy of meme marketing speech acts.	Speech Act Theory, Theory of Consumption Values	699 memes analyzed	The study identifies a taxonomy of meme marketing speech acts, emphasizing the complexity of internet memes and their communicative functions, thus improving understanding of meme marketing as a strategic communication tool.	Not specified
8.	(Yang, 2022)	This study looks at whether humorous memes improve brand recall, post attitudes, and the intention to share content online.	Not specified	258 respondents	Humorous memes in Twitter posts improve consumers' perceptions of humor, resulting in a more positive attitude and greater sharing intent, whereas more serious images are more effective at increasing brand recall.	Brand recall, post attitude and online sharing intention

Contd.

Table 1 *Contd.*

S. No.	*Authors*	*Objective of the Study*	*Theory Used*	*Number of Memes Analyzed/ Survey Respondents Considered*	*Major Findings*	*Branding Impact on Consumer Behavior Through Memes*
9.	(Malodia, Dhir, Bilgihan, Sinha, & Tikoo, 2022)	The study's goal is to look into the factors that influence the creation and impact of viral memes, including their causes, consequences, and boundaries.	Social schema and social contagion theory	Lab experiment ($n = 200$), Online: Quasi experiment ($n = 458$), Event study ($n = 15$), Survey ($n = 127$)	The study discovered that viral memes, which are topical and relatable, increase customer engagement and brand recall. Marketers can use these insights to create relevant, humorous, and shareable content, enhancing consumers' escapism and social gratification levels. This approach organically broadens reach, boosting brand performance through increased recall and engagement.	Brand recall and brand engagement
10.	(Teng, Lo, & Lee, 2022)	To investigate the effect of Internet memes on brand image.	Social contagion and symbolic value theory	348 survey respondents	Meme traits such as brand interaction, prestige, and humor are positively related to brand image, whereas emotions and high spreadability are not.	Brand image

that appear amusing to one person can become serious or hurt another person's feelings (Brubaker, Church, Hansen, Pelham, and Ostler, 2018). Brands must exercise caution when presenting their brand message because people of various cultures and ethnicities hold different ideologies. Brands must respect each of them and avoid controversies. Recently, due to the ease of use of social media and freedom of expression on these platforms, boycott culture has become quite prevalent, with social media users boycotting brands or celebrity endorsers if something hurts their sentiments (Agrawal, et al., 2024). This strategy has both pros and cons like:

Pros of Meme Marketing:

1. **High Engagement:** On social media platforms, engagement can be created via likes, shares, comments, story posts, and memes within your social circle on messages. Memes that catch the audience's attention and resonate with them, tend to have a high engagement rate (Malodia, Dhir, Bilgihan, Sinha, and Tikoo, 2022). The humor, wit, and relatability of the memes make them highly shareable content within their network. Audiences themselves become brand ambassadors for these brand memes and thus create organic growth and engagement without adding any additional costs to the brand.
2. **Increased Virality:** Virality on social media occurs when a piece of content, such as an image, video, or post, spreads quickly across multiple platforms, generating millions of views and impressions from the audience that uses its products and services. Memes are short pieces of content that anyone can understand in a second, and people like to share them because they are witty and creative (Rathi and Jain, 2024). They are shared across multiple platforms, spreading like wildfire. Brands capitalize on current trends that resonate with their target audiences. Brands use memes to focus on entertaining the audience rather than hard selling their products and spare the consumers from ad fatigue, which makes them the organic drivers of brands and has the potential to go viral (Yang, 2022).
3. **Relatability:** It is one of the memes' key strengths because they capitalize on current trends, events, cultural moments, and shared experiences. Audience can easily relate to these topical memes. When an event occurs, such memes are immediately shared online. Their ability to depict real-life situations or current events makes them instantly recognizable and relatable, encouraging viewers to engage with the content (Agrawal, et al., 2024). Memes can also bridge generational gaps by appealing to a wide range of demographics, thereby improving brand communication.
4. **Connect with Audience:** Memes use humor and cultural references to connect with audiences, particularly younger demographics such as Gen Z and Millennials. These audiences value relatable, light-hearted content that reflects their own experiences and online culture. Memes give people a sense of belonging by aligning with their interests, building community, and establishing stronger, more authentic connections with the brand (Vardeman, 2024).

5. **Brand Personality:** Brands that use memes in their marketing strategy position themselves as young and cool, understanding the pulse of the younger generation (Agrawal, et al., 2024). Their brand messages are written in a language that current generations use frequently and understand immediately. Gen Z prefers to associate themselves with brands that reflect their personality. This makes brands more human and approachable, and it encourages users to visit their pages more frequently.
6. **Increases Social Media Presence:** Memes are characterized as shareable content and due to their high like, comments, and shares, they get high visibility. Consistent use of memes over time makes brands highlight and gives them engagement (Chugh, 2024). Their cross-platform adaptability ensures a wide reach and regular meme posting keeps the brand relevant, increasing recognition and building a vibrant online community.
7. **Encourages User-Generated Content:** Brands like Gucci and Ola Electric have laid out campaigns where they asked their audiences to prepare memes for them (Dixit, 2023; Yang, 2022). Gucci started with a campaign #TFWGucci where they invited audiences to create memes at the launch of one of their new watch collections. This created a buzz and showcased that brands love their audience to engage with them, and the audience also feels a part of the community as their created memes were shared on the brand's story and social media pages.

Cons of Meme Marketing:

1. **Limited Appeal:** Memes are more appealing to the younger generation comprising Gen Z and millennials. They are more familiar with online humor and trends. Something that resonates with them might not resonate with other generational cohorts. This may lead to a reduction in engagement rates. This limits the appeal of memes when dealing with diverse groups online.
2. **Not Suitable for all brands:** Meme marketing is not suitable for all brands, particularly those in the traditional, conservative, or luxury sectors, where a playful tone may clash with their image of sophistication or professionalism (Vardeman, 2024). These brands frequently prioritize exclusivity and formal messaging, so the casual, humorous nature of memes is incompatible with their identity, potentially alienating their core audience.
3. **Risk of misinterpretation:** A poorly constructed meme runs the risk of being misinterpreted, especially if the humor or cultural references are unclear or inappropriate (Faustino, 2023). Audiences may interpret it as offensive, tone-deaf, or irrelevant, resulting in negative reactions. Such errors can harm the brand's reputation, cause backlash on social media, and necessitate significant effort to manage the fallout and rebuild trust.
4. **Limited life span:** Memes are inherently linked to fleeting trends and cultural moments, resulting in a very short lifespan. Their relevance frequently peaks and fades as audiences move on to newer topics or viral content (Faustino, 2023). This transient nature presents a challenge for brands, as

memes that were interesting one day may appear outdated or irrelevant the next. To maintain audience interest and remain competitive, brands must constantly monitor trends and create new, timely memes, which can be time-consuming. Furthermore, relying too heavily on rapidly changing memes can make a brand's content appear inconsistent or disconnected if trends shift before execution.

5. **Constant Monitoring:** Brands must stay on top of the latest trends, pop culture moments, and shifting audience preferences when it comes to meme marketing. Staying relevant entails identifying and leveraging trending memes quickly, as timing is critical for maximizing impact. This requires ongoing social media monitoring, cultural awareness, and an agile content creation process. The time and effort can strain resources, particularly for smaller teams or businesses. Furthermore, the need for quick turnaround increases the likelihood of errors or misjudgments, which can result in poorly received content. Over time, this level of vigilance and adaptability can become demanding, making meme marketing a difficult strategy to maintain.
6. **Risk of Offence:** Humor, the key component of most memes, is highly subjective and varies across cultures, demographics, and personal experiences. A meme intended to be funny or relatable may inadvertently offend certain groups or individuals due to misunderstanding, insensitivity, or cultural differences (Rathi and Jain, 2023). For example, jokes about sensitive topics or stereotypes may backfire, resulting in criticism or social media backlash. In today's connected world, such controversies can quickly escalate, resulting in widespread negative publicity. This not only tarnishes the brand's reputation, but it may also drive away loyal customers. To avoid unintentional harm, meme creators must have a thorough understanding of audience sensitivities and exercise caution.
7. **Oversaturation:** As more brands use meme marketing, the digital space becomes saturated with similar formats and humor styles, resulting in decreased novelty and impact. Audiences, particularly younger demographics, who are frequently the primary users of memes, may become accustomed to overused templates and repetitive content. When memes become predictable or appear forced, they lose their ability to surprise, entertain, and engage (Faustino, 2023). Memes that are oversaturated may be ignored or dismissed as generic marketing ploys rather than authentic, creative expressions. To remain relevant, brands must constantly innovate and strike a balance between capitalizing on popular trends and providing fresh, unique content.

4.2 Brands Capitalizing on Meme Marketing

Meme marketing has transformed the digital marketing landscape, allowing brands to connect with their audiences in a fun, relatable, and highly engaging manner. Memes are effective in more ways than just making people laugh; they also produce measurable results. Businesses that use memes in their marketing strategies are

60% more likely to generate sales, demonstrating the direct impact of this approach on consumer behavior (Persuasion Nation, 2024). This is not surprising given that memes have approximately ten times more reach and sixty percent more organic interaction than traditional marketing graphics, making them an effective tool for increasing visibility and engagement (Forbes, 2018).

Meme marketing's economic value reflects its growing popularity. The global meme industry was worth $2.3 billion in 2020 and is expected to reach $6.1 billion by 2025 (Persuasion Nation, 2024). This exponential growth demonstrates how memes evolved from a niche online phenomenon to a mainstream marketing strategy across industries. Their viral nature enables brands to gain massive exposure with minimal investment, making them an effective alternative to traditional advertising methods. Memes' ability to evoke shared cultural moments and emotions makes them highly relatable and appealing. They enable brands to humanize their image and build stronger relationships with audiences, particularly younger demographics such as Millennials and Gen Z (Vardeman, 2024). This demonstrates how memes not only boost engagement but also influence purchasing decisions, thereby increasing brand loyalty and conversion rates.

Memes are highly shareable, making them perfect for spreading quickly on platforms like Instagram, X, TikTok, and Meta. Their straightforward and humorous nature fits well with today's fast-moving, content-heavy world. However, brands must keep up with current trends and ensure their memes are relevant to both their audience and their brand image.

With their increasing popularity and success in influencing consumer behavior, memes have become more than just a trend—they are a valuable strategy for businesses aiming to stay relevant and impactful in the digital era. Many brands have started using memes as part of their marketing efforts:

A. Food and Beverage Industry

1. **Zomato (Food Delivery Application):** Zomato, widely recognized as the pioneer of meme marketing in India, has become a global phenomenon in the food delivery industry (Forbes India, 2022). Recently, it acquired Uber Eats in India, further strengthening its market presence (The Economic Times, 2020). Known for its humorous and relatable memes, Zomato consistently shares witty content on its social media platforms to engage with its audience. With over one million followers on Instagram, the brand has built a strong connection with the younger generation, making it one of the most popular and engaging brands (Social Nation, 2023).
2. **Swiggy (Food Delivery Application):** Swiggy, another leading food delivery app in India, is a fierce competitor to Zomato and also excels in meme marketing (Statista, 2023). Known for its creative integration of pop culture trends and relatable food moments, Swiggy's content often achieves high virality, driving organic growth and engagement (ET Brand Equity, 2024). Some of its most popular memes, such as #CravingDuringWorkHours, festival specials, and humorous takes on *Game of Thrones* and Bollywood

culture, have not only captured the audience's attention but also become trending topics across social media platforms.

3. **McDonalds (Fast Food Chain):** McDonald's has mastered meme marketing by blending its iconic imagery and memorable slogans with contemporary pop culture trends (Kao, Hong, Perusse, and Sheng, 2020). This approach creates content that resonates deeply with its target audience. One of its most notable campaigns, the "Travis Scott Meal," capitalized on the rapper's popularity by showcasing his favorite McDonald's meal, sparking massive engagement and sales (Adobe Express, 2023). Additionally, McDonald's effectively uses memes to spotlight its seasonal menu offerings, ensuring they stay top-of-mind for customers. These creative strategies highlight the brand's ability to connect with consumers in a fun and relatable way.
4. **Wendy's (Fast Food Chain):** Wendy's has become a meme marketing pioneer, known for its sharp wit and humor that connects with social media users. The fast-food chain has mastered the art of creating viral moments by combining relatable memes with playful yet scathing responses, which frequently target competitors or address current events (Lind, 2019; Razzaq, Shao, and Quach, 2023). Wendy's X account, in particular, has grown in popularity due to its bold, snarky comebacks and interactive campaigns such as #NationalRoastDay, in which fans request to be humorously "roasted" by the brand. This unfiltered, edgy approach not only entertains but also strengthens Wendy's distinct brand voice, setting it apart in the competitive fast-food industry. Wendy's has successfully connected with younger audiences by cleverly using memes and real-time engagement, transforming everyday interactions into marketing gold.

B. Entertainment and Media

1. **Netflix (Streaming Services):** Netflix's meme marketing strategy is a masterclass in using humor, relatability, and pop culture to engage viewers. The streaming service employs memes to spark conversations, generate buzz, and build emotional connections with its audience (Ziegler, 2018). By focusing on trending topics and scenes from its shows and movies, Netflix creates content that is easy to share and deeply relatable, ensuring that its message reaches a large audience organically. Netflix's social media accounts frequently feature playful, witty memes that appeal to a wide range of audience segments, whether through humor, nostalgia, or clever commentary on current events. Campaigns that use iconic moments from Stranger Things, Squid Game, or The Witcher as meme templates increase user engagement and the reach of their content (Forbes, 2022).
2. **Spotify (Music Streaming):** Spotify has also adopted this marketing strategy by relating memes with music habits. The streaming platform cleverly uses memes to highlight quirky user behaviors, such as overplaying favourite songs or creating oddly specific playlists (Phillips, 2020). Its annual *Spotify Wrapped* campaign is a standout example, sparking a wave of user-generated

memes as listeners share their personalized statistics online (Ramnani, 2023). Spotify also collaborates with trending topics, artists, and pop culture to create humorous and shareable content. By blending data-driven insights with creativity, Spotify's meme marketing creates community engagement and reinforces its position as a culturally attuned, user-centric brand.

5. Implications

5.1 Measuring Meme Success with Analytics

Memes are no longer just a source of entertainment; they have evolved into a powerful marketing tool capable of driving successful campaigns and delivering impressive results. These simple yet engaging visuals, often infused with humor, relatability, or cultural references, can capture attention and evoke emotional responses within seconds. Brands are increasingly leveraging memes to connect with audiences on a deeper level, particularly with younger, tech-savvy demographics who thrive on quick, shareable content (Yang, 2022).

In addition to their creative appeal, memes can be thoroughly analysed using various analytics measures available on social media platforms.

Metrics such as likes, shares, comments, and reach provide a clear picture of how well a meme resonates with its audience (Malodia, Dhir, Bilgihan, Sinha, and Tikoo, 2022). Sentiment analysis tools help brands gauge the emotional response generated by memes—whether positive, negative, or neutral (Sevenois, 2020). Furthermore, insights into demographic data, click-through rates, and follower growth allow marketers to refine their strategies and ensure their content remains impactful.

When paired with social media analytics, meme marketing evolves from a playful experiment to a data-driven strategy, maximizing its impact and delivering measurable results (Adobe Express, 2023). This unique blend of creativity and analysis ensures that memes are not just for fun—they are a serious and cost-effective business tool in the digital age, capable of driving engagement, increasing brand awareness, and boosting conversions. These metrics that can be used to measure a meme's success are explained in detail:

I. *Reach and Impressions:* Reach is the number of unique users who have seen the meme. It indicates that the meme is reaching a broad audience, which is significant for brand awareness. Impressions track how many times the meme has been displayed on the user's screen, regardless of unique views (Kenan, 2024a). A high number of impressions indicates repeated exposure, reinforcing brand messaging.
II. *Engagement metrics:* Likes, shares, and comments are the direct indicators of how well a meme has been comprehended by the audience. A high number of likes shows appreciation, shares indicate that the content is valuable or entertaining enough to be passed on, and comments reflect active participation, sparking conversations or debates (TIS Digitising Brands, 2024). Analyzing

the tone and nature of comments provides further insights into audience sentiment.

III. *Engagement Rate:* This measures the level of audience interaction relative to the meme's reach or impressions (Keyhole, 2024). It is calculated as:

$$\text{Engagement Rate} = \frac{\text{Total Engagement (Likes + Comments + Shares)}}{\text{Total Followers or Impressions}} \times 100$$

IV. *Click-Through Rate (CTR):* On any meme, if the brand adds the link to its product page, website page, social media page, or any landing page, CTR measures the percentage of viewers who have clicked on the link (Chugh, 2024). This can be calculated as:

$$\text{CTR} = \frac{\text{Number of Clicks}}{\text{Total Impressions}} \times 100$$

V. *Tracking through Hashtags:* A hashtag is a keyword or phrase preceded by the "#" symbol (e.g., #MarketingAnalytics). It acts as a label or tag that organizes and categorizes content on social media platforms, making it easily discoverable by users who search for or click on that specific tag. Hashtags are widely used on platforms like Instagram, X, LinkedIn, and TikTok to link conversations, trends, and content around a common theme or topic. If a brand meme includes a branded or trending hashtag, analytics tools can track its usage across platforms (Kenan, 2024b). Metrics like the number of posts using the hashtag, impressions, and reach help gauge the meme's adoption and visibility within broader trends.

VI. *Time Spent on Content:* Social media platforms like Instagram, and TikTok can track how long any audience spends time on memes (Hill, 2024). A high viewership reflects that the meme has captured the attention of the audience and engaged enough to hold their interest, which is significant for storytelling or brand messaging.

VII. *Sentiment Analysis:* Sentiment analysis tools evaluate the emotional tone of user interactions—comments, mentions, or reactions. They classify responses as positive, negative, or neutral (Sevenois, 2020). For example, a humorous meme with overwhelmingly positive comments suggests alignment with audience expectations. On the other hand, negative sentiment could highlight areas for improvement or cultural missteps.

VIII. *Conversion Metrics:* If the meme is part of a campaign, conversion metrics (such as purchases, sign-ups, and downloads) provide tangible proof of its success. For example, a meme promoting a discount code can be directly linked to sales made with that code.

Conclusion

With the onset of COVID-19 and its continuous duration, memes became more prevalent in online conversation (Muirhead-Hernandez, 2020; Priyadarshini, et al., 2022). This further led to the rise of meme marketing. Memes have become one of the top three most shared content types online (Enberg, 2021). They have become a universal language of the younger generation and are often discussed in the physical environment as well (Vardeman, 2024). Memes have created a fear of missing out quite significantly, as one feels an outsider if they are not able to contribute to their social circle discussions. This makes youth highly updated on ongoing trends. This has further led brands to use memes as they have found the pulse of the younger generation who are hesitant to miss out on memes. Brands use social media to connect with global audiences (Razzaq et al., 2024). A generation who are characterized as digital natives and has short attention spans, brands had to consider memes as a marketing strategy as they aptly cater to this category (Shen, Lee, and Lin, 2023). While memes have proven extremely effective in driving engagement, virality, and visibility, brand managers must keep the following key points in mind when implementing this strategy:

1. Stay Authentic: Memes should align with the brand's personality. If a brand presents itself as a serious brand, offering similar services like medical, health, or financial related, meme marketing may not be the right fit (Vardeman, 2024). Brands that adopt this marketing strategy should also avoid over-promotional content; memes should feel natural and fit within the brand's language.
2. Knowing About the Audience: Understanding the target audience is a must. Companies should be aware of the age, gender, culture, location, and behavior of the people that visit their social media pages. (Adobe Express, 2023). Brands can then easily customize their meme accordingly, as what resonates with one group may not resonate with another.
3. Be Relevant: Timeliness is essential in meme marketing. Trends shift quickly, so brands must stay on top of the latest trends to keep their content fresh and relevant (Malodia, Dhir, Bilgihan, Sinha, and Tikoo, 2022).
4. Memes Should be Simple: Simplicity is key as memes should convey the intended message immediately. Brands should avoid complexity because it reduces the effectiveness of this medium.
5. Creativity with Trends: While jumping on trends is effective, brands can add a creative twist to memes that eventually reflect brand values and showcase brand image.
6. Adding Humor: Brands should use humor that is light-hearted, positive, and suitable for the audience (Agrawal, et al., 2024). One should avoid controversial or offensive content that can harm a brand's reputation.
7. Knowing about Platform Dynamics: Brands that are available on different platforms like Meta, Instagram, X, Snapchat, LinkedIn, and Pinterest should create memes differently for each platform as every platform has a different purpose and audience (Adobe Express, 2023). Each audience is from a different generation, and each generation has a distinct way of thinking about and processing humor. Hence, for every platform, a brand can have a different set of memes.

8. Monitor Engagement: When incorporating so much effort, it becomes vital for brands to track and monitor engagement (Malodia, Dhir, Bilgihan, Sinha, and Tikoo, 2022). Brands can track the likes, comments, and shares, to evaluate the success of their marketing campaigns and can refine their approaches accordingly.

In conclusion, meme marketing is an effective tool for brands to connect with younger audiences; however, its success is dependent on strategic execution, creativity, and authenticity. By preserving those ideas in mind, brands can effectively use memes to drive meaningful engagement and strengthen connections with their target audience.

References

Adobe Express. (2023). Meme marketing: how to make memes part of your marketing strategy. Retrieved from https://www.adobe.com/express/learn/blog/meme-marketing

Agrawal, A., Sharma, D., Mishra, M.K., Shah, M.A., Alam, M. and Khandelwal, V. (2024). How to cook a meme: exploring content strategies in brand and user-generated memes on Instagram. *Cogent Business & Management, 11*(1). doi:https://doi.org/10.1080/23311975.2024.2366001

Benaim, M. (2018). From symbolic values to symbolic innovation: Internet-memes and innovation. *Research Policy, 47*(5), 901-910. doi:https://doi.org/10.1016/j.respol.2018.02.014

Beresford Research. (2024). *Beresford Research: Making Smart Companies Smarter since 1991.* Retrieved from https://www.beresfordresearch.com/age-range-by-generation/

Brinkman, C.S., Gabriel, S. and Paravati, E. (2020). Social achievement goals and social media. *Computers in Human Behavior, 111*, 106427. doi:https://doi.org/10.1016/j.chb.2020.106427

Brubaker, P.J., Church, S.H., Hansen, J., Pelham, S. and Ostler, A. (2018). One does not simply meme about organizations: Exploring the content creation strategies of user-generated memes on Imgur. *Public Relations Review, 44*(5), 741–751. doi:https://doi.org/10.1016/j.pubrev.2018.06.004

Chan, F.Y. (2011). Selling through entertaining: The effect of humour in television advertising in Hong Kong. *Journal of Marketing Communications, 17*(5), 319–336. doi:https://doi.org/10.1080/13527261003729220

Chou, H.-Y. and Lien, N.-H. (2010). Advertising effects of songs' nostalgia and lyrics' relevance. *Asia Pacific Journal of Marketing and Logistics, 22*(3), 314–329. doi:https://doi.org/10.1108/13555851011062278

Chugh, S. (2024, February 26). What is Meme Marketing and How to Use Memes to Boost Brand Voice. Retrieved from https://emeritus.org/blog/sales-and-marketing-meme-marketing/

Dix, S. and Marchegiani, C. (2013). Advertising Appeals. *Journal of Promotion Management, 19*(4), 393–394.

Dixit, P. (2023). *Send memes and win Ola S1 Pro special edition.* Retrieved May 27, 2023, from Business Today: https://www.businesstoday.in/amp/auto/story/send-memes-and-win-ola-s1-pro-special-edition-says-ola-ceo-bhavish-aggarwal-383083-2023-05-27

Duarte, F. (2024). *X (Formerly Twitter) User Age, Gender, & Demographic Stats (2024).* Exploding Topics. Retrieved from https://explodingtopics.com/blog/x-user-stats

El-Murad, J. and West, D.C. (2004). The Definition and Measurement of Creativity: What Do We Know? *Journal of Advertising Research, 44*(2), 188–201. doi:https://doi.org/10.1017/S0021849904040097

Enberg, J. (2021). *To Meme or Not to Meme? A Marketer's Guide to Memetic Media in 2021 and Beyond.* eMarketer. Retrieved from https://www.insiderintelligence.com/content/to-meme-or-not-to-meme

ET Brand Equity. (2024). Swiggy's new ad results in the multiverse of madness. Retrieved from https://brandequity.economictimes.indiatimes.com/news/advertising/swiggys-new-ad-results-in-the-multiverse-of-madness/112992347

Faustino, A. (2023, September 08). How Meme Marketing is Taking Over Social Media. Retrieved from https://capforge.com/how-meme-marketing-is-taking-over-social-media/

Flanagan, C. (2020). *Early Socialisation: Sociability and Attachment.* London: Routledge.

Forbes. (2018). *Memes: A Digital Marketing Tool For Every Industry.* Retrieved August 10, 2018, from Forbes: https://www.forbes.com/sites/forbescommunicationscouncil/2018/08/10/memes-a-digital-marketing-tool-for-every-industry/?sh=3e32ce682664

Forbes. (2022). How Netflix is Resonating with 'Stranger Things' Fans through Experiential Marketing. Retrieved from https://www.forbes.com/sites/charlesrtaylor/2022/08/31/how-netflix-is-resonanating-with-stranger-things-fans-through-experiential-marketing/

Forbes India. (2022). *Engineering the viral: The science behind meme marketing.* Retrieved from https://www.forbesindia.com/article/storyboard18/engineering-the-viral-the-science-behind-meme-marketing/78877/1

Forsgren, E. and Byström, K. (2018). Multiple social media in the workplace: Contradictions and congruencies. *Information Systems Journal, 28*(3), 442–464. doi:https://doi.org/10.1111/isj.12156

Gavilanes, J.M., Flatten, T.C. and Brettel, M. (2018). Content Strategies for Digital Consumer Engagement in Social Networks: Why Advertising is an Antecedent of Engagement. *Journal of Advertising, 47*(1), 4–23. doi:https://doi.org/10.1080/00913367.2017.1405751

Hill, C. (2024). The social media metrics to track in 2024 (and why). Retrieved from https://sproutsocial.com/insights/social-media-metrics/

Instagram. (2020). Instagram Year in Review: How Memes were the Mood of 2020. Instagram from Meta.

Joa, C.Y., Kim, K. and Ha, L. (2018). What Makes People Watch Online In-Stream Video Advertisements? *Journal of Interactive Advertising, 18*(1), 1–14. doi:https://doi.org/10.1080/15252019.2018.1437853

Jung, A.-R. and Heo, J. (2021). Does cluttered social media environment hurt advertising effectiveness? The moderation of ad types and personalization. *Journal of Research in Interactive Marketing, 15*(4), 592–606. doi:https://doi.org/10.1108/JRIM-11-2020-0238

Kang, M. (2005). *A Study on the Effect of Features of Brand Community Using One-person Media on Consumers.* Seoul: Seoul National University.

Kantar. (2024). Time to get serious about humour in advertising. London. Retrieved from https://www.kantar.com/inspiration/advertising-media/time-to-get-serious-about-humour-in-advertising

Kao, G., Hong, J., Perusse, M. and Sheng, W. (2020). Memes as Marketing. In *Turning Silicon into Gold: The Strategies, Failures, and Evolution of the Tech Industry* (pp. 99-107). doi:10.1007/978-1-4842-5629-9_15

Kemp, S. (2024). *Digital 2024 October Global Statshot Report.* Datareportal. Retrieved from https://datareportal.com/reports/digital-2024-october-global-statshot

Kenan, J. (2024a, February 20). Reach vs. impressions vs. engagement: What's the difference? Retrieved from https://sproutsocial.com/insights/reach-vs-impressions/#:~:text=Impressions%20are%20the%20total%20number,%2C%20comments%2C%20shares%20and%20clicks.

Kenan, J. (2024b). Hashtag analytics: Your complete guide to tracking hashtag performance. Retrieved from https://sproutsocial.com/insights/hashtag-analytics/

Keyhole. (2024). How to Calculate Engagement Rate in 2024 + Free Tool Included. Retrieved from https://keyhole.co/blog/calculate-engagement-rate/

Kim, J. and Kim, M. (2024). Viral Dynamics on Social Media: Enhancing Brand Engagement through Meme Marketing Strategies. *Journal of Current Issues & Research in Advertising*, 1–18. doi:https://doi.org/10.1080/10641734.2024.2390850

Kitchen, P.J. (2023). Digital does not stand alone… the road to further critical analysis lies open. *Journal of Marketing Communications, 29*(6), 533–534. doi:https://doi.org/10.1080/13527266.2023.2244799

Kusumasondjaja, S. (2018). The roles of message appeals and orientation on social media brand communication effectiveness: An evidence from Indonesia. *Asia Pacific Journal of Marketing and Logistics, 30*(4), 1135–1158. doi:https://doi.org/10.1108/APJML-10-2017-0267

Laskey, H.A., Day, E. and Crask, M.R. (1989). Typology of Main Message Strategies for Television Commercials. *Journal of Advertising, 18*, 36–41. doi:https://doi.org/10.1080/00913367.1989.10673141

Lind, W. (2019). Meme Marketing to Fellow Kids.

Liu, S.S. and Stout, P.A. (1987). Effects of Message Modality and Appeal on Advertising Acceptance. *Psychology & Marketing, 4*(3), 167–187. doi:https://doi.org/10.1002/mar.4220040303

Malodia, S., Dhir, A., Bilgihan, A., Sinha, P. and Tikoo, T. (2022). Meme marketing: How can marketers drive better engagement using viral memes? *Psychology & Marketing, 33*(9), 1775–1801. doi:https://doi.org/10.1002/mar.21702

McKinsey. (2023). Gen Z mental health: The impact of tech and social media. Retrieved from https://www.mckinsey.com/mhi/our-insights/gen-z-mental-health-the-impact-of-tech-and-social-media

Muirhead-Hernandez, C. (2020, April 06). Memes: More popular than ever thanks to COVID-19? doi:https://www.linkedin.com/pulse/memes-more-popular-than-ever-thanks-covid-19-chris-muirhead-hernandez/

Pelsmacker, P.D., Geuens, M. and Anckaert, P. (2002). Media Context and Advertising Effectiveness: The Role of Context Appreciation and Context/Ad Similarity. *Journal of Advertising, 31*(2), 49–61. doi:https://doi.org/10.1080/00913367.2002.10673666

Persuasion Nation. (2024). 25+ Latest Meme Marketing Statistics You Should Know. Retrieved from https://persuasion-nation.com/meme-marketing-statistics/

Phillips, M. (2020). How Spotify Memed Their Way to 130 Million Subscribers: A showcase in knowing your market. Retrieved from https://medium.com/swlh/how-spotify-memed-their-way-to-130-million-subscribers-f298d59fe2d4

Priyadarshini, I., Chatterjee, J.M., Sujatha, R., Jhanjhi, N., Karime, A. and Masud, M. (2022). Exploring Internet Meme Activity during COVID-19 Lockdown Using Artificial Intelligence Techniques. *Applied Artificial Intelligence, 36*(1), 2014218. doi:https://doi.org/10.1080/08839514.2021.2014218

Ramnani, K. (2023). These Spotify Wrapped Memes Have Me Cackling. Retrieved from https://www.hercampus.com/culture/spotify-wrapped-2023-memes/

Rathi, N. and Jain, P. (2023). The Power and Pitfalls of Internet Memes in promoting brand awareness and engagement on social media. *World Scientific News, 184*, 51–62.

Rathi, N. and Jain, P. (2024). Impact of meme marketing on consumer purchase intention: Examining the mediating role of consumer engagement. *Innovative Marketing, 20*(1). doi:10.21511/im.20(1).2024.01

Razzaq, A., Shao, W. and Quach, S. (2023). Towards an understanding of meme marketing: conceptualisation and empirical evidence. *Journal of Marketing Management*, 1–32. doi:https://doi.org/10.1080/0267257X.2022.2158906

Razzaq, A., Shao, W. and Quach, S. (2024). Meme marketing effectiveness: A moderated-mediation model. *Journal of Retailing and Consumer Services, 78*, 103702. doi:https://doi.org/10.1016/j.jretconser.2023.103702

Redondo, I. (2012). The effectiveness of casual advergames on adolescents' brand attitudes. *European Journal of Marketing, 46*(11/12), 1671–1688. doi:https://doi.org/10.1108/03090561211260031

Sevenois, E. (2020). Sentiment Analysis of Internet Memes on Social Media Platforms;. Ghent University. Retrieved from chrome-extension://efaidnbmnnnibpcajpcglclefindmkaj/https://libstore.ugent.be/fulltxt/RUG01/003/006/973/RUG01-003006973_2021_0001_AC.pdf

Shen, Y.-C., Lee, C.T. and Lin, W.-Y. (2023). Meme marketing on social media: the role of informational cues of brand memes in shaping consumers' brand relationship. *Journal of Research in Interactive Marketing*. doi:https://doi.org/10.1108/JRIM-01-2023-0029

Shewale, R. (2024). *LinkedIn Statistics In 2024 (Global Demographics)*. Demand Sage. Retrieved from https://www.demandsage.com/linkedin-statistics/

Shifman, L. (2013). Memes in a Digital World: Reconciling with a Conceptual Troublemaker. *Journal of Computer-Mediated Communication, 18*(3), 362–377. https://doi.org/10.1111/jcc4.12013

Smith, A. (2024). How Gen Z uses social media and what that means for brands. Retrieved from https://sproutsocial.com/insights/gen-z-social-media/

Social Nation. (2023). Zomato's mouthwatering meme marketing magic! Retrieved from https://www.socialnationnow.com/zomatos-mouthwatering-meme-marketing-magic

Statista. (2023). *Most popular food delivery companies across India as of April 2023*. Retrieved from https://www.statista.com/statistics/1149293/india-popular-food-delivery-apps/

Statista. (2024). *Distribution of Instagram users worldwide as of April 2024, by age group*. Statista. Retrieved from https://www.statista.com/statistics/325587/instagram-global-age-group/

Stemler, S. (2001). An Overview of Content Analysis. *Practical Assessment, Research, and Evaluation, 7*(17), 17. doi:https://doi.org/10.7275/z6fm-2e34

Taecharungroj, V. and Nueangjamnong, P. (2015). Humour 2.0: Styles and Types of Humour and Virality of Memes on Facebook. *Journal of Creative Communications, 10*(3), 288–302. doi:https://doi.org/10.1177/0973258615614420

Teng, H., Lo, C.-F. and Lee, H.-H. (2022). How do internet memes affect brand image? *Online Information Review, 46*(2), 304–318. doi:https://doi.org/10.1108/OIR-05-2020-0192

The Economic Times. (2020). Zomato acquires Uber Eats in an all-stock transaction. Retrieved from https://economictimes.indiatimes.com/small-biz/startups/newsbuzz/zomato-acquires-uber-eats-in-an-all-stock-transaction/articleshow/73465982.cms?from=mdr

TIS Digitising Brands. (2024). Memes as a Marketing Strategy: How to Turn Viral Culture into Profit. Retrieved from https://www.tisdigitech.com/blog/memes-as-a-marketing-strategy/

Vardeman, C. (2024). Advertising to Gen-Z college students with memes? A focus group study. *Qualitative Market Research, 27*(1), 1–18. doi:https://doi.org/10.1108/QMR-05-2023-0073

Vries, L.D., Peluso, A.M., Romani, S., Leeflang, P.S. and Marcati, A. (2017). Explaining consumer brand-related activities on social media: An investigation of the different roles of self-expression and socializing motivations. *Computers in Human Behavior, 75*, 272–282. doi:272–282

Wang, Z., Morey, A.C. and Srivastava, J. (2014). Motivated Selective Attention During Political Ad Processing: The Dynamic Interplay Between Emotional Ad Content and Candidate Evaluation. *Communication Research, 41*(1), 119–156. doi:https://doi.org/10.1177/0093650212441793

Waqas, M., Hamzah, Z.L. and Salleh, N.A. (2021). Customer experience with the branded content: a social media perspective. *Online Information Review, 45*(5), 964–982. doi:https://doi.org/10.1108/OIR-10-2019-0333

Weber, R.P. (1990). *Basic Content Analysis* (Vol. 49). Sage. doi:10.4135/9781412983488

Weinberger, M.G. and Gulas, C.S. (2021). The emergence of a half-century of research on humour in advertising: what have we learned? What do we still need to learn? In *Humor in Advertising* (p. 46). Routledge.

Yang, G. (2022). Using Funny Memes for Social Media Marketing: The Moderating Role of Bandwagon Cues. *Journal of Promotion Management, 17*(1), 944–960. doi:https://doi.org/10.1080/10496491.2022.2054904

Yoo, C. and MacInnis, D. (2005). The brand attitude formation process of emotional and informational ads. *Journal of Business Research, 58*(10), 1397–1406. doi:https://doi.org/10.1016/j.jbusres.2005.03.011

Zhang, H., Sun, J., Liu, F. and Knight, J.G. (2014). Be rational or be emotional: advertising appeals, service types and consumer responses. *European Journal of Marketing, 48*(11/12), 2105–2126. doi:https://doi.org/10.1108/EJM-10-2012-0613

Ziegler, J. (2018). *Medium*. Retrieved from https://medium.com/rta902/from-movies-to-memes-examining-netflixs-social-media-stardom-cf6987d5d79e

15 Ethical and Legal Challenges in Business Analytics and Digital Technologies

Aditi Sharma[1*], Arun Bhatia[2], Imali N. Fernando[3] and Mridul[4]

The growing reliance of modern societies on technology is creating new ethical and legal issues, necessitating the creation of new laws and ethical standards to guard against the possible harm that careless technology usage may cause to society. Emphasising algorithmic bias, data security, and cybersecurity breaches, this chapter explores digital technology's and business analytics' moral ramifications. Thus, practitioners and data scientists need to employ ethics as a moral compass while creating applications that help resolve issues that people still cannot resolve. Providing ethical and legal guidelines to the technologists can ensure the development of technologies that mitigate the human suffering, like using them for finding a cure to deadly diseases like AIDS and cancer. The developments in the field of AI/ML have outpaced the development of legal frameworks and the increase in fraudulent activities and compromise of individual privacy. This chapter also examines the intersections and tensions between ethics and law, and it is important to recognize that ethical guidelines may help in resolving the moral dilemma of the technologist/software developers, it is the law that prevents, controls, and sanctions the violations.

1. Introduction

Business analytics is becoming increasingly ubiquitous in our daily lives and humans are increasingly relying upon algorithmic decision-making, yet there is a dark side to the excessive reliance on AI, like its use in the development of drone technology by the US military (Vidgen et al., 2019) in cyberattacks worldwide. The importance of ethical considerations in business analytics is growing as organisations increasingly rely on data for decision-making. The generation and sharing of personal and impersonal data by information brokers pose a significant challenge to the long-held cherished societal beliefs of privacy and autonomy of humans. The growing use of data in corporate analytics presents serious ethical questions about its collection and use, creating new moral dilemmas for the data

[1] 0000-0001-5774-528X
[2] 0000-0003-3464-6128
[3] 0000-0001-9487-5569
[4] 0000-0003-2453-7024
* Corresponding author: draditisharmacuhp@hpcu.ac.in

DOI: 10.1201/9781003514282-15

scientists who collect and analyze the massive amounts of data produced globally. (Weinhardt, 2020). The usage of this data, if done for the benefit of society, is a great service to humanity. At the same time, data security and confidentiality are the major ethical questions that the firms need to address at both moral and legal levels. Data breaches in many technological giants like Facebook and LinkedIn put user data in jeopardy and compelled them to limit their exposure to public access (Sayed, 2023).

Philosophical frameworks can provide utilitarian and deontological ethics guidance for ethical decision-making in data analytics. Developing and integrating clear ethical guidelines into data governance strategies ensures all data-related activities are conducted ethically (Adaga et al., 2024). The ethical considerations surrounding big data include human rights ethics, moral ethics, information ethics, management ethics, and responsibility ethics. Human rights ethics focus on fairness, autonomy, and holistic development of individuals, while the moral ethics of society keep evolving and depend on the limits set by the moral codes. Civic engagement, ethics, and laws all play a vital role in regulating the responsible handling of enormous amounts of data for business analytics. Big data presents an ethical dilemma as morality is a fluid idea that changes as society evolves. The ethical standards about the creation, sharing, and usage of information are known as information ethics. Examination of the notion of responsibility ethics in contemporary society involves analysis of the distribution of responsibility, and an exploration of responsibility connections (Sayed, 2023).

1.1 Chapter Organization

The chapter is divided into the following sections: The introduction highlights the pervasiveness of technology in almost every sphere of human activity and how it poses ethical challenges for AI/ML developers in the absence of any proper ethical guidelines. At the same time, the legal frameworks may not alone be sufficient to deal with the rapid technological disruptions leading to damaging consequences for the public at large. The next section reviews the philosophical and ethical foundation in merging areas like technoethics, roboethics, and computational ethics. Ethical frameworks developed by researchers for resolving ethical dilemmas and ethical decision-making have been discussed. The legal aspects focus on challenges such as data sharing, informed consent, and governance issues. The intersectionality between ethics and AI, particularly in the field of medicine and robotics, emphasizes the need for continuous ethical reflection and legal compliance. The conclusion section underscores the importance of embedding ethical principles in designing and deploying technologies that are fair, inclusive, responsible, and accountable.

1.2 Objective of the Study

This chapter examines the ethical and legal difficulties arising from increased human-machine interaction and the ethical challenges it poses for developers.

Data scientists, technologists, coders, and developers face the ethical dilemma of reconciling the organization's desire to remain competitive and profitable with the societal expectation of dignity, privacy, and autonomy. It provides insights into the moral and legal frameworks for democratic decision-making, leading to the development and implementation of sustainable and socially responsible technologies.

2. Methodology

The authors reviewed the existing studies on ethical and legal challenges in the context of digital technologies. Scopus and Google Scholar were the two major databases used for literature searches.

3. Philosophical and Ethical Frameworks

Vidgen et al. (2019) proposed an ethical toolkit required for OR and data science practice based on the following principles: (1) transcend broad standards for professional and ethical behaviour; (2) embody a toolkit that does not necessitate a thorough philosophical background; (3) represent the normative status of ethical reasoning; and (4) be practically applicable to ethical decisions in the real world.

Conventional ethical theories, whether deontological, consequentialist, or virtue-based, are often anthropocentric, emphasising human behaviour and how it affects other people or, in some situations, sentient beings. Applied ethics focuses on applying moral principles to real-world situations, a departure from traditional ethics, as it integrates philosophical inquiry into real human issues, occurrences, and behaviors. In this information age, knowledge creation and dissemination has become as fast as the speed of light due to the rapid strides made in information and communication technologies (Luppicini, 2010). Aspects of technology such as its development and societal ramifications are included in the ethical study fields of AI and digital technologies (DT). A few of these include:

3.1 Technoethics

Technoethics is the study of the moral and social implications of how science and technology are created, used, regulated, and assessed in society. It examines the ethics of developing new technologies and their impact on individual authority (Luppicini, R., 2010). The development of ethically and socially acceptable technical standards would be greatly supported by adopting a technoethics approach, which incorporates objective valuation and factual information to aid ethical decision-making in technology-related endeavors (Bunge, 1977).

3.2 Computer Ethics

This subfield of philosophy examines the ethical use and societal effects of computer technology. Applied ethics encompasses ideas on computer system work ethics and user behaviour (Han, 2022). Computer ethics focuses on the societal impact of computer technology and the ethical policies governing its use.

Initially viewed as advanced calculating machines, computers sparked debates about their potential to replicate human reasoning and their implications for humanity. Over time, concerns shifted to government power centralization, extensive data collection, and emerging technologies like software systems, video games, and virtual reality. Each innovation brought moral uncertainty, raising questions about its effects on human values and relationships, including those between governments and citizens, employers and employees, and businesses and consumers (Johnson, D.,1994). Utilitarian, deontological, and social contract theories have often been used to frame ethical dilemmas about computers. According to Moore, computer ethics is the study of properties and social impacts of computer technology that leads to the creation of laws governing its moral use. The absence of moral guidelines creates a policy vacuum as there are no set rules for moral conduct amid the new opportunities that technology has provided (Moor, 1985), which gives rise to several moral dilemmas and crimes such as breach of computer security, program manipulation, computer addiction, etc (Wiener, 1950).

3.3 Information Ethics

Hauptman coined the term "information ethics" to refer to any moral concerns about the creation, preservation, availability, and sharing of information, including media and literacy (Han, J., 2022). "Information Ethics" (IE) is a new ethical paradigm for the information era. The "infosphere," which includes all information entities and their interactions, is the centre of IE, which maintains that each thing is deserving of dignity and respect just because of its existence. It addresses the creation, organization, dissemination, and use of information alongside ethical standards for human conduct. In contact with the infosphere, it encourages objectivity, universality, and consideration, guaranteeing the welfare of all information beings (Floridi, 1999). Information ethics needs to go beyond concerns related to the internet to encompass the broader digital ontology and the ethical implications of digital technology. advocating for an **intercultural dialogue** to address the ethical challenges of the digital age, moving beyond an "ethics *in* the infosphere" to an "ethics *of* the infosphere". According to IE, moral attention should be extended to all information entities, including digital objects, algorithms, and the larger "infosphere." This change highlights how all entities in the information environment are interrelated and how agents must take the health of the infosphere as a whole into account" (Capurro, R. 2006).

3.4 Roboethics

Roboethics is a subfield of technoethics that studies the ethical implications of robotics, emphasizing the moral design and use of robots for humanity's benefit. The study of the ethical ramifications and effects of robotics technology, especially intelligent/autonomous robots, on human society is the focus of the contemporary multidisciplinary research area of "roboethics," which sits at the nexus of applied ethics and robotics. Motivating the ethical design, development, and application of robots for the good of humans is the main goal of roboethics (Tzafestas, 2018).

3.5 Machine Ethics

Machine ethics is an emerging field focused on implementing moral decision-making faculties in computers and robots, addressing key ethical questions in the context of autonomous systems. The study of machine ethics broadens the scope of computer ethics to include enquiries into the actions of the machines themselves, in addition to concerns about human behaviour. It explores the ethical behavior of machines toward humans and other machines. The development of Artificial Moral Agents (AMAs) raises challenges related to ethical decision-making capabilities, moral psychology, and the simulation of moral behavior in machines. Thus there is a need for a collaborative effort from philosophers, engineers, and social scientists to address the boundaries of autonomous moral beings, their design methodologies, and their applicability (Allen, et al., 2006).

3.6 Computational Ethics

Computational ethics is an essential subfield of AI ethics that focuses on the practical implementation of ethical decision-making in AI systems. It moves beyond theoretical debates to develop and test algorithms, frameworks, and procedures that allow AI to operate ethically in the real world. Computational ethicists collaborate with engineers and computer scientists to develop ethically conscious AIS. Although ethics is the foundation of computational ethics, its application mostly depends on other fields. Knowledge Representation and Reasoning assist in putting abstract ethical ideas into a calculable form, robotics/embodiment gives AI a physical shape, and logic gives ethics a formal framework (Segun, 2021).

3.7 ICT Ethics

The widespread use of ICT in global society necessitates the adoption of a communicative, democratic approach to global ICT ethics that addresses the issues of freedom, intellectual property, and social justice (Górniak-Kocikowska, 2007). It investigates how information and communication technologies affect human values, attitudes, and behavior. Information ethics is a branch of ethics that examines the link between the creation, organisation, distribution, and use of information. It is described as the moral principles and ethical standards that guide human behaviour (Han, 2022).

3.8 AI and Machine Learning Ethics

AI can transform several industries, including healthcare and national security, but there have been worries about safety, discrimination, and racial bias. The employment of artificial intelligence in "narrow" applications, such as self-driving vehicles, deadly autonomous weapon systems, and surveillance systems, has raised concerns. In response, companies, governments, civil society groups, and multi-stakeholder organisations have issued high-level AI ethical standards, such as the OECD AI standards accepted in May 2019 and the G20 Human-centered AI Principles agreed in June 2019. These guidelines seek to address ethical issues about AI application in military and government situations (Zhang, et al., 2021). The AI/ML ethics deal with issues like responsibility, transparency, auditability, and the morality of intelligent systems. These domains span the physical, cognitive, and informational dimensions of ethical considerations in DTs (Madan et al., 2022).

4. Digital Ethics

As digital technology becomes more and more integrated into our daily lives, digital ethics become an existential issue worldwide (Hanna and Kazim, 2021).

Defined as "the systems of values and moral principles for electronic interactions," digital ethics integrates technological (Utilitarian) and societal (Kantian) moral values, encompassing both hard ethics (regulations) and soft ethics (beyond compliance) (Madan et al., 2022). Digital ethics, as a macroethic, takes into account the ethical implications of the entire digital environment. It investigates the interplay of data, algorithms, practices, and infrastructures while advocating for a shift from an information-centric to a digital-centric perspective (Floridi et al., 2019). Digital ethics encompasses the study and evaluation of moral problems related to data, algorithms, and practices in the digital age, aiming to formulate and support morally good solutions. The field of digital ethics is multidisciplinary and addresses ethical challenges in areas such as internet infrastructure and cyber conflicts, emphasizing responsibilities, risks, and rights. Decision-making in data science is a complex process as data scientists have to make decisions that require a trade-off between margin of error and fairness. The Professional Code of Conduct may not completely suffice, hence the ethical frameworks can be valuable guidebooks in these scenarios (Schwarz, 2005).

4.1 Business Ethics Canvas (Vidgen et al., 2019)

Adopting an action research approach, Business Ethics Canvas was developed by following the guidelines of Business Analytics Methodology. The framework begins with the recognition of an ethical problem, subsequent identification of used cases, and creation of an analytics portfolio. Thereafter, an Opportunity Canvas and BEC are placed jointly. Markkula Centre's ethical principles serve as the basis for the choice of its implementation after pertinent data has been gathered and framed in a corporate ethics canvas. The final choice is made after considering the unforeseen implications and meeting the legal standards outlined in the GDPR.

4.2 *Digital Ethics Framework (Madan et al., 2022)*

The authors conceptualised the ethical implications of AI on seven DT archetypes using an ontological framework. The authors placed digital ethics at the centre of eight ethical study areas and included a critical domain of Governance within the context of Digital Transformation. The authors' conceptual map offers managers a morally guided method for carrying out an ethical effect study of AI on a particular DT typology.

4.3 *Ethics as a Process Approach (Hanna and Kazim, 2021)*

The ethical framework for digital and AI ethics is based on the Kantian theory of human dignity, which recognizes the transcendental value and moral standing of human real persons. The authors emphasize the metaphysical nature of human dignity and its intricate connection with moral and legal obligations, ultimately asserting the importance of a well-grounded and generally accepted broadly Kantian dignitarian digital and AI ethics for humanity.

4.4 *Computational Reflective Equilibrium (Awad et al., 2022)*

The framework involves "algorithmic characterization of normative and descriptive ethics" requiring designers to make forced choices to enable the development of stress-handling models by testing a wide array of used cases. At the implementation stage, the behaviour that emerges out of interactions between people and AI is observed, and it is the human interaction that establishes whether the AI system has a positive or negative impact on humans and society.

4.5 *Value Theory of Means, Goals, and Side Effect (Bungee, 1977)*

The author developed a set of rules for technologists that serve as ethical guidelines for making decisions. Before making any decision, the technologists must assess the overall cost or input by combining the gain and side effects. They must refrain from taking any decision that is practically and morally inefficient. Thus, moral principles need to be moulded based on objective assessment and factual information rather than embracing them as general principles.

These frameworks provide an action-oriented approach to addressing ethical and legal challenges in business analytics and digital technologies. The common underlying assumption is that each of these frameworks is the continuous reflection, adaption, and alignment with the changing business environments and legal standards for developing technologies that are not only innovative but morally and socially responsible.

5. Legal Issues in Business Analytics

Formalities and legalities defined in the business domain for effective use of business analytics set the tone and pattern for organisations to use it efficiently.

Most business is built on futuristic business strategies and processes using business analytics. On the other hand, some organisations operate in an environment created around the legal issues of using business analytics. Legal concerns in digital analytics must also be addressed since consumers must be aware of how, when, and where businesses utilize their data. Maintaining surveillance of the many agencies with which data is exchanged and changing account passwords regularly are two ways to fight this.

5.1 Overview of Legal Frameworks

Guan and Zhou (2017) provide a critical overview of multiple problems that need to be taken care of in the information society for business analytics, raising concerns for information rights and obligations leading to quality of life. In addition to the same, the researchers do address issues like accountability and control along with information brokers. One of the major concerns in contemporary society originates from customer-driven data and needs to be taken care of using web analytics where data reidentification is found to be the right kind of society in modern, driven information society (Berendt, Preibusch, and Teltzrow, 2018). Adaga et al. (2024) highlighted that the current theme of the data is not limited to the better use of data only but other issues also need a critical understanding, like the impact of more generation of data on the environment and energy consumption and last but not the least understanding the digital life cycle technologies. In a developing nation like India, governing the legal framework in business analytics has been outlined in two of the most important acts, the Digital Personal Data Protection Act (DPDP Act) 2023 and the Information Technology Act, 2000. Both acts are of main importance in digital framework overcoming problems such as interpretation and implementation of the act, data silos, and last but not least digital illiteracy. Chowcat, Kay, and Korn (2015) do admit the legal issues and challenges where it has been a greater concern that how the data is to be shared among the various departments, various organisations, and among various individuals to understand that data is a resource for the organisation and asset for an individual to be shared publicly. In the same article, informed consent and transparency have been put at the top of the legal framework in the United Kingdom. Numerous legal and ethical frameworks have emerged, enabling various agencies worldwide to consider the problem and potential remedies for both macro and micro data analytics.

5.2 Key Legal Challenges

The key challenges as addressed in the past by (Birckan et al., 2020) are highly reciprocal to the benefits business analytics accrue to society, along with the non-availability of any regulation in data sharing features in the online society among the government and the corporations. Hard press success in business analytics is highlighted in the five domain areas, namely descriptive analytics, real-time analytics, diagnostic analytics, predictive analytics, and prescriptive analytics (Wolniak and Grebski, 2023a), which is also a challenge in contemporary society.

One of the challenges outlined by Vidgen, Shaw, and Grant (2017) is to oversee how value can be created in the organisation where modern businesses are more data and information-driven. Liu, Han, and De Bello (2018) in a research article titled "The Challenges of Business Analytics: Successes and Failures" provided that application of business analytics is a challenge for several organisations where organisations are not able to capture the true benefits of decision making from business analytics power thus leading to ineffective businesses. Omar, Minoufekr, and Plapper (2019) within the key benefits gained from business analytics, the other highlighted challenges are isolated data along with the noninteroperability capabilities among the departments, challenges coming from nonacceptance to digital analytics from the senior management, etc. One of the other key challenges outlined in the business analytics domain is that most organizations are not able to define the data policy within the organisation (Soldić-Aleksić, Krasavac, and Karamata, 2020). The only remedy to the above-mentioned problem lies in defining the data structures to be created in the organisation, and the data protection centres, and limiting logical access to the data. Ajah and Nweke (2019) define challenges in the three interlinked areas of business analytics, ranging from access to the right kind of data, data accuracy, and lastly consistency. One of the modern challenges that emerge from data analytics is the consistent use of social media influence on the buyers of the products and services, and vice-versa. The same has been a critical area of research as a social media paradox and business analytics. Adaptive social media strategy is the key to overcoming such a challenge (Kaisler et al., 2014). According to Reddy and Prasad (2021), the usage of data analytics for e-commerce and commerce activities does align with the idea of protecting the data of many customers online, where e-commerce and commerce activities are outpacing enterprises' offline operations. Bhatia (2020) provides the key challenge is not only on the data protection for the business analysis, but it is also important to look into all the 3Cs in data-driven business models coming out of content, community, and commerce. The workplace issues with data analytics are presented by Wolniak and Grebski (2023b). Data has emerged as a key decision-making factor in organizations, and the majority of employees base their effective decision-making on making better use of data.

6. Ethical-Legal Overlap and Conflicts

There exists a close relationship between ethical and legal frameworks governing the development and deployment of AI. While **ethics** provide moral guidelines for responsible AI development and use, the **law** establishes formal rules and regulations to ensure compliance and minimize potential harm. The ethical ramifications and difficulties related to the use of robots in fields such as medical, assistive, sociorobot, war, autonomous car, and cyborg ethics, Tzafestas (2018) examined the moral and legal issues surrounding the employment of robotic technology in military applications. The authors cautioned that when it comes to the creation and application of robots, especially in assistive and medical technology, autonomy, privacy, and moral responsibility must be carefully considered. Theragnostic technologies are a new area of personalised medicine that combines diagnosis and therapy. Nuclear medicine, interventional radiology, and nanomedicine are a few examples of uses. For example, in interventional radiology, the interventionist takes complete responsibility for diagnosis and treatment by performing a minimally invasive procedure in place of surgery. This might result in several situations where the AI/ML model can be used simply for diagnosis, just for treatment, or for both diagnosis and therapy. On the flip side as AI becomes increasingly integrated into these technologies, the potential risks to patient safety and liability increase (Drabiak, et al., 2023).

Conclusion

"The emerging topic of 'theory of machine' investigates lay theories of how machines work. People's 'theory of machine' is likely to influence how they behave around machines, how they treat machines, and the extent to which they are willing to trust machines. The dynamic exchange between a machine's theory of the human mind and a human's theory of machine mind is particularly important for the challenge of formalizing the ethics of care" (p. 399, Awad et al., 2022)

Law, ethics, and society are intricately related. The rules, regulations, and policies formed in the organizations are based on the legal framework. The adherence to ethical principles enables these firms to build a strong brand image. The degree of contemporary society's reliance on technology is indisputable, but from an ethical perspective, this over-reliance must be carefully examined. Given the close interaction between people and technology, there are ethical questions regarding the potential for technology to significantly alter human potential. However, it's crucial to recognise that technology cannot function independently without human involvement. For example, computer virus transmission, cyberattacks, and abuse are human-mediated activities (Johnson, 1994). Ellul (1964) warned of the perils of a technological society that enslaves individuals by depriving them of their autonomy in the name of freedom and ensuring the pointless submission to technology by humans. To protect digital privacy by explicit or implicit rational permission, Bunge (1977) argued that human dignity should serve as the foundation for AI ethics and should not result in damage or prejudice. Jonas (1985) highlighted the ethical dilemmas presented by genetic engineering in his paper "Ethics and biogenetic art," particularly about the potential for unforeseen ecological consequences and the control of human inheritance through gene-splicing and cloning. He questioned if mankind is capable of using such power sensibly. To arrive at solutions that enhance the moral worth of digital innovation for the benefit of our societies,digital ethics must navigate the difficult terrain between societal rejection and legislative ban (Floridi et al., 2019). The effective exploration of ethical issues is possible only if they are embedded in the analytics creation process in the organization (Vidgen et al., 2019).

Beyond maximising profits, ethical business analytics also includes a dedication to reducing any possible harm to society. To make sure that their data practices lead to favourable social outcomes, organizations should proactively identify and resolve any potential harms resulting from data-driven choices. Transparency and accountability, embedding ethical standards into data governance plans, and incorporating philosophical ideas into corporate data governance plans are some ways to solve these ethical issues. The ethical ramifications of gathering and using data for business analytics necessitate serious thought and prompt action. embracing ethical norms, implementing open and responsible data practices, and embracing philosophical ideas may help organizations responsibly traverse this complicated terrain and contribute to a more sustainable and fair society (Adaga et al., 2024).

As seen in the well-known Satyam case in India, a strong legal framework can discourage corporate entities from engaging in fraud such as misleading shareholders and investors by disclosing manipulated data. At the same time, it is also important to realize that laws alone cannot control the unlawful activities arising out of the huge chunk of data and information that is created and generated by the data firms. Thus, ethics play a vital role in developing self-regulatory mechanisms that may prevent the misuse of data and information by anti-social elements. The moral precepts of digital ethics should guide the legal decisions made about the use of digital technology, even if this implies questioning current laws or customs. Legal frameworks in the digital sphere should be developed and used by the moral need to safeguard individual autonomy and dignity (Hanna, & Kazim, 2021).

The dovetailing of ethics with the law is critical for safeguarding fundamental human rights. While ethics ensure that the moral compass of technologists is set in the right direction, the legal frameworks protect the basic human rights of dignity, autonomy, privacy, etc. by ensuring compliance with the laws and regulations made by the governments. While the law ensures that business firms comply with regulations like GDPR (applicable in the European Union) and DPDP Act (for firms operating in India), ethical guidelines help in mitigating the potential harm that may occur due to misuse of technology.

References

Adaga, E.M., Egieya, Z.E., Ewuga, S.K., Abdul, A.A. and Abrahams, T.O. (2024). Philosophy in business analytics: a review of sustainable and ethical approaches. *International Journal of Management & Entrepreneurship Research*, *6*(1), 69–86.

Ajah, I.A. and Nweke, H.F. (2019). Big data and business analytics: Trends, platforms, success factors and applications. *Big Data and Cognitive Computing*, *3*(2), 32.

Allen, C., Wallach, W. and Smit, I. (2006). Why machine ethics?. *IEEE Intelligent Systems*, *21*(4), 12–17.

Awad, E., Levine, S., Anderson, M., Anderson, S.L., Conitzer, V., Crockett, M.J.... and Tenenbaum, J. B. (2022). Computational ethics. *Trends in Cognitive Sciences*, *26*(5), 388–405.

Berendt, B., Preibusch, S. and Teltzrow, M. (2008). A privacy-protecting business-analytics service for on-line transactions. *International Journal of Electronic Commerce*, *12*(3), 115–150.

Bhatia, A. (2020). Extent and Determinants of ICT adoption for digital inclusiveness in rural tourism a study in Himachal Pradesh. (PhD Unpublished thesis). Department of Tourism and Travel Management. Central University of Himachal Pradesh, India. http://hdl.handle.net/10603/333381

Birckan, G., Dutra, M., de Macedo, D. and Viera, A. (2020). Effects of data protection laws on data brokerage businesses. *EAI Endorsed Transactions on Scalable Information Systems*, *7*(27).

Bunge, M. (1977). Towards a technoethics. *The Monist*, 96–107.

Capurro, R. (2006). Towards an ontological foundation of information ethics. Ethics Inf Technol 8, 175–186. https://doi.org/10.1007/s10676-006-9108-0

Chowcat, I., Kay, D. and Korn, N. (2015). The legal, risk and ethical aspects of analytics. *Library analytics and metrics: Using data to drive decisions and services*, 157–165.

Drabiak, K., Kyzer, S., Nemov, V. and El Naqa, I. (2023). AI and machine learning ethics, law, diversity, and global impact. *The British Journal of Radiology*, *96*(1150), 20220934.

Ellul J. (1967). The Technological Society. New York: Random House.

Floridi, L. (1999). Information ethics: On the philosophical foundation of computer ethics. *Ethics and Information Technology*, *1*(1), 33–52.

Floridi, L., Cath, C. and Taddeo, M. (2019). Digital ethics: its nature and scope. *The 2018 yearbook of the digital ethics lab*, 9–17.

Górniak-Kocikowska, K. (2007). From computer ethics to the ethics of global ICT society. *Library Hi Tech*, *25*(1), 47–57.

Guan, P. and Zhou, W. (2017). Business Analytics Generated Data Brokerage: Law, Ethical and Social Issues. In: Doss, R., Piramuthu, S., Zhou, W. (eds) Future Network Systems and Security. FNSS 2017. Communications in Computer and Information Science, vol 759. Springer, Cham. https://doi.org/10.1007/978-3-319-65548-2_13

Han, J. (2022). An information ethics framework based on ICT platforms. *Information*, *13*(9), 440.

Hanna, R. and Kazim, E. (2021). Philosophical foundations for digital ethics and AI Ethics: a dignitarian approach. *AI and Ethics*, *1*(4), 405–423.

Jonas, H. (2004). Ethics and biogenetic art. *Social Research: An International Quarterly*, *71*(3), 569–582.

Johnson, D.G. (1994). *Computer ethics*. Prentice-Hall, Inc.

Kaisler, S.H., Espinosa, J.A., Armour, F. and Money, W.–H. (2014, January). Advanced Analytics–Issues and Challenges in a Global Environment. In *2014 47th Hawaii international conference on system sciences* (pp. 729-738). IEEE.

Liu, Y., Han, H. and DeBello, J. (2018). The challenges of business analytics: Successes and failures.

Luppicini, R. (Ed.). (2010). *Technoethics and the Evolving Knowledge Society: Ethical Issues in Technological Design, Research, Development, and Innovation: Ethical Issues in Technological Design, Research, Development, and Innovation*. IGI Global.

Madan, R., Joha, A. and Sivarajah, U. (2022) Ethical framework for artificial intelligence and digital technologies. *International Journal of Information Management*, 62. 102433. ISSN 0268-4012 doi: https://doi.org/10.1016/j.ijinfomgt.2021.102433

Moor, J. (1985). "What is computer ethics?" *Metaphilosophy* 16(4): 266–275.

Omar, Y.M., Minoufekr, M. and Plapper, P. (2019). Business analytics in manufacturing: Current trends, challenges and pathway to market leadership. *Operations Research Perspectives*, *6*, 100127.

Reddy, J.M. and Prasad, S.V.A.V. (2021). The Role of Business Analytics in E-Commerce. *Journal of Interdisciplinary Research (ISSN: 2408–1906)*, *6*(1).

Sayed, R. (2023). Ethical considerations in the use of big data for business analytics: a cross-industry perspective. *Journal of Innovations and Sustainability*, *7*(4), 08. https://doi.org/10.51599/is.2023.07.04.08

Schwarz, T. (2005). Teaching Ethics and Computer Forensics: The Markkula Center for Applied Ethics Approach. Proceedings of the 2nd annual conference on Information security curriculum development, September 23–24, Kennesaw State University, Georgia, US. pp. 66–71.

Segun, S.T. (2021). From machine ethics to computational ethics. *Ai & Society*, *36*(1), 263–276.

Soldić-Aleksić, J., Krasavac, B.C. and Karamata, E. (2020). Business analytics: new concepts and trends. *Management: Journal of Sustainable Business and Management Solutions in Emerging Economies*, *25*(2), 15–29.

Tzafestas, S.G. (2018). Roboethics: Fundamental concepts and future prospects. *Information*, *9*(6), 148.

Vidgen, R., Shaw, S. and Grant, D.B. (2017). Management challenges in creating value from business analytics. *European Journal of Operational Research*, *261*(2), 626–639.

Vidgen, R., Hindle, G. and Randolph, I. (2019) Exploring the ethical implications of business analytics with a business ethics canvas. *European Journal of Operational Research* 281 (3), pp. 491–501. ISSN 0377-2217.

Weinhardt, M. (2020). Ethical issues in the use of big data for social research. *Historical Social Research/HistorischeSozialforschung*, *45*(3), 342–368.

Wiener, N. God & Golem, Inc (1950) A Comment on Certain Points Where Cybernetics Impinges on Religion, 7th ed.; MIT Press: Cambridge, MA, USA, 1966. ISBN 978-026-273-011-2.

Wolniak, R. and Grebski, W. (2023a). The five stages of business analytics. *Silesian University of Technology Scientific Papers. Organization and Management Series*, *178*, 735–752.

Wolniak, R. and Grebski, W. (2023b). Work Safety and Health – The Business Analytics Usage In Industry 4.0 Conditions. *Scientific Papers of Silesian University of Technology. Organization & Management/ZeszytyNaukowePolitechnikiSlaskiej. Seria OrganizacjiiZarzadzanie*, (186).

Zhang, B., Anderljung, M., Kahn, L., Dreksler, N., Horowitz, M.C. and Dafoe, A. (2021). Ethics and governance of artificial intelligence: Evidence from a survey of machine learning researchers. *Journal of Artificial Intelligence Research*, *71*, 591–666.

Index

A

Advanced Analytics 5, 9, 14, 19, 66, 107, 108, 113, 114, 196-200, 205, 207-212, 215, 216, 226

AHP Analysis 134, 139

AI 9, 10, 12, 22, 25 273

AI ethics 273, 275, 278

Applications 1-3, 18, 25, 37, 58, 60, 61, 63, 65, 67, 69-74, 81-83, 94, 99, 101, 104, 106, 111, 122, 129, 147, 167-173, 180, 182, 199, 207, 211, 233, 269, 274, 277

Artificial Intelligence 9, 10, 22, 55, 59, 66, 67, 82, 85, 101, 113, 116, 119, 120, 128, 130, 131, 138, 151, 153, 159, 180, 196-200, 205, 208, 210, 212, 219-221, 224, 226, 274

Artificial Intelligence in Supply chain 200

B

Big Data 2, 5, 9, 10, 14, 16, 25, 36, 58, 61, 65, 68, 72, 81, 83, 85, 86, 90, 94, 152, 159, 165, 167, 208, 219, 223, 226, 270

Blended Education 176

Brands 81, 82, 90, 92, 244, 245, 248-250, 254, 257, 260, 262, 264, 265

Business Analytics 1-6, 8, 10, 15, 17, 21, 24, 151-157, 219-227, 269, 274

Business Intelligence 5, 8, 66, 88, 94, 154, 165, 168, 174, 219, 221, 227

C

Challenges 4, 9, 10, 24, 25, 34, 41, 53, 58, 60, 67, 70, 72, 74, 76, 85, 99, 100, 104, 112, 115, 120, 129, 135, 147, 152, 154, 169, 199, 228, 270, 277

Computer Vision 219

CPI 230-232, 236, 238, 240

Customer Experience 19, 59, 62, 63, 82, 84, 90, 91, 214

Customer touchpoints 81, 82, 84, 87, 90, 94-96

D

Data Analytics 1, 5, 6, 8, 24, 39, 58, 59, 67, 89, 100, 103, 106-109, 111, 114, 115, 153, 165, 167, 173, 180, 184-188, 191, 192, 207, 208, 212, 221, 223, 226, 270, 276

Data mining 2, 6, 9, 11, 14, 16, 25, 58, 81, 83, 85, 87, 88, 93, 94, 167, 219

Data Science 2, 58-65, 67, 70, 71, 73, 75, 77, 83, 99, 113, 235, 271

Data-driven 1, 2, 8, 9, 19, 25, 34, 40, 55, 59, 61, 70, 74, 76, 81, 82, 89, 90, 99, 100, 106, 109, 112, 114, 120, 122, 154, 166, 168, 171, 173, 202, 209, 210, 219, 220, 226, 228, 262, 277, 279

Data-driven Decision-making 8, 36, 61, 62, 82, 166, 168, 210

Decision making 1, 3, 5, 8, 10, 12, 17, 19, 21, 34, 36, 38, 39, 41, 45, 55, 59, 61, 64, 68, 82, 83, 86, 95, 100, 101, 104, 106, 113, 115, 151, 152, 166, 174, 221, 223, 270, 277

Deep Learning 58, 59, 61, 66, 70, 101, 109, 116, 223

Digital Ethics 274, 275, 278, 279

Digital Transformation 151-154, 159-161, 275

Dimensions 9, 34, 38, 51, 60, 136-138, 156, 274

DPDP Act 2023 276

E

Economic Growth 73, 134-136, 139, 140, 143, 144, 146, 204, 206, 230

Education 20, 43, 90, 125, 137, 147, 165-173, 176, 177, 183, 189, 223

Employee onboarding 120, 123, 129, 130

Engagement 12, 18, 42, 43, 48, 50, 53, 63, 82, 92, 110, 112, 122, 125, 137, 138, 151, 156, 161, 166, 168, 170, 173, 175, 177, 182, 183, 192, 197, 244, 250, 252-254, 256, 257, 258, 260-264, 270

Environmental analytics 99-104, 106, 110, 111, 113-116

Environmental monitoring 102, 103, 111, 113, 114, 116

F

Family-Owned Enterprises 151, 155, 157
Financial Inclusion 71, 73, 74, 134-144, 147
FinTech 58-60, 63, 66, 68, 71, 76, 134, 136, 147
Fintech Adoption 147
Forecasting 12, 13, 18, 20, 59, 82, 104, 105, 108, 196, 200, 203, 205, 207-209, 232, 233, 235, 241
Future of Business 219

H

Healthcare industry 19, 119
Healthcare worker 119
HR Analytics 34, 35, 37-39, 42, 44, 45, 49, 52, 54, 55

I

Indian Economy 230
Innovation 2, 21, 22, 58, 60, 61, 66, 67, 70, 73, 75, 76, 93, 100, 120, 136, 138, 151, 154, 157, 160, 161, 182, 184, 186, 190, 192, 195, 196, 204, 207, 209, 220, 227, 272, 278
IoT 4, 22, 25, 59, 66, 69, 85, 90, 95, 100, 102, 107, 108, 110, 111, 113, 115, 165, 209, 219, 223, 224, 227

M

Machine Learning 2, 3, 9, 13, 15, 22, 58, 59, 61, 66, 67, 74, 82, 93, 100, 104, 105, 111, 113, 115, 119, 120, 123, 125, 127, 129, 196, 197, 209, 212, 215, 219, 222, 223, 226, 228, 231, 233, 235, 273
Marketing analytics 81-85, 87, 91, 94, 96
Meme marketing 247, 248, 251-253, 255, 257-262, 264, 265
Metrics 4, 20, 35-38, 40-42, 45, 53, 65, 70, 80, 89, 121, 122, 124, 129, 134, 155, 166, 172, 209, 212, 225, 230, 231, 235, 241, 262, 263
MLR 105, 230, 232, 234, 238, 239, 241, 242

N

Natural Language Processing 10, 13, 22, 23, 25, 67, 69, 72, 82, 93, 119, 121, 129, 182, 193, 224

P

Pareto Chart 140, 144
Performance Metrics 123, 129, 155, 231, 235
Predictive Modeling 67, 103, 104, 106, 110, 111, 122, 209, 222, 230

R

Robotic Process Automation 61, 69, 129, 212, 215, 219

S

Sentiment analysis 11, 20, 72, 81-87, 92-95, 262, 263
Social media 8-11, 15, 20, 23, 41, 59, 66, 69, 71, 81, 82, 84, 87, 89, 90, 92-96, 103, 154, 156, 157, 165, 168, 172, 176, 183, 244-251, 254, 257-261, 263, 264, 277
Sustainability 35, 39, 41, 52, 53, 55, 76, 99, 100, 101, 102, 107, 108, 110, 112, 115, 117, 129, 153, 184, 196, 200, 203, 208, 212, 216

T

Technological Advancements 1, 4, 9, 10, 59, 159, 165, 192, 220
Technology Adoption 154, 161, 183, 184, 186, 191
Tools and Techniques 14, 25, 37, 60, 111, 211, 216
Tourism Supply Chain 196-212, 215, 216

V

Virality 252, 257, 260, 264

Editors Biography

Dr. Ravindra Sharma is an Assistant Professor at the School of Management Studies at Swami Rama Himalayan University, Dehradun, India. He earned his Ph.D. in Management from Uttarakhand Technical University, India. He holds master's degrees in business administration (MBA) and computer applications (MCA). He has qualified for the University Grant Commission National Eligibility Test in Management (UGC-NET). Dr. Sharma has more than 18 years of corporate and academic experience. He has conducted many faculty development programs for national and international audiences. He has published research papers in refereed journals. He has published books in the area of employer branding, HR 4.0, Internet of Things (IoT), artificial intelligence (AI), and green innovations for businesses with reputed publishers. Dr. Sharma has contributed chapters in books from major publishing houses. He has presented research papers at national and international conferences. He has also attended a paper development workshop in the Indian Institute of Management in Rohtak, India. He was honored as a session chair and keynote speaker at international conferences at various universities. His research interests include employer branding, entrepreneurship, Industry 4.0, talent management, and business innovations.

Dr. Geeta Rana is an Associate Professor at the School of Management Studies at Swami Rama Himalayan University, Jolly Grant, Dehradun, India. She has earned her Ph.D. from the Indian Institute of Technology (IIT, Roorkee) in Management. She has also done a certification course in HR Analytics from the Indian Institute of Management Rohtak (IIMR). She is engaged in teaching, research, and consultancy assignments. She has more than 20 years of experience in teaching and in handling various administrative as well as academic positions. She has to her credit over 65 papers published in refereed journals. She has presented several research papers at national and international conferences and has contributed to several chapters in books published by major publishing houses. She has authored a book titled Counseling Skills for Managers. She has conducted and attended various workshops, FDPs, and MDPs. Dr. Rana is a recipient of many awards. Her research interests include Data Analytics, Human Resource Management 4.0, Knowledge Management, Managerial Effectiveness, Justice, Values, Employer Branding, and Innovation.